ENVIRONMENTAL IMPACT ASSESSMENT

TITLES OF RELATED INTEREST

Agriculture: people and policies
G. Cox *et al.* (eds)

Cost-benefit analysis in urban and regional planning
J. Schofield

Countryside conservation
B. Green

Development and the landowner
R. Goodchild & R. Munton

Engineering planning and design
R. Warner & G. Dandy

Environment and development
P. Bartelmus

Environmental groups in politics
P. Lowe & J. Goyder

Floods and drainage
E. C. Penning-Rowsell *et al.*

Geomorphological hazards in Los Angeles
R. Cooke

The Green Revolution revisited
B. Glaeser (ed.)

Greenprints for the countryside?
A. & M. MacEwen

Hedgerows and verges
W. Dowdeswell

Intelligent planning
R. Wyatt

Interpretations of calamity
K. Hewitt (ed.)

Landscape meanings and values
E. C. Penning-Rowsell & D. Lowenthal (eds)

Learning from China?
B. Glaeser (ed.)

London's green belt
R. Munton

The management of toxic chemicals
R. Coté & P. Wells

National parks: conservation or cosmetics?
A. & M. MacEwen

Nature's place
W. Adams

Policies and plans for rural people
P. Cloke (ed.)

Regional development and settlement policy
R. Dewar *et al.*

Risk and society
L. Sjoberg (ed.)

Town and country planning in Britain
B. Cullingworth

Urban and regional planning
P. Hall

ENVIRONMENTAL IMPACT ASSESSMENT

Theory and practice

Peter Wathern
Internationales Institut für Umwelt und Gesellschaft, Wissenschaftszentrum, Berlin

London
UNWIN HYMAN
Boston Sydney Wellington

Published by the Academic Division of
Unwin Hyman Ltd
15/17 Broadwick Street, London W1V 1FP, UK

Allen & Unwin Inc.,
8 Winchester Place, Winchester, Mass. 01890, USA

Allen & Unwin (Australia) Ltd,
8 Napier Street, North Sydney, NSW 2060, Australia

Allen & Unwin (New Zealand) Ltd in association with the Port Nicholson Press Ltd,
60 Cambridge Terrace, Wellington, New Zealand

First published in 1988

British Library Cataloguing in Publication Data

Environmental impact assessment: theory and practice.
1. Environmental impact analysis
I. Wathern, Peter
333.7′1 TD194.6

ISBN 0–04–445042–7
ISBN 0–04–445052–4 Pbk

Library of Congress Cataloging-in-Publication Data

Environmental impact assessement: theory and practice / (edited by) Peter Wathern.
p. cm.
Bibliography: p.
Includes index.
ISBN 0–04–445042–7. ISBN 0–04–445052–4 (pbk.)
1. Environmental impact analysis. I. Wathern, Peter.
TD194.6.E595 1988
333.7′1 – dc19 87–33336 CIP

Typeset in 10 on 12 point Bembo by
Nene Phototypesetters Ltd, Northampton
and printed in Great Britain by
Biddles of Guildford

To Julie, Andrea and Matthew

Foreword

Environmental impact assessment (EIA) has been a significant development in environmental management over the last two decades. Consequently, it has been an important element of the research of the International Institute for Environment and Society (Internationales Institut für Umwelt und Gesellschaft – IIUG), a constituent research unit of the Wissenschaftszentrum Berlin (WZB), almost since its inception in 1977. During that time, a number of IIUG research fellows, visiting research workers and outside organizations involved in collaborative ventures with the IIUG have investigated various aspects of EIA.

The present volume draws together much of the experience of past and present research fellows of the IIUG, augmented with contributions from outside experts. Collectively, the work represents the views of a distinguished group of authors widely acknowledged to be amongst the leading experts in their respective fields. The result is an authoritative text which I am pleased to commend to what I hope will be a wide audience. The book should especially appeal to those involved in the day-to-day application of EIA, as well as to those currently undertaking courses in the environmental sciences, planning, resource management and political science at undergraduate and postgraduate level.

Berlin,
December 1986

UDO E. SIMONIS
Director, IIUG

Acknowledgements

This venture has provided me with an opportunity to renew some old friendships and hopefully to cultivate some new ones. The primary acknowledgement, of course, must be to the contributors for the way they have responded with characteristic good humour and indulgence to my alarmingly short deadlines and unreasonable pedantry. I hope they feel that the final result does justice to their efforts. I should like to thank my former colleagues at CEMP, Aberdeen University, who have collaborated in the production of this volume, namely Brian Clark, Paul Tomlinson and especially Ron Bisset who provided some invaluable introductions at the outset when the structure of the book was being discussed. Special thanks should also go to the Director of IIUG, Prof. Udo Simonis, for his support and encouragement during the preparation of this volume.

Ros Laidlaw provided the graphics and I duly acknowledge her skills. Not least, I should like to thank Julie Wathern for her invaluable assistance during the preparation and editing of the final manuscript, and Andrea Wathern for compiling the index.

PETER WATHERN
Berlin, December 1986

We are grateful to the following individuals and organizations who have kindly given permission for the reproduction of copyright material (figure numbers in parentheses):

Figures 1.3 & 1.7 reproduced from *Landscape and urban planning* by P. Wathern *et al.*, by permission of Elsevier; Figure 1.8 reproduced from *Environmental policy analysis: operational methods and models*, by P. Nijkamp, by permission of John Wiley & Sons Ltd copyright © 1980; Figure 3.4 reproduced from *Ecological Modelling*, 3(3), Gilliland & Risser (eds), by permission of Elsevier; Tables 9.1, 9.2, 9.3, 9.4, 9.5, 9.6, 9.7, 9.8, reproduced by permission of Academic Press.

Contents

List of Tables

List of Contributors

Richard Andrews is the Director of the Institute for Environmental Studies, University of North Carolina at Chapel Hill, Chapel Hill, North Carolina, USA.

Gordon Beanlands is the Director of Research of the Canadian Federal Environmental Assessment Review Office and Adjunct Professor at the School for Resource and Environmental Studies at Dalhousie University, Halifax, Nova Scotia, Canada.

Ronald Bisset, when these papers were written, was the Deputy Executive Director of the Centre for Environmental Management and Planning, University of Aberdeen and is now an Associate of Cobham Resource Consultants, Edinburgh, Scotland, UK.

Eric Giroult is the Regional Officer for Environmental Health Planning and Management of the World Health Organization Regional Office for Europe, Copenhagen, Denmark.

Paul de Jongh is a senior adviser at the Directorate General for Environment within the Ministrie van Volkshuisvesting, Ruimtelijke Ordening en Milieubeheer, Leidschendam, the Netherlands.

John Horberry is a senior consultant with Environmental Resources Ltd., London, England.

William Kennedy, when this paper was written, was a senior adviser within the Ministrie van Volkshuisvesting, Ruimtelijke Ordening en Milieubeheer, Leidschendam, the Netherlands, and has since taken up a post at the Organization for Economic Co-operation and Development, Paris, France.

Joanne Kerbavaz currently manages the Hazardous Waste Management Unit of the California Department of Transportation, Sacramento, California, USA.

Norman Lee is a senior lecturer in the Department of Economics, Manchester University, England.

Iara Verocai Moreira is an architect working in the Fundação Estadual de Engennaria do Meio Ambiente, Rio de Janeiro, Brazil.

Nay Htun is the Regional Director and Representative for Asia and the Pacific Region of the United Nations Environment Programme, Bangkok, Thailand.

Barry Sadler is the Director of the Institute of the North American West, Victoria, British Columbia and a consulting associate of the Banff Centre School of Management, Banff, Alberta, Canada.

Anna Starzewska is a research team leader at the Environmental Pollution Abatement Centre, Katowice, Poland.

Paul Tomlinson, when the paper was written, was Senior Environmental Scientist at the Centre for Environmental Management and Planning, University of Aberdeen, Scotland. He is now an independent consultant working in the UK.

Geoffrey Wandesforde-Smith is in the Department of Political Science and the Division of Environmental Studies, University of California at Davis, Davis, California, USA.

Peter Wathern is a senior research fellow at the Internationales Institut für Umwelt und Gesellschaft, Wissenschaftszentrum, Berlin, Federal Republic of Germany and lecturer in the Department of Botany and Microbiology, University College of Wales, Aberystwyth, Wales.

Christopher Wood is a senior lecturer in the Department of Town and Country Planning, Manchester University, Manchester, England.

Part I

INTRODUCTION

1 *An introductory guide to EIA*

P. WATHERN

In the sixteen years since its inception, environmental impact analysis (EIA), a procedure for assessing the environmental implications of a decision to enact legislation, to implement policies and plans, or to initiate development projects, has become a widely accepted tool in environmental management. EIA has been adopted in many countries with different degrees of enthusiasm where it has evolved to varying levels of sophistication.

In the United States, EIA required under the National Environmental Policy Act of 1969 (NEPA) has given a federal dimension to land-use planning which existed in only rudimentary form prior to 1970 and has created a situation where decisions on major federal activities can only be taken with foreknowledge of their likely environmental consequences. The influence of these federal measures can be gauged from the rapidity with which they have been echoed in state and local statutes. A host of other industrialized countries have since implemented EIA procedures. Canada, Australia, the Netherlands and Japan, for example, adopted legislation in 1973, 1974, 1981 and 1984 respectively, while in July 1985 the European Community (EC) finally adopted a directive making environmental assessments mandatory for certain categories of projects after nearly a decade of deliberation.

Countries in the developed world have not been alone in realizing the potential of EIA. Many less developed countries (LDCs) have been quick to appreciate that the procedures offer a means of introducing some aspects of environmental planning, often in the absence of any formal land-use planning control system. Colombia became the first Latin American country to institute a system of EIA when procedures were adopted in 1974. In Asia and the Pacific region, Thailand and the Philippines now have long-established procedures for EIA. There is a dearth of information on the general situation in Africa, although a number of nations including Rwanda, Botswana and the Sudan have experience of EIA (Klennert 1984).

In the centrally planned economies of Eastern Europe, it is increasingly realized that EIA should be an integral component of state planning, although Marxist theory places another perspective on the interrelationships between development and the environment. Hungary has made the consideration of environmental issues one of the elements upon which investment decisions are based and Poland has initiated studies on the application of EIA which will probably lead to its formal adoption.

Bilateral and multilateral agencies have also become interested in the potential of EIA. The Organization for Economic Co-operation and Development (OECD) adopted recommendations concerning EIA within its constituent states in 1974 and 1979 and for development aid projects in 1985. The United Nations Environment Programme (UNEP) has provided guidance on the assessment of development proposals (UNEP 1980) and supported research on EIA in developing countries (Ahmad & Sammy 1985). The World Health Organization (WHO) has become concerned with the need to assess not only the opportunities to improve the quality of life presented by development, but also consequent adverse effects upon human health mediated through environmental change.

In recent years, the breadth of EIA has expanded perhaps even more rapidly than its rate of geographical spread. Thus, it now comprises a number of discrete specialisms and has spawned related disciplines concerned with other effects of development, particularly social impact assessment (SIA). SIA is now well established in its own right and, consequently, is not covered in this book; a guide to the large literature on SIA can be found in Leistritz & Ekstrom (1986). Environmental health impact analysis (EHIA), assessing the health implications of development, is becoming increasingly important and appears to be on the point of developing into an independent discipline comparable to SIA. Thus, in a few short years a new subject area has emerged, generated considerable controversy, stimulated the development of new technical and administrative skills, become established and gained widespread acceptance.

NEPA was a timely piece of legislation in that it emerged at a point when there was a growing environmental constituency prepared to ensure that its provisions, unlike previous attempts to reform the federal decision-making process, would be assiduously applied by federal agencies. The almost instantaneous success of environmental groups in using litigation to force EIA upon federal agencies proved an important spur to its early evolution. The risk of a court case over an inadequate environmental impact statement (EIS) created the demand for guidance on EIS preparation from administrators and technical experts alike. This preoccupation resulted in the early EIA literature being dominated by methodological issues. The intensity of this type of research during the early 1970s can be gauged from the fact that many of the main conceptual innovations in EIA methodology were established at that time. There were also many attempts to produce complete handbooks for preparing EISs, encompassing not only EIA methods, but also techniques for determining individual impacts, such as dispersion models for pollutants and economic multipliers. Typical examples produced in the USA include Canter (1977), Cheremisinoff & Morresi (1977) and Jain *et al*. (1977). Clark *et al*. (1976, 1981b) produced comparable handbooks for the UK, while UNEP (1980) represents the response of an international agency to the need for guidance on preparing environmental assessments for industrial projects. An alternative approach to the problem of guidance is presented in Carpenter (1983) where the functioning of natural systems prone to disruption by development is described to help planners identify potential impacts. The main emphasis is on systems which are relevant to LDCs.

After this initial concentration on technical aspects, attention has focused increasingly upon EIA within the overall decision-making process. The major deliberation has been the relevance of material included in EIAs to decision making. The utility of the EIA system as a means of effecting a reappraisal of the decision-making process has been called into question and, at an individual level, the content of EISs has frequently borne little relationship to the needs of the decision maker.

EIA has been regarded as both 'science' and 'art', reflecting the concern both with technical aspects of appraisal and the effects of EIA upon the decision-making process (Kennedy 1984). This distinction is useful, indeed it underpins much of the organization of material within this book. It is, however, somewhat artificial for they are inexorably linked. Thus, for example, certain EIA methods developed to deal with technical problems impinge directly upon decision making (Bisset 1978). In the extreme case of adaptive environmental assessment and management (AEAM), these two facets have become firmly intermeshed.

This volume is not intended to be a handbook of 'EIA as science'; currently, a number of textbooks admirably fill this role. Nor can it be a complete review of the ramifications of EIA for decision making throughout the world, 'EIA as art'; the subject has become too vast to make this feasible.

Authors of the succeeding chapters were invited to review recent developments within their own area of EIA specialism to give an up-to-date overview of EIA, a task which is now clearly beyond the competence of one person. Throughout, the intention has been to temper theoretical considerations of EIA with experience of the process as it operates in practice. As part of this remit, authors were requested to ignore early EIA material irrespective of its intrinsic interest, with 1978 a suggested cut-off point. Although an arbitrary suggestion, on reflection it appears defensible as 1978 represents a pivotal year in certain respects. For example, the Council on Environmental Quality (CEQ) consolidated US experience up to that point in sweeping new federal EIS regulations (CEQ 1978); Fairfax (1978) called the efficacy of NEPA into question; a radically new approach to EIA mainly developed in Canada was suggested by Holling (1978); in Europe, the views of the EC's EIA consultants became widely available (Lee & Wood 1978).

Without some idea of this early period, however, the significance of recent developments may appear obscure. Consequently, to make the post-1978 material accessible to those with only a perfunctory knowledge of the subject, this chapter is intended to provide a short synopsis of EIA rather than the weighty conceptualizing which might otherwise be expected of an introduction. The remainder of the chapter is divided into three main components. First, some simple definitions are unavoidable. Secondly, an attempt to project a unified model of the EIA process is made. Although EIA procedures differ in detail around the world, they are united in being designed to deal with particular issues. Rather than knowing the minute details of EIA systems in different countries, it is much more important to understand how the need for certain activities at certain points dictates the nature of the process. These individual issues and the procedural needs which they generate are discussed and, subsequently, fitted together to show the whole EIA process. Finally, some aspects of the debate concerning EIA in the decision-making process are reviewed.

Some terms

ENVIRONMENTAL IMPACT ANALYSIS (EIA)

Largely following the definition of Munn (1979), EIA can be described as a process for identifying the likely consequences for the biogeophysical environment and for man's health and welfare of implementing particular activities and for conveying this information, at a stage when it can materially affect their decision, to those responsible for sanctioning the proposals. Davies & Muller (1983) argue for an extension of this definition to cover socioeconomic effects to provide for a unified appraisal.

Thus, EIA is a process having the ultimate objective of providing decision makers with an indication of the likely consequences of their actions. Over the years, it has become increasingly evident that the authorization of proposals is not the sole decision point. There are many decision makers involved in the evolution of a set of development proposals and the influence of most of them is exerted long before the submission of an application for formal project authorization. The definition adopted above is equally applicable to this expanded view of decision making in the planning of development proposals. In the past, attention has tended to focus on the most spectacular decision point, authorization, and the importance of a well-integrated appraisal in the refinement of development proposals has largely been undervalued. EIA is no longer seen as an 'add-on' process. Indeed, the greatest contribution of EIA to environmental management may well be in reducing adverse impacts before proposals come through to the authorization phase.

Although generally considered a tool of project management, EIA is equally applicable at other levels of planning. Little experience, however, yet exists of the use of EIA for assessing legislation, programmes, policies and plans.

ENVIRONMENTAL IMPACT STATEMENT (EIS)

The outcome of an EIA is usually some formal document. This report has a variety of names throughout the world, although the term 'environmental impact statement' (usually abbreviated to EIS) is most widely known and carries the least scope for confusion. 'Environmental assessment' and 'environmental appraisal' are commonly adopted synonyms. Despite minor differences throughout the world, there is a general consensus on the content of an EIS. Table 1.1, for example, details the content of an EIS for US federal proposals

Table 1.1 Content of an EIS for US federal proposals as required by CEQ (1978).

Summary

Statement of purpose and need

Alternatives including proposed action
- Discussion of all options considered
- Discussion of 'no-action' option
- Identification of agency-preferred alternative
- Discussion of mitigation measures

Affected environment
- Baseline environmental description of area affected by each alternative

Environmental consequences
- Environmental impact of each alternative
- Unavoidable effects
- Relationship between local short-term use of environment and enhancement of long-term productivity
- Irreversible and irretrievable commitment of resources

List of preparers

---

as required by the Council on Environmental Quality (CEQ). The EC EIA directive, on the other hand, also requires proponents to highlight areas of uncertainty by indicating 'technical deficiencies or lack of know-how' encountered in compiling information included in an environmental assessment (Council of the European Communities 1985).

ENVIRONMENTAL IMPACT

The terms 'impact' and 'effect' are frequently used synonymously, although some have advocated differentiating between natural or man-induced changes in the biogeophysical environment, effects, from the consequences of these changes, namely impacts (see, for example, Catlow & Thirlwall 1976 and Munn 1979). An impact has both spatial and temporal components and can be described as the change in an environmental parameter, over a specified period and within a defined area, resulting from a particular activity compared with the situation which would have occurred had the activity not been initiated. It is most easily envisaged graphically (Fig. 1.1).

Environmental systems are not static, but change over the course of time even without the influence of man. Some are very dynamic, while others only change imperceptibly. In order to make predictions about impacts, assumptions have to be made about natural change. In order to assess the impact of a development project, for example, it would be necessary also to analyse natural changes in the rate of sedimentation in an estuarine system over the same period. In contrast, a description of the present state would probably suffice if the proposed development was situated on a stable hard-rock coastline.

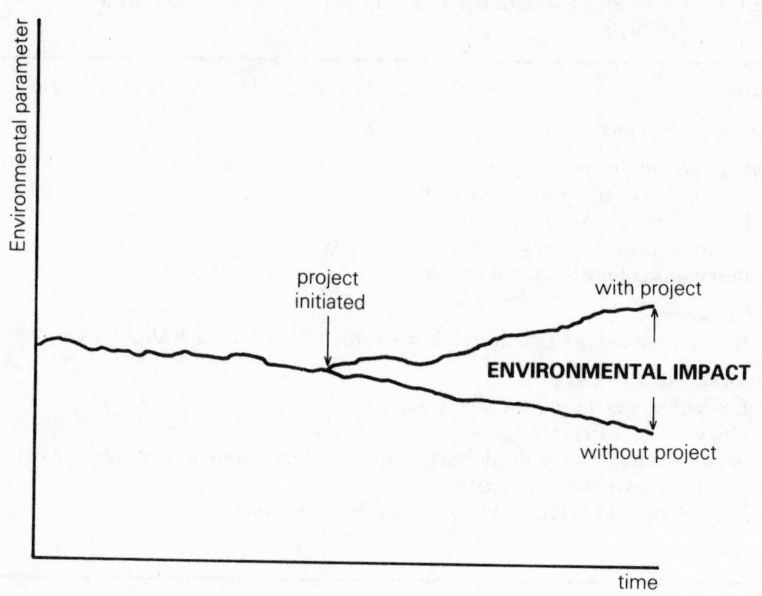

Figure 1.1 An impact.

A major deficiency of many EISs has been the failure to establish a time frame indicating when impacts are likely to be manifest. Impacts are also site-specific and determination of their spatial distribution is also important. Spatial aspects are usually considered more adequately than temporal ones.

It is useful to distinguish between direct (primary) and indirect (secondary, tertiary and higher order) impacts. Some impacts are a direct consequence of a particular activity. Thus, without adequate mitigating measures, construction of a dam on a river will prevent the upward movement of migratory fish. This would be a direct impact of the project. Other impacts, however, occur as a result of changes in a chain of environmental parameters. Thus, to continue this example, there would also be indirect impacts upon fish populations. Reductions in streamflow and turbulence would lower the oxygen tension and affect survival. Reduced water flow would also affect the nature of the streambed, the consequent siltation making conditions unsuitable for migratory fish to breed.

Major issues in the EIA process

The design of effective EIA procedures can be envisaged as the search for mechanisms to deal with issues generated by the need to juxtapose the planning and authorization of proposals. Some issues deal with technical matters such as impact identification and prediction, 'EIA as science'. Most, however, relate to the management of information within and between the two processes and, as such, are issues of 'EIA as art'.

IDENTIFICATION OF PROJECTS REQUIRING EIA

Land-use planners have long argued that all development proposals should be subject to appraisal, the level of analysis being commensurate with the significance of the issues raised. EIA, however, implies a special type of analysis involving a careful, thorough and detailed analysis of the likely implications of a development. This indicates the need for some threshold of 'significance' being exceeded in order to trigger the full EIA process, a procedure commonly referred to as screening.

Many countries have developed lists of projects which should be subject to EIA (Table 1.2). The main considerations in drawing up such lists are project type, size and the consequence of likely impacts. The converse, a list of 'categorical exclusions' exempted from the EIA requirement, has also been adopted. Project location is also a determinant of impact, as a development in one area may be far more severe than if it were located elsewhere. Thailand, for example, has identified such environmentally sensitive areas. Novelty is also an issue which should be addressed, as some small installations may represent unknown quantities in terms of impact.

Establishing rigid screening criteria may be unsatisfactory as it is the combination of project and location which determines the magnitude and significance of impacts. In Canada, for example, screening is a phased process (Federal Environmental Assessment Review Office 1978). Matrices are recommended for determining whether more stringent investigations are required. If major uncertainty still exists a more detailed study called an initial environmental evaluation is undertaken. Depending upon the outcome of this study a full EIA may be required.

IDENTIFICATION OF IMPACTS TO BE ASSESSED

Many of the impacts of a proposed development may be trivial or of no significance to the decisions which have to be taken. In practice, a decision will generally turn upon only a small subset of issues of overwhelming importance. Scoping is the process for determining which issues are likely to be important. Several groups, particularly decision makers, the local population and the scientific community have an interest in helping to delineate the issues which should be considered, and scoping is designed to canvass their views.

ASSESSMENT OF IMPACTS

A clear distinction should be drawn between techniques for predicting individual changes, such as Gaussian dispersion models with which likely ground-level concentrations of atmospheric pollutants can be calculated, and EIA methods used in assessments. EIA methods are used for various activities, namely: impact identification; prediction; interpretation; and communication; and in devising monitoring schemes. A particular method may not be equally useful for each activity.

There were many methodological developments in the early 1970s. Some methods lean heavily upon approaches used in other spheres of environmental

Table 1.2 Example of a list of projects for mandatory EIA.

1 *Extractive industry*
Extraction and briquetting of solid fuel
Extraction of bituminous shale
Extraction of ores containing fissionable and fertile material
Extraction and preparation of metalliferous ores

2 *Energy industry*
Coke ovens
Petroleum refining
Production and processing of fissionable material
Generating of electricity from nuclear power
Coal gasification plants
Disposal facilities for radioactive waste

3 *Production and preliminary processing of metals*
Iron and steel industry, excluding integrated coke ovens
Cold rolling of steel
Production and primary processing of non-ferrous metals and ferro-alloys

4 *Manufacturing of non-metallic mineral products*
Manufacture of cement
Manufacture of asbestos-cement products
Manufacture of blue asbestos

5 *Chemical industry*
Petrochemical complexes for the production of olefins, olefin derivative, bulk monomers and polymers
Chemical complexes for the production of organic intermediates
Complexes for the production of basic inorganic chemicals

6 *Metal manufacture*
Foundries
Forging
Treatment and coating of metals
Manufacture of aeroplane and helicopter engines

7 *Food industry*
Slaughter houses
Manufacture and refining of sugar
Manufacture of starch and starch products

8 *Processing of rubber*
Factories for the primary production of rubber
Manufacture of rubber tyres

9 *Building and civil engineering*
Construction of motorways
Intercity railways, including high-speed tracks
Airports
Commercial harbours
Construction of waterways for inland navigation
Permanent motor and motorcycle racing tracks
Installation of surface pipelines for long-distance transport.

Note: In fact, this list of projects was not adopted.
Source: Commission of the European Communities.

management, but there has also been much innovation. No attempt will be made to catalogue all early EIA methods, they are adequately described elsewhere (see, for example, Clark *et al*. 1980), only the main approaches and a few of the available methods will be highlighted here.

The simplest approach is a checklist of potential impacts which should be considered. An example is shown in summary form in Table 1.3. The main disadvantage of checklists is that they must be exhaustive if no serious impact is to be overlooked. An exhaustive checklist is likely to be unwieldy and may stifle initiative during assessment. The checklist summarized in Table 1.3, for example, has 47 subcategories of potential impacts.

Table 1.3 Checklist of impact categories for land development projects (summarized from Schaenam 1976).

1	*Local economy*
	Public fiscal balance
	Employment
	Wealth
2	*Natural environment*
	Air quality
	Water quality
	Noise
	Wildlife and vegetation
	Natural disasters
3	*Aesthetics and cultural values*
	Attractiveness
	View opportunities
	Landmarks
4	*Public and private services*
	Drinking water
	Hospital care
	Crime control
	Feeling of security
	Fire protection
	Recreation – public facilities
	Recreation – informal settings
	Education
	Transportation – mass transit
	Transportation – pedestrian
	Transportation – private vehicles
	Shopping
	Energy services
	Housing
5	*Other social impacts*
	People displacement
	Special hazards
	Sociability/friendliness
	Privacy
	Overall contentment with neighbourhood

Instructions

1. Identify all actions (located across the top of the matrix) that are part of the proposed project
2. Under each of the proposed actions, place a slash at the intersection with each item on the side of the matrix if an impact is possible
3. Having completed the matrix, in the upper left-hand corner of each box with a slash, place a number from 1 to 10 which indicates the MAGNITUDE of the possible impact; 10 represents the greatest magnitude of impact and 1, the least, (no zeros). Before each number place + (if the impact would be beneficial). In the lower right-hand corner of the box place a number from 1 to 10 which indicates the IMPORTANCE of the possible impact (e.g. regional vs. local); 10 represents the greatest importance and 1, the least (no zeros)
4. The text which accompanies the matrix should be a discussion of the significant impacts, those columns and rows with large numbers of boxes marked and individual boxes with the larger numbers

Sample matrix

	a	b	c	d	e
a		2/1			8/5
b		7/2	8/8	3/1	9/7

		Proposed actions	A. Modification of regime													B. Land transformation and construction																			C. Resource extraction						
			a. Exotic flora or fauna introduction	b. Biological controls	c. Modification of habitat	d. Alteration of ground cover	e. Alteration of ground water hydrology	f. Alteration of drainage	g. River control and flow modification	h. Canalization	i. Irrigation	j. Weather modification	k. Burning	l. Surface or paving	m. Noise and vibration	a. Urbanization	b. Industrial sites and buildings	c. Airports	d. Highways and bridges	e. Roads and trails	f. Railroads	g. Cables and lifts	h. Transmission lines, pipelines and corridors	i. Barriers including fencing	j. Channel dredging and straightening	k. Channel revetments	l. Canals	m. Dams and impoundments	n. Piers, seawalls, marinas and sea terminals	o. Offshore structures	p. Recreational structures	q. Blasting and drilling	r. Cut and fill	s. Tunnels and underground structures	a. Blasting and drilling	b. Surface excavation	c. Subsurface excavation and retorting	d. Well drilling and fluid removal	e. Dredging	f. Clear cutting and other lumbering	g. Commercial fishing and hunting
CHEMICAL CHARACTERISTICS	1 Earth	a. Mineral resources																																							
		b. Construction material																																							
		c. Soils																																							
		d. Land form																																							
		e. Force fields and background radiation																																							
		f. Unique physical features																																							
	2 Water	a. Surface																																							
		b. Ocean																																							
		c. Underground																																							
		d. Quality																																							
		e. Temperature																																							
		f. Recharge																																							
		g. Snow, ice and permafrost																																							

Figure 1.2 A section of the Leopold matrix (Courtesy US Geological Survey).

Leopold *et al.* (1971) were the first to suggest the use of a matrix method for EIA. Matrices are particularly useful for EIA as they reflect the fact that impacts result from the interaction of development activities and the environment. The Leopold matrix (Fig. 1.2) is complex. The 8800 cells result from ranging 88 environmental parameters along one axis and 100 development characteristics along the other. The matrix format is ideally suited for impact identification, although the ability of the Leopold matrix to identify indirect impacts has been questioned. Environment Canada (1974) contains an alternative matrix approach which can be used to identify such effects and a quantified modification is given in Wathern (1984). The Leopold matrix is also used to present the results of an appraisal. Numbers representing magnitude and significance, expressed on a 10-point scale, are included in each cell indicating where a likely impact is anticipated.

Sorensen (1971) developed a hybrid approach which reviewers have generally included amongst network approaches (Fig. 1.3). Networks are relatively effective at revealing indirect impacts as the ramifications of a change can be followed through chains of intermediaries.

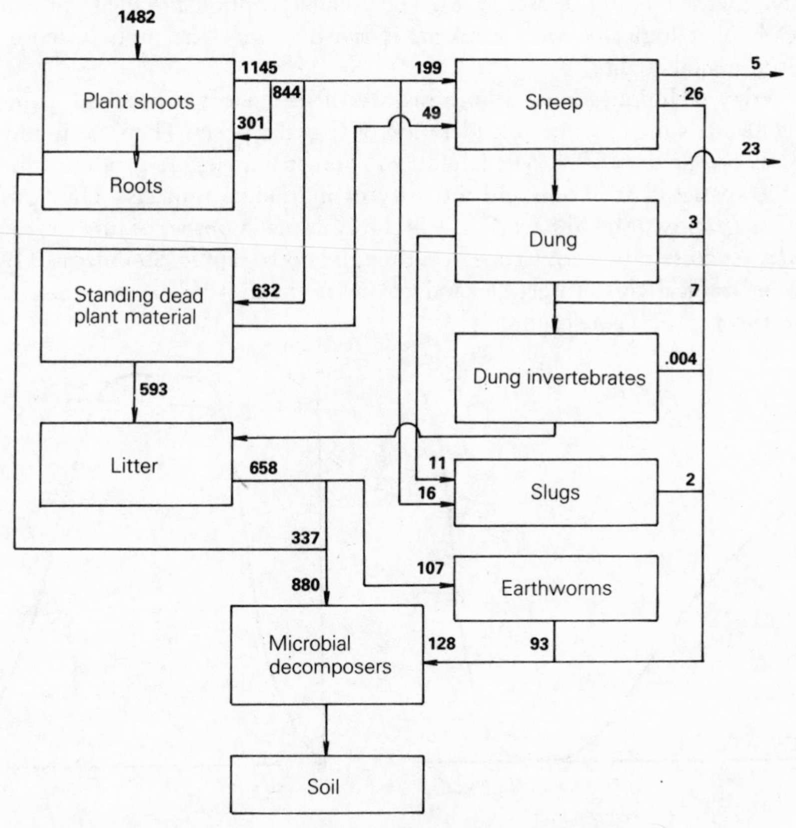

Figure 1.3 Network showing interrelationships in an upland ecosystem. *Source:* Wathern *et al.* (1987).

Condensing information on complex environmental variables into some manageable form is a recurrent problem of assessment. It occurs in various facets of environmental management, for example, in ecological evaluation (Wathern *et al.* 1986). Aggregation, sometimes called weighting and scaling, methods represent a technical fix to this problem (Dee *et al.* 1973, Solomon *et al.* 1977). Advocates favour combining numerical values indicative of individual impacts into a surrogate reflecting overall impact. These methods have two elements, scaling and weighting. In the Environmental Evaluation System (EES), value functions have been concocted to translate the state of individual environmental parameters into arbitrary, environmental quality indices all expressed on the same scale (Dee *et al.* 1973). Figure 1.4 shows two hypothetical curves for landscape features, based upon conventional theories of perception. Each parameter is also ascribed a weight according to its putative importance. Environmental quality scores are multiplied by the appropriate weightings and added to give a total score of environmental quality for each option under consideration. Many subjective elements are subsumed within both the weighting scheme and the value functions. In effect, the basis for a decision is created by the method (Bisset 1978). The preferred option and, hence, the only decision that logically can be taken, is the one with the highest score for environmental quality.

Overlay techniques have a long history of use in environmental planning being ideally suited for the consideration of spatial aspects. Their use in impact analysis pre-dates NEPA (McHarg 1968). Transparencies are produced showing the spatial distribution and intensity of individual impacts. They can be overlain to show total impact (Fig. 1.5). Only a small number of impacts can be overlain successfully, about a dozen, although photographic (Steinitz *et al.* 1976) and hierarchical clustering (Alexander & Mannheim 1962) approaches overcome this practical constraint.

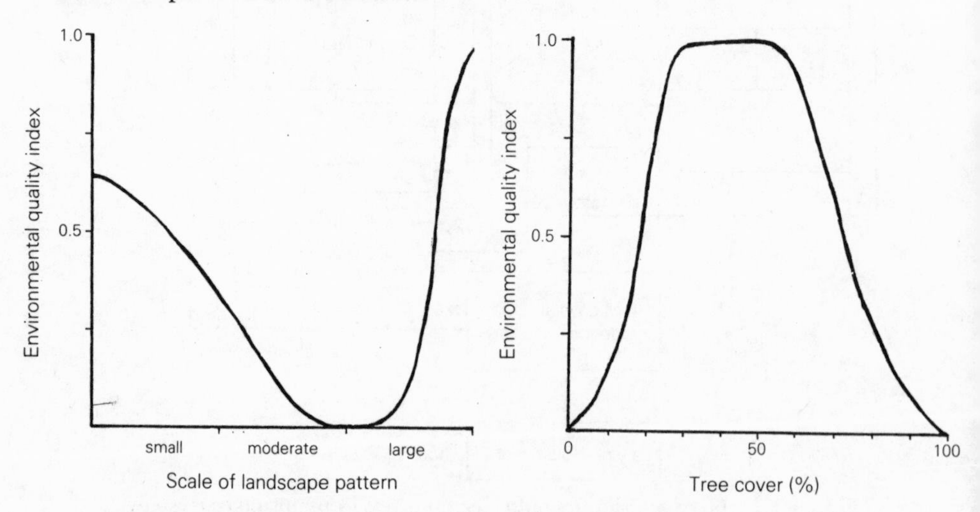

Figure 1.4 Hypothetical environmental quality indices for landscape features.

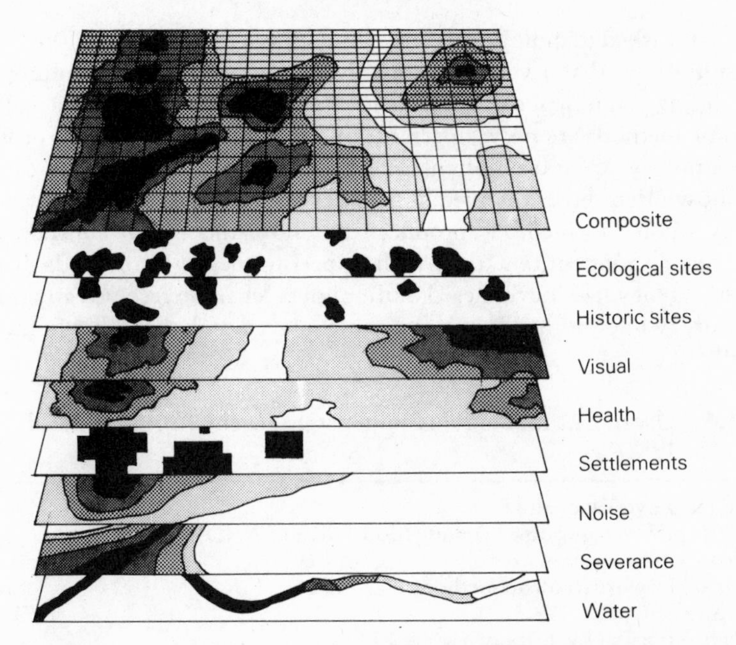

Figure 1.5 The use of overlays to show environmental impacts.

Computer developments have revealed the full potential of overlay approaches. The raw data files can be manipulated, for example by changing weighting values or by aggregating types of impact in various combinations. This is impracticable with manual overlays. In addition, the data can be used with computer-aided design software to select locations with specific siting criteria such as minimum environmental impact. Examples include Krauskopf & Bunde (1972) and Dooley & Newkirk (1976). Potential sources of technical error in using overlays are reviewed in MacDougall (1975).

Adaptive environmental assessment and management (AEAM) is a simulation-based approach to EIA (Holling 1978). Although it is generally described as an EIA method, it differs from all others in that the whole EIA process is subsumed within it. Workshops are used to establish the scope of an appraisal, to identify the key components of environmental systems which may be affected and to determine how these might respond to perturbations. Computer simulations are used to determine the likely outcome of the proposals based upon certain assumptions. Generally, decision makers can be shown these outcomes in a graphical form. One advantage of the approach is that the assumptions can be varied and the simulation rerun repeatedly to show the implications of a range of decisions. The approach has been most widely used for natural resource management programmes and for proposals to exploit natural resources (see, for example, Holling 1978, Walker *et al.* 1978 and Everitt 1983).

Staff from the Corps of Engineers, the Bureau of Land Management, the Forest Service and the Soil Conservation Service were given a list of EIA methods and asked to indicate their use within the agency (Caldwell *et al*. 1982). The results show that a variety of methods are used in EIA, including some long-standing planning techniques such as the goals achievement matrix (Table 1.4). Some methods are not transferable outside a restricted range of proposals and are unlikely to be used often. The large number in the 'other' category reflects how often those involved in EIA develop their own approaches rather than rely upon the methods produced on more theoretical considerations. Indeed, one respondent remarked with respect to the list of methods that staff within his agency had 'never heard of the above, let alone received any training or exposure' while another owned to having 'played with them, but found none that really fit'.

Table 1.4 Use of EIA methods in four federal agencies (from Caldwell *et al*. 1982).

Method	Number
McHarg overlay method	38
Metropolitan Landscape Planning Model (METLAND)	3
Goals achievement matrix	10
Surrogate worth tradeoff method	1
Leopold matrix	25
Environmental Evaluation System (EES)	7
Environmental Quality Assessment (EQA)	8
Environmental Quality Evaluation Procedure (EQEP)	6
Water Resources Assessment Methodology (WRAM)	15
Wetland Evaluation System (WES)	19
Sorensen network analysis	21
Adaptive Environmental Assessment and Management (AEAM)	4
Habitat Evaluation Procedure	75
Decision analysis	9
Kane Simulation Model (KSIM)	1
Other	106

COMPLETION OF AN APPRAISAL

The final decision with respect to project authorization may appear a logical point at which to terminate an appraisal. This, all too frequently, has been the situation in the past. If appraisal is halted at this stage, however, there is no way of knowing whether predicted impacts actually occur. EIA should be characterized by a stream of data collection and analysis running from information on environmental status at the outset, baseline data, through a gradual process of refinement and augmentation during impact prediction to the collection of data on actual impacts. Post-implementation, that is, monitoring, data can be used either to refine the proposal, perhaps by the inclusion of additional remedial measures and the relaxation of constraints found to be unnecessarily restrictive, or to modify the decision. In the most extreme case, it may be necessary to rescind authorization if predictions severely underestimate adverse impacts.

Data collection after proposals have been implemented can also be used to assess the accuracy of EIA. Such audits involve a comparison of the predicted situation with that which actually occurs. The success of an audit, paradoxically, depends upon the thoroughness with which the EIA was originally carried out. In an EIA, potential impacts must be described adequately in terms of their anticipated magnitude, spatial distribution and timing so that accuracy can be assessed.

THE OBJECTIVITY OF APPRAISAL

There is a dilemma in determining who should prepare an EIA. Obviously, a proponent will have more information on the characteristics of a proposal than any of the others involved in EIA. In addition, if EIA is to be fully integrated into project formulation, as is almost universally advocated, responsibility for EIA preparation must lie with the proponent. However, a developer cannot be expected to view proposals completely dispassionately. Without adequate safeguards, proponents may be tempted to regard EIA simply as a means of obtaining project authorization and present only those results which show proposals in a favourable light.

Transferring responsibility for an EIA to a public body, such as the authorizing agency, would divorce EIA from project formulation, a retrograde step. In addition, it would present problems concerning detailed knowledge about the proposals and carry major cost implications. In most countries proponents are required to submit an EIS with an application for authorization. The external checks come during a period of review when technical experts, administrators, interest groups and the public are given the opportunity to comment. In some instances, review panels of independent technical experts may be commissioned by the authorizing agency. Some proponents are able to produce an EIS 'in-house', although the use of outside consultants does introduce an additional element of disinterest into appraisal.

The EIA process

The structure of an EIA process is dictated primarily by the need to accommodate each of the key issues discussed above. Although there may be variations in the detailed procedures adopted within a particular country, most systems, in essence, conform to the pattern shown in Figure 1.6.

From a technical point of view, EIA can be thought of as a data management process. It has three components. First, the appropriate information necessary for a particular decision to be taken must be identified and, possibly, collected. Secondly, changes in environmental parameters resulting from implementation must be determined and compared with the situation likely to accrue without the proposal. Finally, actual change must be recorded and analysed.

EIA has been so widely adopted in project planning that there is a danger that its use will be confined to the appraisal of projects. Many countries had pre-existing systems for the appraisal of development projects which could be

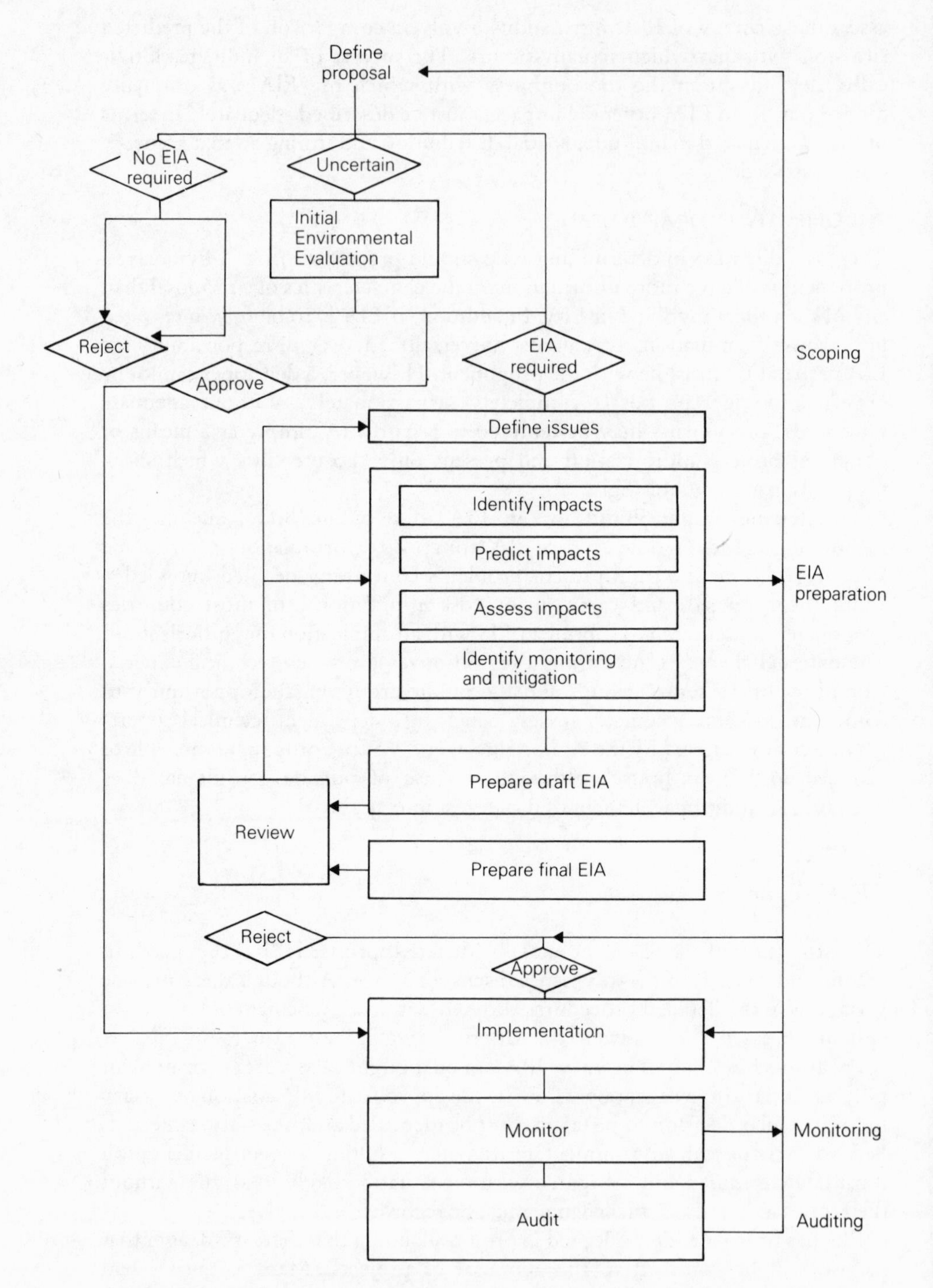

Figure 1.6 Flow diagram showing the main components of an EIA system.

extended into formal procedures. The EIA process summarized in Figure 1.6, however, is equally applicable at other levels of planning.

Experience has shown that certain issues cannot be addressed efficiently at the project level (Clark *et al.* 1981a). Development projects are not generally formulated in isolation. Thus, a proposal to build a nuclear power plant must be set within the context of the policies concerned with future energy supply strategies and the programmes and plans devised to implement them. Similarly, major development proposals often have such profound implications that they dictate the course of future policy. This was the case, for example, with a number of North Sea oil development projects in Scotland.

Lee & Wood (1978) proposed a tiered approach to EIA in order to reflect this structure. However, there seems to be a need for an additional tier between plans and projects in order to accommodate strategic developments which effect change at, at least, a regional scale. Foster (1984) has reviewed examples of such developments.

The nature and practicality of EIA at various levels differ. The policy to project sequence can be regarded as a theoretical hierarchy. Passing up through the hierarchy is characterized by increasing uncertainty and generality. This means, for example, increasing difficulty in determining the informational requirements for appraisal and less precise predictions concerning the consequences of change. These should not be insurmountable problems, because decision makers may have lower expectations regarding precision. Furthermore, progressively longer lead times are subsumed in the hierarchy which will allow feedback between data generation and impact prediction, permitting greater clarification of the areas of uncertainty. This iterative process, for example, will allow policies to be more precisely formulated and their impacts more narrowly defined over time.

If the above sequence can be considered as a hierarchy, application of EIA at the various levels can be regarded as a classic ecological pyramid of numbers. Many thousands of EIAs have been carried out for projects around the world. EIA has been used occasionally in plan preparation, for example, for the preparation of areawide plans in the USA (Department of Housing and Urban Development 1981) and in formulating land-use plans in the Netherlands (in't Anker & Burggraaff 1979) and in the UK (Collins 1986). EIA has been used only rarely in programme development, primarily in the preparation of 'generic' EISs in the USA, such as that concerned with the use of plutonium in mixed oxide fuel in light water reactors (Nuclear Regulatory Commission 1976). With respect to policy appraisal, for example, Wathern *et al.* (1987) were able to suggest only how EIA might be used (Fig. 1.7).

EIA in the decision-making process

The objective of EIA is not to force decision makers to adopt the least environmentally damaging alternative. If this were the case, few developments

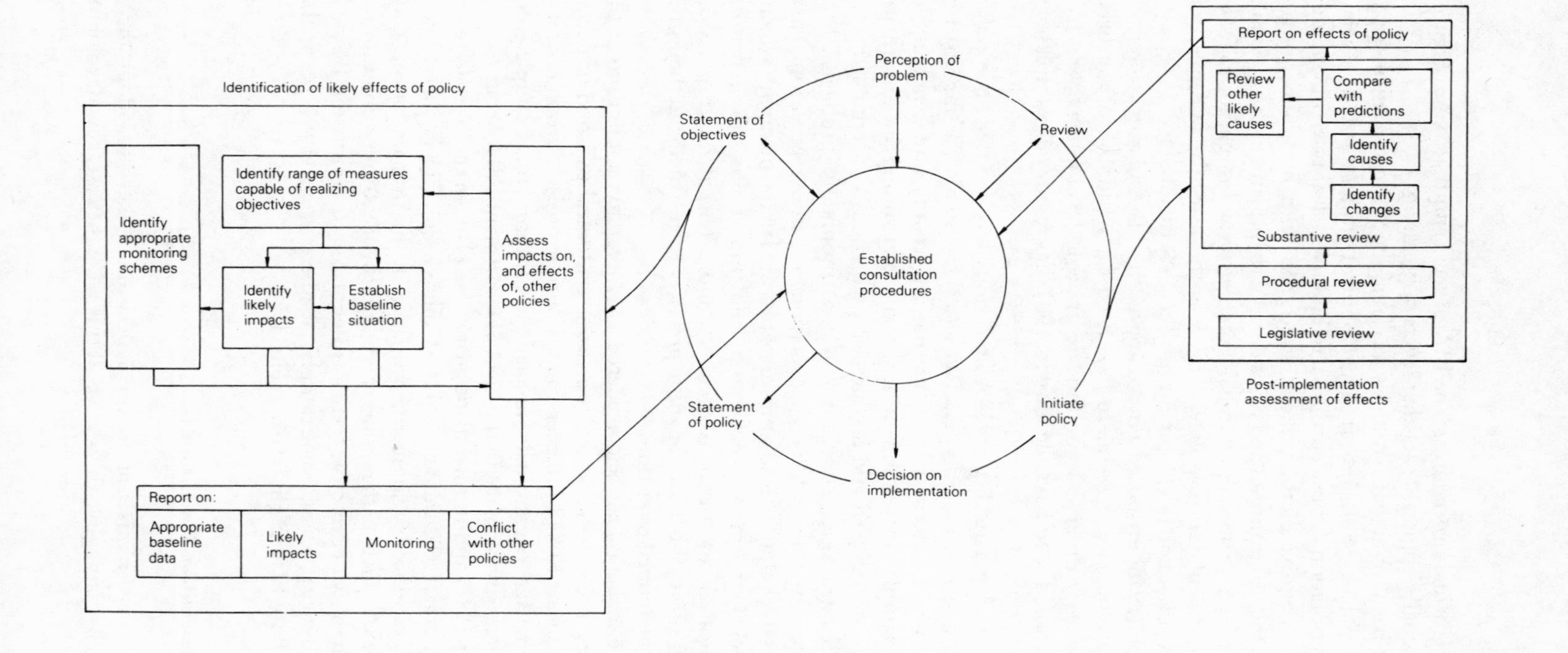

Figure 1.7 EIA in policy appraisal. *Source:* Wathern *et al.* (1987).

would take place. Environmental impact is but one of the issues addressed by decision makers as they seek to balance the often competing demands of development and environmental protection. Social and economic factors may be far more pressing.

Nor is an EIS the only appraisal which decision makers are likely to have at their disposal. Different professions have brought forward other approaches for assessing proposals. In the past, the main consideration of decision makers has been to ensure that the economic benefits accruing from development exceed the costs. The most widely adopted technique for economic appraisal is cost–benefit analysis (CBA). In fact, CBA was the basis for project appraisal prior to the advent of EIA and it is still routinely used. The more sophisticated CBAs go beyond a simple consideration of direct economic issues and also assess indirect aspects. Thus, for example, multiplier effects at a local and regional level figure amongst the benefits, and remedial measures for pollution control amongst the costs. Certain attributes, such as aesthetics, cannot be expressed in monetary terms without some arbitrary manipulation. More comprehensive CBAs carry some consideration of such 'intangibles' and 'unquantifiables'. However, other effects on environmental media which act as pollutant sinks, such as the atmosphere or watercourses, are regarded as externalized costs and are excluded from the analysis.

Risk assessment (RA) is of particular interest as, in many respects, it parallels EIA. Both are concerned with the likely consequences of environmental change. RA is frequently used to assess the probability and likely consequences of a particular catastrophic event, such as an explosion, associated with a hazardous installation. It may also be used to assess policy issues such as the implications of introducing a novel chemical, such as a new pesticide.

Risk assessments tend to be highly numeric appraisals; they are essentially statistical analyses of likely events based upon certain probabilities of occurrence. The use of RA has certain implications particularly for non-numerate decision makers. First, it suffers from the same potential weakness as all quantitative predictions of change. Non-quantified parameters must either be forced into a numerical guise, based upon arbitrary considerations, or ignored. These subjective elements are generally obscured within the analysis. Robustness, that is, the dependence of the outcome of an analysis upon the assumptions built into it, may not be determined. Secondly, decision makers may treat such highly quantified assessments reverentially, affording them greater credence than is warranted and weighting them more highly than more descriptive treatments of likely impacts.

As decision makers may be dealing with a disparate range of information on which to base a decision, integration has been a recurrent problem. Economic and environmental analyses, for example, tend to be treated separately. In fact, they are closely interrelated. Nijkamp (1980) has proposed a framework for relating potential economic, social and environmental change (Fig. 1.8).

Most detailed analyses of the procedural and political implications of EIA are based upon US experience. A number of factors combine to give the United

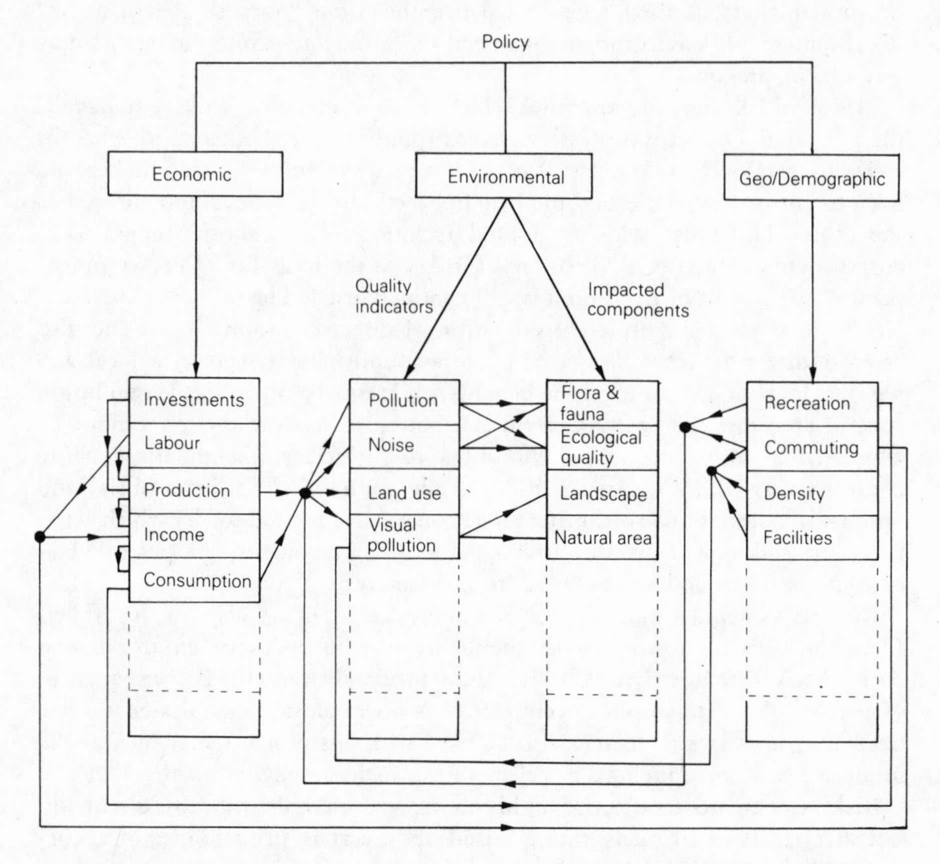

Figure 1.8 An integrated approach to project appraisal. *Source:* Nijkamp (1980).

States this prominent position. The bias results from the fact that EIA is more firmly established, has been implemented for a longer period and has been adopted for a wider range of proposals in the USA than anywhere else. Canter (1984), for example, reports upwards of 15 000 full federal EISs and in excess of 200 000 initial evaluations up to that date. There is also a more sophisticated view of the function of EIA within the decision-making process. It is hardly surprising, therefore, that concern in the USA has focused increasingly on making the system more efficient and more effective, at a time when people in other countries have been more intent on ensuring merely that EIAs are carried out. Recently, experience from other countries, notably Canada and the Netherlands has started to make an impact upon thinking about EIA.

The dominance of US experience is likely to generate three responses. First, US procedures will be considered definitive and transferred into systems ill designed to accommodate them. Secondly, because of the unusual administrative system that exists in the USA, all experience will be deemed nontransferable. Finally, the adoption of some aspects of the US system, which seems the sensible approach, will be rejected because of fears that the unwanted features will also appear. This can be the only possible explanation of, for example, UK paranoia over mandatory EIA. It was assumed that the adoption of formal procedures for the preparation of EISs would lead inexorably to the vast litigation associated with EIA in the United States. As a consequence, much of the valuable experience from the USA has had to be learned anew in other countries.

EIA LITIGATION

In a review of US experience, NEPA has to be the starting point of any analysis. The legislation was enacted by the Congress in recognition of the need for care in the use of the country's natural resources. It is important to appreciate that the federal government, through its various agencies, plays a crucial custodial role in the management of natural resources. Approximately 738 million acres (375 million hectares), about a third of the country, are the responsibility of the federal government and Congress has the authority to determine how these public lands are used. Enacting NEPA sought to reverse a 'clear and intensifying trend toward environmental degradation' and to remedy the lack of 'environmental awareness of many federal agencies whose policies were in conflict with the "general public interest"' with its main function 'to hold the federal government accountable as trustee for the protection of the American environment' (Holland 1985).

The provisions of the Act are now well known and need only be outlined here; good analyses of the evolution of NEPA and its provisions can be found in, for example, Anderson (1973), and Heer & Hagerty (1977). In brief, there are three main elements. First, a general policy for the environment, long on rhetoric and aspiration but short on concrete measures is enunciated. Secondly, Section 102 (2) (C) requires the preparation of an EIS for 'major federal actions significantly affecting the quality of the human environment'. Finally, the Act established the Council on Environmental Quality (CEQ) to administer the provisions of the legislation (*U.S.Publ.L.91–190,42,U.S.C., 4321–4347*). Clearly, the EIS requirements were the action forcing provisions which acted as the 'two by four which got the government mule's attention' (Anderson 1973).

NEPA details the points to be addressed in an EIS. These are: the environmental impact of the proposed action; adverse environmental impacts which cannot be avoided; alternatives to the proposed action; the relationship between short-term use and the maintenance and enhancement of long-term productivity; and any irreversible and irretrievable commitment of resources. The intent is clear. Agencies were required to switch their assessment process from a limited economic accounting review to a broader analysis embracing non-monetary environmental considerations. The ways in which this change was to be brought about were specified more fully in EIS guidelines published by the CEQ in 1971 and 1973.

Understandably in the USA, a country where recourse to law is often the first, rather than the 'last resort', this piece of vaguely worded legislation has been the subject of litigation. Many of these cases can be readily explained. The responsibilities of federal agencies needed clarification. The combination of NEPA, CEQ guidelines and the guidance issued by individual agencies generated considerable regulatory redundancy, conflict and inconsistency, where even some agency officials were uncertain as to what their responsibilities were (Legore 1984).

Primarily, however, it has been the interpretation of the Act's provisions which have needed clarification. Litigation has clearly established that agencies must give appropriate consideration to the environmental consequences of their actions in the decision-making process. The courts have ruled that the intent of the Act is to ensure that the public is provided with complete and accurate information about the environmental consequences of such actions. As such, NEPA has been described as a 'full-disclosure' law, placing a responsibility on federal agencies to investigate fully and to reveal the likely consequences of their actions. The crucial issue, however, remains whether the provisions of the Act are substantive, that is, compelling agencies to adopt the least environmentally disruptive option, or procedural, requiring agencies only to comply with the procedures specified. When addressed in a previous review of US experience written in 1976, included in Clark *et al.* (1978), this issue was considered unresolved. Holland (1985) regards this question as still 'not settled'; the case law quoted by Holland, however, implies a balance of argument in favour of a procedural stance. This would be an important interpretation as it would support the contention that the responsibility for decision making lies with the agencies, not with the courts.

EISs have also been the subject of much litigation. Many decisions not to produce an EIS for a particular development and the adequacy of many completed EISs have been challenged in the courts. Case law concerning individual EISs has helped to clarify the threshold criteria that trigger the preparation of EISs, namely 'major federal actions' and 'significant impact'. The courts have determined that an adequate EIS is one that contains 'sufficient detail to ensure that the agency has acted in good faith, made a full disclosure, and ensured the integrity of the process' (Holland 1985).

Judicial review is an ongoing process in the USA with the courts progressively clarifying NEPA's provisions. Interpretation of the requirement to consider alternatives, for example, is an evolving one, but certain pointers are beginning to emerge. Amongst the most important considerations are that agencies should: consider the option of doing nothing; consider alternatives outside the remit of the agency; and consider achieving only a part of their objectives in order to reduce impact.

There has been much EIA litigation in the United States. Kennedy (1984) quotes a total of 1602 NEPA-related lawsuits up to that time, representing almost 10 per cent of the total number of federal projects for which EISs had been prepared. Almost 40 per cent of suits are filed by environmental groups (Canter 1984). Much of this litigation can be regarded as legitimate in that it has helped clarify definitional issues. Challenging either the lack of an EIS or its adequacy, however, has been seized upon as a means of stalling development. Public interest environmental groups and, to a lesser extent, individuals have successfully delayed development in the courts and have scored notable victories, even with the abandonment of some projects. This may seem an abuse of the right to judicial review. It is clear, however, that those with grievances feel that they are not adequately addressed by the system and are utilizing the only means of expression available. This implies that the review system is inadequate.

Europeans point to the apparent high time and cost penalties of litigation as a fundamental weakness of the US system. It does ensure, however, that decisions are eventually taken. UK experience is no more reassuring. The Sizewell 'B' inquiry, the most recent of a series of protracted public inquiries, sat for over two years hearing evidence and even 18 months later had not reported. Similarly, Kennedy (1986) from a comparative study of highway planning in the USA and the Federal Republic of Germany, concluded that US EIS procedures eventually facilitate development. In contrast, West German environmental groups have been able to delay projects indefinitely in the absence of formal procedures. UK protestors have also been able to impede road projects for long periods at public inquiries.

EFFICIENCY OF THE EIA PROCESS

Criticisms of EIA tend to fall into five categories (Kennedy 1984). They are deemed to have little effect upon the decision-making process, few tangible environmental benefits and inadequate opportunities for public involvement as well as being costly and a source of delay. The available evidence, however, seems to refute some of these assertions. A study of EISs produced in connection with the programme for waste-water treatment facilities of the Environmental Protection Agency (EPA) showed that there were significant changes to projects during the EIA process with marked improvements in environmental protection measures (EPA 1980). This contrasts with early experience. The lack of a 'grandfather clause' exempting projects already under way resulted in 3635 EISs being submitted to CEQ in the first two years of NEPA. Under these circumstances, EIA for many projects became merely an add-on, rather than an integral element of the planning process and could effect little change.

The EPA study also showed that EIA gave net financial benefits. The costs of EIA preparation and any delays were more than covered by the savings accruing from modifications to individual projects identified during the analysis.

Superficially there appears to be ample opportunity for public involvement. More than 95 per cent of EPA projects involved public meetings. The high incidence of litigation discussed above, however, suggests that people consider such meetings an inadequate mechanism for incorporating their concerns into a project.

The consideration of delay is a more complex issue as it depends to a large extent upon the quality of an EIA. Delay is an inescapable consequence of litigation resulting from an inadequate EIS which may halt implementation for more than a year (Liroff 1985). When reduced to an add-on process, the time spent on EIA can only be regarded as a delay directly attributable to the system. The EIA is also likely to be poor. The time spent on EIA should run parallel to other activities and, therefore, should be subsumed within the overall process which, for major proposals, is likely to involve long lead times.

There have been few investigations of the quality of EISs. Since 1975, however, EPA has rated both draft and final EISs submitted to it for review (EPA 1975). Between 1975 and 1982, 91% of all final EISs were in the top two categories devised by EPA. Over the same period, the comparable figure for draft EISs rose from 59% to 76% (Kennedy 1984). Although Kennedy cautions against affording these figures too much significance, they remain amongst the few independent data which are available. Culhane & Armentano (1982) concluded that the coverage of acid rain in power plant EISs had improved, but that the discussions left 'something to be desired'.

The high probability of litigation, approximately a 1 in 10 chance of a suit being filed against an EIS, encouraged agencies to adopt an 'encyclopaedic' approach to EIA, particularly in the early years of EIA. This approach involved putting every available scrap of information concerning the proposals and particularly the local area into an EIS. This device was an attempt to build a defence against the charge of inadequacy into an EIS. Thus it could always be claimed that every aspect included in an EIS had been 'considered'. In the early 1970s, there were some spectacularly vast multi-volume EISs because of this phenomenon.

In 1976, CEQ published the results of an investigation into the operation of the EIS system (CEQ 1976). On the basis of this review, CEQ proposed major amendments to the process, codified in a set of regulations. These were aimed at decreasing the volume of material presented in an EIS, reducing delay and producing more environmentally sensitive decisions (CEQ 1978). The regulations became operational during 1979. The major changes were: the introduction of scoping; limits on the length of EISs; stress on the role of EIA in the decision-making process; and an emphasis on a scientific and interdisciplinary approach to EIA.

Scoping has been discussed earlier in this chapter. Adopting a scoping process should ensure that inconsequential aspects are omitted, while the involvement of the public should reduce the likelihood of litigation. The new procedures should ensure that a succinct document is produced, not only by restricting the range of issues considered, but also by the imposition of a 150-page limit (300 pages for complex proposals), on the length of an EIS.

The modifications effected by the CEQ regulations are aimed at providing decision makers with concise, pertinent information on which to base a decision. The courts have long upheld the view that the content of an EIS is only one of the factors considered in decision making (Holland 1985). The CEQ regulations, however, aim to make decision makers, at least, responsive to an appraisal. A final EIS must be accompanied by a record of decision detailing how the EIS was used in arriving at a decision. In addition, it should indicate which alternative is preferable on environmental grounds and other relevant factors, such as economic issues and national policy considerations, which also influenced the decision.

EFFICACY OF EIA

The last major issue which began to attract attention in the United States at the end of the 1970s concerned the overall effectiveness of NEPA. Fairfax & Ingram (1981) have argued that any success which NEPA has achieved cannot be attributed to the legislation itself, which is poorly and imprecisely worded. Rather, it is a reflection of the preparedness of environmental groups to use the courts to make agencies respond to its provisions. Notwithstanding whether the legislation is the proximal or ultimate cause of reform, many agencies appear to have become more responsive to environmental issues. The situation is not one of unmitigated success. Some agencies such as the United States Agency for International Development were reluctant to comply with the legislation but were compelled to do so by the courts (Horberry 1984). However, when environmental groups successfully delayed the 'development' activities of the Bureau of Land Management pending compliance with NEPA, powerful cattle interests were able to ensure, through the Congress, that improvements on public range lands still took place (Fairfax & Ingram 1981).

Fairfax (1978) has gone beyond this to argue that environmental groups have fought over the wrong issues. The characteristics of the legislation have persuaded environmental groups to argue lack of procedural compliance and not to contest the substantive merits of a particular case. This approach has ensured that the preparation of an EIS defensible in the courts has become the objective of EIA rather than the means for making environmentally sound decisions.

Only qualified support for this view comes from a survey of agency practice conducted by Caldwell *et al.* (1982). This study, commissioned shortly after the CEQ regulations were adopted, provides some insight into the effectiveness of NEPA at that time. The main conclusions from the study were that political and institutional changes were more important than technical improvements in increasing the efficacy of EIA, while the attitudes of personnel largely shaped agency implementation of the provisions of the Act. Staff most directly involved in EIA had a positive attitude towards NEPA and considered its policy elements to be more important than the EIS provisions. The study revealed that NEPA seemed to have broadened the base of technical advice available to decision makers, to have encouraged a more interdisciplinary approach to assessment as a result of the influx of personnel with new skills. In certain agencies there was evidence that the fundamental change in decision making aspired to in NEPA had been achieved. These observations do not concur with Fairfax's view. However, certain participants in the study considered that some agency legal and executive staff still regard the primary function of an EIS to be a means of avoiding litigation. Such staff may exert a disproportionate influence on agency practice, because of their position close to the centre of policy making.

In view of the high incidence of litigation associated with EIA in the USA, alternative means of resolving environmental disputes have been investigated. Since 1974, there has been a marked growth in the use of negotiated resolution of disputes through an independent mediator. The process has even become embodied in law in some states in the USA. Bingham (1986) has established that agreement between the parties is eventually reached in a majority of cases related to both policy and site-specific issues. Although high rates of implementation of negotiated agreements on site-specific issues are generally achieved, these are substantially lower for policies.

The apparent success of environmental mediation does not portend the demise of EIA. Under present statutes there will still be a need for EIA, even for mediated proposals. It may merely eliminate delays resulting from litigation, without necessarily speeding up the overall process. Mediation is time-consuming and implies a considerable commitment of resources over the period of negotiation. Environmental groups may find litigation a comparatively low-cost alternative. In addition, such groups may be reluctant to give up a familiar process which in the past has afforded them a prominent position in the media, as well as some notable successes.

Conclusions

From this brief overview it is apparent that EIA is now fully embodied in the decision-making process in many countries. Having been adopted by the bureaucracy its future seems assured. Experience of EIA application has expanded rapidly, but unequally in the years since the United States adopted EIS provisions in 1970. Consequently, EIA is most highly evolved in the USA, while elsewhere, with the possible exceptions of Canada and the Netherlands, experience lags behind by as much as a decade. Thus, considerable evolution of EIA practice is likely in many countries in the coming years, however, US experience should not be used to predict the future state of EIA around the world as few countries will elect to follow the USA slavishly.

One function of EIA is to provide decision makers with an indication of the environmental consequences of the options open to them. Environmental issues, however, rarely form the sole basis for a decision related to the implementation of a particular set of proposals. Politicians may perceive a pressing need for economic development, jobs and revenue generation or for remedying some social ill as an overriding consideration despite consequent environmental degradation. Thus, the case for development often seems overwhelming. Nor must it be assumed that development and protection of environmental quality are necessarily conflicting. Indeed the converse may be true, as experience from some LDCs clearly points to the serious environmental impact of poverty. Even when sanctioning a development appears the only decision which can be countenanced, applying EIA may still yield benefits. An EIA may reveal other ways of achieving the same objectives, but with less environmental disruption. In addition, there may be economic benefits from using EIA. Mitigating measures identified during EIA may be incorporated more economically at the design stage than subsequently.

EIA can also be an aid to design, but this can only be achieved if it is an integral component of project formulation. As such, EIA is a technical process bound up with the identification, investigation and refinement of alternative options. Further development of this capability seems to be the most logical step in the evolution of EIA. The use of EIA has gone through two stages of development. First, EIA has often highlighted conflict. The formal identification of potential impact tends to focus attention on the unavoidable consequences of a proposal which become the basis for conflict between developers and protestors. Secondly, EIA integrated into project formulation offers a means of resolving potential conflict without the involvement of a mediator. A residual impact which cannot be designed out of a proposal, however, is always likely to remain. Thirdly, therefore, EIA may achieve a more dynamic characteristic in the future by indicating any compensatory measures that could be adopted by the proponent to offset these residual impacts.

In this introductory chapter, the duality of EIA as science and art has been mentioned, but each of these attributes could be given no more than perfunctory discussion. This theme, however, is taken up and explored in more detail in the following chapters of this book. The contributions are grouped into four parts.

Part II contains five chapters concerned with the mechanics of EIA. Gordon Beanlands's chapter on scoping and baseline studies is based primarily on Canadian experience. Ron Bisset describes recent advances in the development of EIA methods. The management of uncertainty is a recurrent problem in EIA and Paul de Jongh reviews the approaches that have been developed to deal with this issue. Risk assessment has many common features to EIA and essentially deals with the same problem, namely the likely consequences of change. Richard Andrews discusses the ways in which the two approaches impinge upon one another and how their future evolution could be aided by the transfer of experience between them or, in some instances, by fusion into a unified process. Chris Wood discusses the application of EIA in plan formulation.

Part III deals with the efficiency of EIA. In the first chapter Ron Bisset and Paul Tomlinson discuss EIA monitoring requirements and how such data can be used to assess the accuracy of impact predictions, a process known as auditing. Process evaluation, effectively an audit on the whole EIA process, is discussed by Barry Sadler, based mainly upon Canadian experience. Finally, the efficacy of EIA ultimately depends upon the availability of adequate numbers of well qualified personnel. Norman Lee reviews how adoption of the EC EIA directive will affect personnel requirements and the likely implications for training.

EIA in different parts of the world is reviewed in Part IV. An attempt has been made to take a broad perspective in these discussions, rather than merely cataloguing EIA legislation and procedures. Geoffrey Wandesforde-Smith and Joanne Kerbavaz review recent US experience at both the federal and state levels. The EC EIA directive provides a nice example of the problems of adopting supra-national policy. The problems resulted in a protracted period being spent agreeing what finally turned out to be quite minimal provisions. The centrally planned economies of Eastern Europe have yet to formally adopt EIA. Anna Starzewska shows that the necessary provisions exist to make its adoption practicable, while some countries are slowly acquiring the experience necessary to do so. Nay Htun's discussion of Asia and the Pacific region shows how diverse this area is. Experience is discussed with respect to the application of various facets of EIA within the region. Finally, in this part Iara Moreira analyses the EIA provisions that exist throughout Latin America. The lack of any discussion on Africa in this regional perspective sadly reflects the dearth of information on a continent currently facing considerable problems which development, based upon proposals adequately assessed through the use of EIA, could play a role in alleviating.

The final part comprises both EIA as science and art. Environmental health impact assessment (EHIA) is becoming increasingly important in EIA primarily at the initiative of WHO. Eric Giroult describes EIA within the context of WHO policy and outlines an overview of EHIA. Bill Kennedy's chapter reviews the efforts made by OECD to formulate guidance on EIA in development assistance for its constituent member states. Finally, John Horberry discusses the ways in which the development assistance programme of the United States Agency for International Development (USAID) was made to comply with the provisions of NEPA and hence ensure that, at least for US aid projects, developments in LDCs would be subject to EIA even in the absence of national provisions.

Part II

METHODOLOGICAL ASPECTS OF EIA

2 *Scoping methods and baseline studies in EIA*

G. BEANLANDS

Introduction

Scoping and baseline studies are activities that are undertaken at early stages in an environmental impact assessment (EIA). It is difficult to overemphasize the importance of these activities since the success of an EIA will depend largely upon how well they are conducted. Scoping refers to the process of identifying, from a broad range of potential problems, a number of priority issues to be addressed by an EIA. In other words, it is an attempt to focus the assessment on a manageable number of important questions. Baseline studies, in turn, are designed to provide information on the issues and questions raised during the scoping exercise.

The importance attached to both scoping and baseline studies arises from the fact that environmental assessments are almost always conducted under serious limitations of time and resources. Any priority-setting activity, therefore, should improve efficiency and provide a more focused product for decision makers. In this paper, scoping and baseline studies are discussed mainly with respect to biological components. It should be stressed, however, that similar considerations relate equally to other environmental attributes.

Scoping in EIA

The term scoping has recently appeared on the environmental impact assessment scene as a result of the 1979 regulations under the US National Environmental Policy Act (NEPA) which require lead agencies to undertake 'an early and open process for determining the scope of issues to be addressed and for identifying the significant issues related to a proposed action' (Council on Environmental Quality 1980). The agencies should achieve this objective through careful consideration of existing information relevant to the assessment as well as the organized involvement of other agencies and consultations with the general public.

This is a somewhat belated recognition of the need to establish clearly the focal point for an assessment at the outset; failure to do so severely limits the probability of obtaining useful and credible results. Scoping, in effect, provides a means whereby the public has a role in translating the policy wording of NEPA, that is, 'restoring and maintaining environmental quality for the overall welfare and development of man', into a tangible specification for each individual impact assessment.

There is no sure way of anticipating the concerns of the general public, if for no other reason than that social values change with time. Although the results can verge on the philosophical, social scoping may help to define the concepts that become formal requirements for impact assessment. For example, consider the following quote from the US Atomic Energy Commission's (USAEC) Directorate of Regulatory Standards:

> A species, whether animal or plant, is 'important' (1) if it is commercially or recreationally valuable, (2) if it is rare or endangered, or (3) if it affects the well-being of some important species within criteria (1) or (2) above, or (4) if it is critical to the structure and function of the ecological system.
>
> (USAEC 1973)

As Eberhardt (1976) noted, it is virtually impossible to translate phrases like 'well-being' into an operational focus for a study. In addition, in most cases the structure and function of natural systems are not understood, so that it would not be possible to determine the critical nature of various components.

With respect to biological components of the environment, social scoping, to be useful as an operational guide, is often expressed as the plant or animal species perceived by society to be important. Thus, among other more ecological criteria, Cairns (1975) used commercial, recreational and aesthetic values as some of the bases for establishing a list of critical species. Similarly, Truett (1978) established the focus for a major impact research programme on 'key species' which were defined on the basis of abundance, as well as their commercial, recreational and food value to man. He considered that there was

> good reason for concentrating research on species considered to be of immediate value to society. The reason relates both to the difficulty of assigning environmental value to species not useful to man and to the fact that species with little value are of little concern to decision-makers. And, lest we forget, the ultimate purpose of an assessment study is to influence decision.

Two publications which provide the most detailed technical direction to those undertaking impact assessment (States *et al.* 1978, Fritz *et al.* 1980) have both treated social and economic values as major factors in narrowing the range of ecosystem components which should be considered. Thus, in summary, scoping might be defined as a 'very early exercise in an EIA in which an attempt is made to identify the attributes of components of the environment for which there is public (including professional) concern and upon which the EIA should be focused' (Beanlands & Duinker 1983).

THE QUESTION OF SIGNIFICANCE

In some respects, adopting a definition for significant impact represents an initial attempt to reduce the scope of assessment studies to the most important potential effects. Any consideration of the significance of environmental effects must acknowledge that environmental impact assessment is inherently an anthropocentric concept. It is centred on the effects of human activities and ultimately involves a value judgement by society concerning the significance or importance of these effects. Such judgements, often based on social and economic criteria, reflect the political reality of impact assessment in which significance is translated into public acceptability and desirability.

In this context, the ecological implications of a proposed development usually get translated into effects on physical and biotic resources valued by man for commercial, recreational or aesthetic purposes. From the perspective of an ecologist, more profound changes to the intrinsic structure and function of natural systems may be involved, but their significance probably will be evaluated by the public in terms of the implications for such resources. In effect, ecologists involved in environmental impact assessment are often required to extend their interpretation of impacts beyond the limits of professional interest and to emphasize those environmental attributes perceived by society to be important.

While the detailed results of a scoping exercise will depend upon the specific nature of the project under consideration, a number of themes seem to recur. The primary concern of the public with respect to environmental matters is human health and safety. All others will be subordinate when man's health is in jeopardy as a result of a proposed development. The public will have a great concern for potential losses of important commercial species or commercially available production. The converse would hold true regarding an increase in the numbers of undesirable species. Society can be expected to place a high priority on species of major recreational or aesthetic importance, whether or not they support commercial activities of any consequence. Special interest groups will usually gain broad support in their concern for rare or endangered species on the basis that mankind has special custodial responsibilities regarding their preservation. Finally, the public can normally be expected to be concerned over habitat losses which represent a foreclosure on future production. In all of these cases, public concern will be heightened in relation to perceived imbalances between supply and demand of species or habitats within a local, regional, or national context.

This approach was adopted in the assessment report for the South Davis Strait offshore exploration programme in Canada (Imperial Oil Ltd. *et al.* 1978) where significance was taken to include reductions in populations of species of subsistence or commercial importance to local users. Likewise, a company representative indicated the regular use of a simple scoping exercise to focus the assessment study effort. This included four categories of species – commercially important, important as indicators, ecologically important and species high in the trophic structure. Some attempt was made to include a few species from each category in the impact assessment studies.

There is growing concern about the need for social scoping very early in the assessment process. Recent hearings to discuss the draft assessment guidelines for the Beaufort Sea hydrocarbon production proposal in Canada can be considered as a scoping exercise. It is not apparent, however, from the final guidelines (Beaufort Sea Environmental Assessment Panel 1982) that the exercise was entirely successful. The document directs the proponents to discuss the biological environment ranging from micro-organisms to mammals. While later sections suggest that studies should be limited to effects 'that are deemed to be significant', it is only at the end of an appendix to the guidelines that the true meaning of this becomes apparent. Significance is defined as affecting 'species that at present are of direct value to society such as those that may be considered rare or endangered or important for subsistence, scientific, commercial or recreation value'.

SOCIAL SCOPING AND SCIENTIFIC ENQUIRY

The compromise struck between the subjectivity of value judgements and the objectivity of the scientific approach is largely a function of the relative importance of the role of science at various stages in the sequence of impact assessment activities (Fig. 2.1). There seems to be a consensus that initially some direction, explicit or implied, must be given to the scientific pursuits. The logic sequence in providing such direction is considered to involve impacts perceived to be socially important; socio-political decisions required; technical questions posed; and scientific answers attempted. Thus, the initial major role of value judgements in establishing a focus for the assessment is gradually replaced with a scientific programme of investigations to address the social concerns.

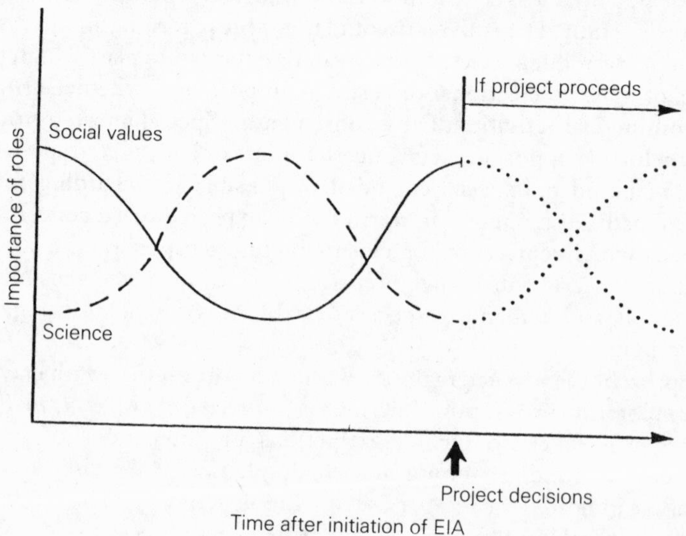

Figure 2.1 Changes in the relative importance of science and social values in EIA.

This translation of social concerns into scientific investigations is fraught with moral, conceptual and operational difficulties for many scientists. It is not surprising that dedicated scientists feel professionally constrained when they are expected to focus their expertise solely on social concerns which often change with time. Furthermore, it is often difficult to conceptualize scientifically the public's perception of an environmental problem; impacts on attributes of aesthetic value are a prime example.

Eventually, the pre-project scientific studies must be concluded and the results presented to those responsible for making project-related decisions. At this stage in the process, the importance of social value judgements may outweigh scientific considerations. Although the implications can be frustrating to scientists involved in environmental impact assessment, the fact remains that project decisions will reflect some compromise between social aspirations and the results of scientific enquiry.

In theory, the role of the scientist will once again dominate in the design and implementation of post-EIS monitoring programmes. The same range of problems is posed as in pre-project studies. However, there is greater opportunity to apply a quantitative approach in measuring changes than in predicting them.

SCOPING METHODS

Determining the priority values of society with respect to the potential effects of a particular development proposal is a major concern. There are various issues which are of concern irrespective of the approach to scoping which is adopted. First, it is extremely important to define clearly the segment of 'society' involved, that is, the target population. Once identified, the target population must be given adequate information about the project and the potential environmental effects in a format which they can understand. In addition, the target population must be given enough time to organize its thoughts and ideas regarding the potential environmental problems. Finally, the mechanism whereby the target population is able to voice its concerns to the decision makers must be clearly understood. Meeting these requirements entails advanced planning, the involvement of competent staff and access to adequate resources.

There are various direct and indirect ways of conducting a scoping exercise. Regardless of which approach is used, however, it always should occur very early in the EIA process. Under NEPA regulations the sponsoring federal agency is required to have meetings involving all parties directly affected or interested in the proposed project. At these sessions, the participants are encouraged to present their concerns about the project and an attempt is made to define the priorities in these perceived problems.

A similar approach based upon public meetings has been used recently under the Canadian Federal Environmental Assessment Review Process (EARP). In this case, prior to formal EIA hearings, smaller community-based meetings are held at which local residents and other interested parties are given the opportunity to discuss their concerns in the presence of the assessment panel and representatives of the industrial proponent. Also in attendance are spokespersons for government agencies and research establishments who also make representations to the panel. All of this interchange takes place in a non-adversarial forum in which the participants do not have to defend their concerns. Ideally, the sessions should lead to increased understanding about the potential environmental effects and clarify the issues perceived by the community at large.

The Canadian scoping sessions are preceded by the distribution of written material on the project as well as small informal sessions where people are encouraged to 'drop in' and learn more about the project. In one case, the scoping programme for a proposed nuclear power plant, a scientific advisory group was also formed which added to the development of priority concerns from a technical perspective. In complex projects, this scientific contribution may be an important aspect of the overall scoping process since many of the potential impacts may be beyond the understanding of the general public.

The advantage of the scoping meeting is that it affords an opportunity for an open dialogue between those responsible for the EIA and the public whose interests they are supposed to represent. Such open discussions can often lead to a resolution of perceived problems which are based upon misunderstandings. The disadvantages are that it is time-consuming, requires financial and manpower resources and needs the full co-operation of the industrial proponent.

If it is not possible to hold scoping meetings, questionnaires or surveys may be used to assess public concern. This indirect approach to scoping is less desirable for a variety of reasons. First, the return rate on surveys is normally quite low and it may be biased towards the more vocal segments of the community. Secondly, proper survey design requires experts in the field who may not be available at the time required. Third, the analysis and interpretation of survey results are subject to professional disagreement which may further confuse the issues rather than clarifying them.

When dealing with isolated or reasonably well-defined target populations, it may be possible to assist the community itself to conduct its own scoping programme. By providing financial support or organizational skills, those responsible for conducting an EIA may be able to encourage local people to use existing communication mechanisms to determine community concerns. Where this can be done, it is highly recommended since it increases the credibility of the resulting advice to government and proponents.

In the normal sequence of events a scoping exercise, by whatever method, would provide a list of priority concerns. These, in turn, would be incorporated into guidelines for the preparation of an environmental impact statement (EIS). Depending upon the nature of the priority issues identified, the baseline study programme undertaken as part of the EIS should be structured around the results of the scoping exercise. Thus, the scoping programme may have a major influence on the focus of the entire EIA and, therefore, upon the advice given to decision makers.

Baseline studies in EIA

Baseline studies are perhaps the most commonly recognized, and yet least understood, element of EIA. The term usually refers to the collection of background information on the environmental and socioeconomic setting for a proposed development project and it is normally one of the first activities undertaken in an EIA. From the above discussion it can be seen that a baseline studies programme may be designed around the results of a scoping exercise. Whether it involves the collation of existing information or requires the acquisition of new data through field sampling, baseline studies frequently account for a large part of the overall cost of an EIA.

Environmental impact assessments are often conducted under severe time restrictions and, since it is relatively easy to collect information and data, there is a tendency to give too much emphasis to baseline studies early in the assessment process. The result is that there is often a great deal of information made available on the environmental setting of a particular project, but it may be irrelevant to the resolution of certain critical questions raised at later stages in the EIA.

There is an almost universal problem with baseline studies as applied in EIA; that is, they are undertaken without clearly defined objectives. Seldom is there an understanding of why data are being collected or to what problem they will be applied. In order to cover all potential requirements an effort is made to gather some information on all aspects of the environment. This inevitably leads to superficial surveys which provide only reconnaissance-level information. In the end, much of the investment in time and resources is wasted.

Perhaps the most glaring inadequacy of many baseline studies is that they do not reflect the ultimate needs of the decision maker involved in project planning. During the planning of development projects there are key decision points for which important environmental and socioeconomic data should be available from baseline studies. If these critical stages and the related information needs are not clearly defined at the beginning of an EIA, it is unlikely that the prime needs of the decision maker will be met.

A DEFINITION FOR BASELINE STUDIES

The term 'baseline studies' entered the environmental literature about the same time as the concept of EIA, about 15 years ago. Although the phrase does not appear in the wording of NEPA it quickly became standard terminology in EISs prepared under that legislation.

There does not appear to be any universally accepted definition for environmental baseline studies. Walsh (1983), in a major dissertation on the subject, records more than 15 formal written definitions of the term. In general, it is taken to refer to a description of some aspects of the physical, biological and social environments which could be affected by the development project under consideration.

It is this vague definition which causes most difficulties with baseline studies.

An attempt is made to describe 'the environment' and this usually means accumulating any information which is available on the general topics of land, water, air and people. The focus is often on information and data which are readily available rather than on what is needed.

A more operative definition is given by Hirsch (1980) who defines a baseline study as a 'description of conditions existing at a point in time against which subsequent changes can be detected through monitoring'. Using the same approach, a group of research scientists reviewing assessment requirements in the offshore marine environment suggested that baseline studies be designed 'to provide insight into the normal variability of phenomena such that appropriate monitoring programs can be designed' (Anon. 1975). From these definitions it can be seen that baseline studies are closely linked to environmental monitoring. In other words, if the practical objective of EIA is to predict changes in the environmental and social systems resulting from a proposed project, baseline studies provide the before-project records whilst monitoring gives the after-project measurements from which changes over space and time can be assessed.

An example may help to stress the operational implications of this definition. Suppose that a coal-fired thermal electric generating plant is to be constructed and there are concerns about the effects on air and water quality, namely increased levels of sulphur dioxide (SO_2) and raised water temperatures resulting from the discharge of cooling water. If there are other industries in the area it is likely that local SO_2 concentrations are already elevated above 'normal' levels and there will be daily and seasonal variations. It is also probable that the water body scheduled to receive the cooling water is already under stress from other industrial sources and its quality and temperature will vary, particularly on a seasonal basis.

In this case, the objective of baseline studies would be to determine the existing levels of SO_2 in the atmosphere in the vicinity of the proposed plant and hourly, daily and monthly variations in the levels. Baseline studies on the aquatic environment would determine the existing species composition, levels of pollutants and the normal water temperature changes throughout the season.

Once these background values were determined for the area in question, they could be compared with the predicted SO_2 emissions and the expected water temperature changes as a result of the cooling water discharge. If the expected changes were considered acceptable (that is, within prescribed standards) and the plant constructed, the actual changes from baseline conditions would be determined through monitoring. Operational procedures at the plant could be altered if the actual environmental changes were greater than anticipated.

THE ROLE OF BASELINE STUDIES IN EIA

Beanlands & Duinker (1983) envisage EIA as a series of basic, sequential steps, Figure 2.2a. Thus, an initial baseline data collection programme would be used to characterize the pre-project state. Cause and effect studies would then be undertaken to predict how stated variables would change as a result of projected

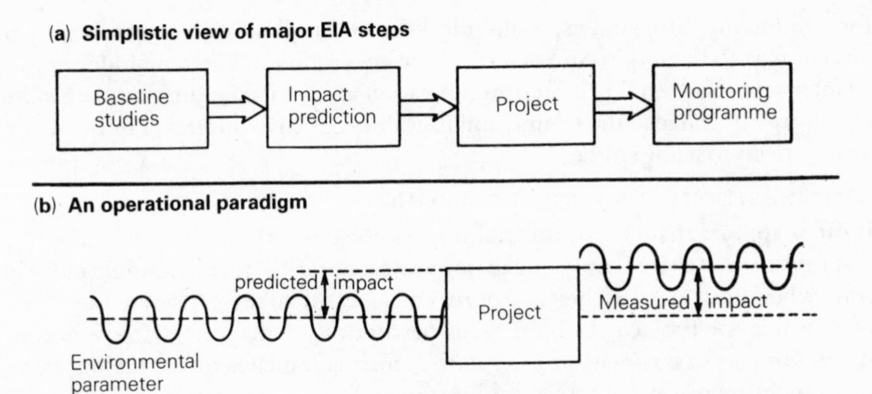

Figure 2.2 EIA showing: (a) the basic steps, and (b) processes translated into the roles of baseline studies and monitoring.

activities. Subsequently, following start-up of the approved project, monitoring would be used to determine actual impact conditions.

These simple steps can be translated into a diagram which clearly shows the relationship between baseline studies and monitoring (Fig. 2.2b). Thus, baseline studies would be directed towards establishing statistically valid descriptions of selected environmental components prior to the onset of the project under consideration. Subsequently, an effort is made to predict the extent to which the values would change as a result of the project. The project may or may not proceed, in its original or altered form, depending upon the reliability and acceptability of the predicted changes. In the event that the project proceeded, baseline variables would be remeasured during project construction and operation to determine the extent to which the predicted changes had occurred. In Figure 2.2b it is important to note the continuity of selected variables from baseline studies through to the monitoring programme.

The above description adequately portrays the role of baseline studies in EIA in a technical sense. However, EIA is fundamentally a planning tool. It should be undertaken to provide environmental and social input into the project decision-making process. In this context baseline studies must do more than provide a statistically valid description of specified environmental components prior to project initiation. They must be linked to critical decision points throughout all phases of project planning. The details of this will become clear in the following section; however, it is important to realize that baseline studies have a key role to play from the inception of a project through to final design and the setting of operational standards. A concentrated effort on baseline studies at the beginning of an EIA fails to optimize the potential use of such information in other stages of the development process.

Of course, one of the prerequisites for such use of baseline studies is a strong working relationship between the project planning team and the people responsible for the EIA. Although this seems to be logical, in reality, many EIAs are considered to be extraneous to the planning process and those responsible for conducting EIA studies, including baseline studies, are not given proper direction as to the environmental information required. The remainder of this chapter will show how better communication and understanding on both sides could greatly enhance the timing and usefulness of environmental information arising from baseline studies.

BASELINE STUDIES AND PROJECT PLANNING

It often appears that environmental impact assessments are undertaken on the assumption that there is only one major project decision, that is, a single point in time when the results of the environmental assessment are considered by those responsible for project planning. This perception is enforced by the environmental impact assessment process itself which culminates in the tabling of an assessment report or EIS. The implication is that the concerns presented in an assessment could not have been taken into account at earlier stages in the project planning cycle. Unfortunately, in most cases, by this time the major elements of a project have been decided, such as need, site and design criteria. Under such circumstances, baseline studies amount to little more than an attempt to rationalize decisions which have already been taken.

As pointed out by McMichael (1975), the reality in project planning is that there is a multitude of decision points shared among various agencies in the public and private sectors. The sequence of decisions is not always recognizable or predictable and many of the decisions are retractable. This decision-making network, spread out in time and among various interested parties, is even more complex when it involves international aid projects since another level of bureaucracy and decision making is added.

Baseline studies must be designed with these realities in mind if they are to be an effective means of bringing the environmental perspective to bear on industrial development. Specific information and data are required by different decision makers at various stages in the project cycle which should be accommodated within the concept and practice of baseline studies.

The project decision network In a very generalized sense, there are four critical stages in the sequence of project-related decisions, with different responsibilities assigned for each. These are shown in Table 2.1. The nature and extent of baseline information required at each of these decision points is quite different, yet equally important from an environmental perspective. Thus, in taking the initial project approval step, the proponent would need information on environmental legislation and policies which could seriously affect the economic viability of the project. If the decision is made to proceed with the project, the choice of a site may reflect concerns over various environmental sensitivities such as endangered species or resource use conflicts, some of which may be directly linked to the resource management responsibilities of government agencies. In considering the design and operation of a project, the proponent needs specific information on the resources potentially at risk at the selected site and on the relationship between projected effects and standards set by regulations or public acceptability.

Table 2.1 Critical decision stages in project planning and associated responsibilities.

	Decision	*Responsibilities*
1	A primary decision on whether the project will be undertaken – an 'approval in principle'	Ostensibly the decision is taken by the project proponent based on economic criteria; actually governments often decide indirectly through the provision of incentive grants, loan guarantees, licences, permits, etc.
2	A decision on where the project will be built – sometimes within a national or regional context	Normally the responsibility of the proponent, but often greatly influenced by government policies respecting political, economic and social development strategies.
3	A decision on how the project is to be built – a consideration of basic design options	Proponent's responsibility, but influenced by standards and limits set by government agencies through legislation or policies.
4	A decision on how the project will be operated – to some extent this depends on the degree of flexibility inherent in the design	(see 3)

In reality, these decisions would not be taken in the simplified sequence suggested above. Nevertheless, the point is that the information needs are quite different at various stages in the planning process. It should also be clear that limited information on the biological and physical characteristics of a development site will not ensure an adequate consideration of environmental concerns in the development planning process.

BASELINE STUDIES FOR A HYDROELECTRIC DEVELOPMENT

In parts of Labrador in eastern Canada there are major rivers which have potential for hydroelectric development. The most promising of these is the Eagle River where preliminary investigations have suggested the possibility of a 600 MW development. This example will be used to demonstrate how important it is to have environmental baseline studies phased in with the major stages in project decision making. For reasons of brevity, the focus will be on a rather narrow set of environmental concerns. However, the principles demonstrated by the example are equally valid for a broad range of environmental concerns.

Should the project be approved in principle? For this proposed development, one of the major environmental concerns is the possible effects of the dam on populations of the Atlantic salmon (*Salmo salar*) for which the Eagle River is a major spawning area. There are various reasons why effects on salmon would be a key factor at this initial decision stage. First, salmon populations throughout eastern Canada are depressed compared with historical levels and the long-term viability of the species is in question. This concern is widespread amongst resource managers, fishermen, and the general public. Secondly, salmon is a valuable species for both recreational and commercial fisheries. For example, it is known that the commercial catch along the Labrador coast makes up about one-third of the total provincial catch. Thirdly, anadromous species, such as salmon, are sensitive to hydroelectric developments and over the years many of the best salmon spawning rivers have been lost due to dam construction. Finally, the management of salmon stocks is a responsibility of the federal government and the proponent could not proceed with the project without a permit issued under the Fisheries Act.

Environmental baseline studies at this stage should be directed towards determining whether it is possible to proceed with the project and meet the requirements of the Fisheries Act. An important consideration is the intention of the federal government to adopt a policy preventing any net loss of salmonid habitat resulting from hydroelectric developments.

Three specific studies would have to be undertaken. First, it would be necessary to determine what percentage of the Labrador coastal fishery is dependent upon production from the Eagle River. A review of historical catch records and recent population surveys conducted by government agencies, along with a limited tagging programme, would provide an approximate answer. Secondly, a study would be required to determine the feasibility of building and operating a salmon hatchery in the area of sufficient size to replace the maximum possible loss of production from the project. Finally, a reconnaissance survey of nearby salmon-producing rivers (of which there are three) could indicate the potential for opening up salmon habitat which is currently inaccessible due to physical obstructions. This information could provide some idea of the possibility of meeting the requirements of the policy regarding no net loss of habitat and its financial implications. The information provided by these studies would be crucial in deciding whether the measures necessary to meet the requirements of the Fisheries Act were of such a magnitude as to affect the economic viability of the project. The most significant point is that such studies would be required before a commitment in principle to the project.

Where should the project be built? Depending in part on the results of the first baseline studies, the proponents might attempt to negotiate an agreement in principle with the fishery management agency on the basis that they could meet the requirements of the Fisheries Act. If such efforts were successful, the next major decision to be addressed would be the location of the dam(s) on the river. In reality, the proponent already would have conducted engineering feasibility studies which would have identified priorities from a list of potential sites. In this case, the proponent had looked at eleven potential sites and had decided that three were economically viable, with a combined potential output of 600 MW.

At this stage, more detailed and focused baseline studies would have to be conducted with the objective of assessing the percentage of the total Eagle River salmon population which would be prevented from reaching their spawning habitat under the proposed dam regime; the distribution of currently available salmon habitat of various classes within the entire Eagle River system; the extent of habitat of various classes throughout the drainage area currently unavailable to salmon and the nature of the blockage; and the losses of habitat of various classes and how much could be made accessible above each of the preferred dam sites. This information would be necessary in deciding the potential trade-offs between the engineering costs associated with different dam sites and the financial implications of maintaining salmon populations and of reducing the loss of spawning habitat. It could be that some of the other eight original sites surveyed would be reconsidered. In any event, the results of the baseline studies undertaken at this site-selection stage would influence the benefit:cost ratio of the project and further define the best strategy to pursue with respect to the construction of a fish hatchery and development of new spawning habitat.

How should the project be designed? In spite of the environmental planning implications arising from the two previous stages in the decision-making process, this is the stage at which conventional baseline studies are undertaken. In most cases the economic viability of the project has been determined, and the site chosen, before baseline studies are undertaken as the first step in the environmental impact assessment process.

Unfortunately, at this stage in the project planning cycle there are only a few options available for mitigating environmental problems. Information would be required on the maximum and minimum downstream water flows necessary to protect migrating adult fish as well as the viability of juvenile stages. The dam design would have to take into account the water flows required at various times throughout the year. This could mean that all of the dams in the system have to be regulated in a co-ordinated manner to maintain the required water flows in the lower reaches of the river, particularly in the estuary. Normally, by this stage, the hydrological regime is well understood, although it is likely that the biological parameters are less well defined, particularly since different rivers have their own salmon production characteristics.

Another important consideration would be whether a fish ladder should be incorporated into the design of the dam. If the production and habitat studies showed only limited projected losses, the proponent may be able to argue effectively for a less expensive programme to net the adults and move them above the dam by truck or other such means.

How should the project be operated? Studies at this stage in the project cycle would be directed towards establishing monitoring and operational feedback systems to ensure that the design features built into the project are properly implemented. A number of objectives could be established. These include, for example, determining the long-term survival of salmon moving past the dams; establishing the survival rates above the dams compared with that in unaffected parts of the river; assessing whether newly developed habitat is being used to the extent predicted; and the relationship between salmon production in the river and commercial salmon catches.

Most would be continuations of studies initiated during previous decision stages. In effect, environmental monitoring during the operational phase of a project is a continuation of pre-construction baseline studies. The overall objective is to ensure that the project is operated in accordance with its design specifications and to determine whether the mitigation measures applied were as effective in protecting the resource base as predicted.

Conclusion

Three basic themes have been pursued in this article. First, environmental baseline studies as conducted in the past have been only marginally effective in influencing key project decisions. Normally, they are undertaken at a stage in the project planning cycle when important opportunities for mitigation are no longer available.

Secondly, the concept of baseline studies as a statistical characterization of the state of the physical and biological environments is far too limited. It must be broadened to include information of a strategic nature, useful in early project decisions with respect to approval in principle and the choice of site.

Finally, no single, key, project decision point exists. A broader definition of baseline studies is needed so that such studies can be integrated with the multitude of decisions and decision makers which represents the real world of project planning.

3 *Developments in EIA methods*

R. BISSET

Introduction

Since the National Environmental Policy Act (NEPA) became law in the United States, much effort has been expended by agency personnel, consultants and academics in devising methods to aid preparation of environmental impact statements (EISs). Most of this work has emanated from the USA, but with the introduction of EIA procedures into more and more countries, the ingenuity devoted to developing EIA methods has increased correspondingly. EIA methods are formulated throughout the world although the USA is still the main source.

For the purposes of this discussion a 'recent' method is one which has appeared in the literature since 1978. This literature consists of articles in journals, EISs, unpublished conference papers and items from the 'grey literature' which contain so much of the thinking and writing devoted to EIA. Pre-1978 methodological developments have been widely reviewed in the literature and comprehensive descriptions can be found in, for example, Clark *et al*. (1980), Canter (1983), Bisset (1984a) and Wathern (1984).

The term 'method' deserves some elucidation. A distinction must be made between methods and techniques used in EIA. EIA techniques are concerned with predicting future states of specific environmental parameters such as noise levels. In any single EIA study a number of techniques may be used. Together, they provide data which are then collated, arranged, presented and sometimes interpreted according to the organizational principles of the EIA method being used. EIA methods have been described alternatively as methodologies, technologies, approaches, manuals, guidelines and even procedures in the literature.

This chapter cannot contain a description or discussion of every method which has been put forward since 1978 nor can it ignore totally the pre-1978 literature, upon which many of the more recent developments are based. Rather it will try to identify common themes or trends which appear to characterize the methods developed since 1978. Attention will be paid to recent and current thinking which has resulted in the development of a particular type of method or its variants. The types of method described here are index approaches, systems diagrams, simulation modelling and the 'sound ecological principles' approach.

Index approaches

Index approaches were developed early in the evolution of EIA practice. Some of the earlier methods continue to attract attention and have seen some further elaboration, although there have been some major conceptual innovations. These methods can be divided into checklists and approaches based on multi-attribute utility theory.

CHECKLISTS

The checklist probably vies with the Leopold matrix (Leopold *et al.* 1971) as the oldest EIA method. It exists in an enormous number of forms from a simple list of environmental factors to be considered in EIA, through variants providing additional guidance on data requirements and relevant predictive techniques. The apotheosis of the checklist has been the development of complex, quasi-mathematical forms in which impacts are transformed into units on a common, notional scale, weighted in terms of relative importance and finally manipulated mathematically to form various indices of 'total' impact. Examples include the Environmental Evaluation System (EES) (Dee *et al.* 1973) and the method developed by Sondheim (1978). Such methods, however, can be stopped short of full integration. For example, the Water Resources Assessment Methodology (WRAM) does not add all impacts, instead total scores for 'sectoral' impacts are derived (Solomon *et al.* 1977). Thus, a score for all social impacts is derived which can be compared with a score for environmental impacts. This approach enables trade-offs between 'sectoral' impacts to be shown explicitly and made by decision makers.

Thailand has been the source for another variant of the EES concept (Mongkol 1982). Mongkol takes as the starting point some important impact characteristics which are not considered *in toto* in EIA methods. These characteristics are impact magnitude (severity of an impact); prevalence (extent of an impact, and the network of causes and effects); duration and frequency (the time elements); risk (probability of an impact occurring); importance (significance of an impact at the time of EIA work); and mitigation (action taken to reduce or eliminate adverse impacts). Mongkol's major omission is the failure to consider 'reversibility', but this attempt to improve the conceptual base of scaling–weighting checklists has considerable merit. This variant of EES involves use of a matrix to take account of mitigation and modified value functions to make them flexible in order to allow for local conditions (Fig. 3.1). The modified value function incorporates maximum and minimum allowable levels for selected environmental parameters, for example, noise levels. Further, Mongkol advocates the use of an error term to accommodate the risk of wrong decisions being taken. Finally, instead of using concepts such as 'net environmental cost' or 'net environmental benefit', Mongkol borrows an idea from cost–benefit analysis and uses an 'environmental benefit–cost' ratio obtained from

$$\text{environmental benefit–cost} = \text{beneficial impact/adverse impact}$$

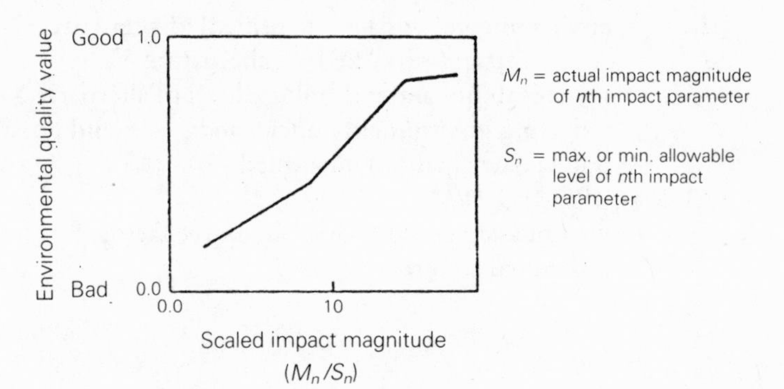

Figure 3.1 Flexible composite value functions.

One of the main difficulties with scaling–weighting checklists concerns the derivation of the weights. Sondheim (1978) has attempted to broaden the basis of weight allocation by removing it from the experts who either develop a method or are involved directly with specific EIA studies. In the Sondheim method, a weighting panel is constituted for each EIA. The members of the panel, chosen by the organization responsible for the study, can be from government, industry, community organizations, interest groups and other parties affected by a proposal. Each member of the panel produces an individual weighting scheme for environmental components. These schemes are amalgamated to produce a single weighting scheme representative of the panel's views.

Yapijakis (1983) has produced an adaptation of the Sondheim method in which an even broader base for deriving weights is advocated. Yapijakis is concerned that major projects likely to have transnational effects should be assessed in an acceptable manner by all of the countries likely to be affected. A scheme for weighting impacts which includes selected nationals from academia, industry and government departments is used to produce national weights, which can be combined in a 'regional' if not 'global' manner to produce an index of 'total' project impacts. In a trial run Yapijakis discovered that nationals of Yugoslavia and Greece produced similar weighting schemes despite socioeconomic and political differences between the two countries.

As well as incorporating environmental and social impacts, Yapijakis also included economic aspects and a new dimension which he termed 'manageability and technology level'. This is a facet of a project which reflects government policies and goals. Amalgamating all of these factors produces the following comprehensive equation to aid international decision making on projects having transnational impacts.

$$PRO = WB(BCR) + WC(CO) + WE(EI) + WS(SI) + WM(MTL)$$

where PRO = project alternative ranking order
BCR = project alternative benefit:cost ratio
CO = capital outlay for project alternative

EI = environmental impact (quantified) of alternative
SI = social impact (quantified) of alternative
MTL = manageability and technology level of alternative, reflecting government policies and goals (and possibly public participation) quantified
WB, WC, WE, WS, WM
= weight assigned to each of the above factors by individual country

Multi-attribute utility theory

Recent literature on EIA methods indicates considerable interest in the application of multi-attribute utility theory. Virtually all of the published studies deal with energy projects and in particular with site selection, yet there is no inherent reason why this method cannot be applied to other types of projects and policies. Although this method was applied initially before 1978, it is only recently that interest seems to have increased.

The method has its theoretical base in the writings of von Neumann & Morgenstern (1953) and associated developments by Keeney & Raiffa (1976). It is a means whereby possible environmental consequences can be 'traded off'. Alternative projects can have many different environmental impacts and also exhibit different 'levels' of the same impact. For example, one alternative might increase ambient noise level by 10 dB (A) whereas another might only increase it by 5 dB (A). This method provides a logical basis for comparing the impacts of alternatives to aid decision making.

Utility theory in EIA has been applied most often to site selection for major power stations (for example, Keeney & Robilliard 1977, Kirkwood 1982). In addition, Uys (1982) has used this method to assess alternative energy policies for South Africa. However, it can be applied to the assessment of the environmental impacts of alternative projects. Collins & Glysson (1980) used this method to assess two alternative solid waste disposal systems. The organizing principles of the multi-attribute utility framework are basically similar irrespective of the specific objectives of a particular application.

The first step is to determine environmental attributes which can be measured, for example, particulates can be measured in $\mu g/m^3$ or dissolved oxygen in mg/l. A number of such attributes which provide a comprehensive picture of likely environmental impact are selected. For each attribute different measures or 'states' may exist which have to be calculated using predictive techniques such as air pollution dispersion models. Once attribute levels have been determined, the principles of this method enable their desirability or undesirability to be established.

It is recognized that this operation relies on the subjective opinion of experts. However, utility theory provides a logically consistent framework for establishing the preference structure of experts (individually or in combination) regarding the relative merits of different levels of each attribute. Systematic comparison of these levels by the decision maker results in the formulation of utility functions (see Fig. 3.2). Utility is measured on a scale of 0 to 1 where 1 is the highest utility. In the example shown in Figure 3.2 the loss of 100 per cent of salmonids leaving natal waters (due to power station impacts) would be very serious and be given a utility value of 0. An 80 per cent loss has a utility value of 0.5

An important feature of this method is the ability to deal with probability, that is, the likelihood that specific levels of environmental attributes will occur. This aspect of utility theory distinguishes it from all other EIA methods. Usually, the probability of an impact occurring is ignored or omitted from the structure of an EIA method. If recognized at all, it is usually relegated to a qualitative commentary.

Once utility functions have been established for individual attributes it is possible to combine them. The first step in this stage of the analysis is to calculate a scaling value (k) for each attribute. Such scaling values reflect the relative importance as perceived by decision makers of the different attributes. Total utility or a composite environmental quality index (EQI) can be obtained from the following equation:

$$EQI = U(x) = \sum_{i=1}^{n} k_i \, U_i \, (x_i)$$

where k_i is scaling factor of attribute x_i, and U_i is utility function. This equation assumes environmental independence between the attributes. Usually this is not the case as attributes are interrelated. A more complex formula is required to deal with this situation.

The results from an analysis can be presented as EQIs for a number of sites or

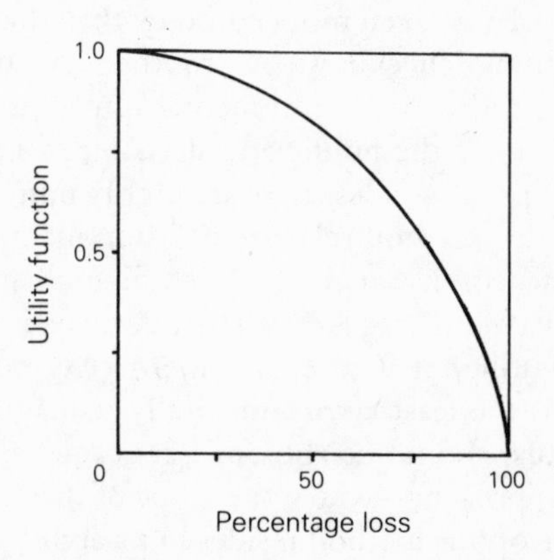

Figure 3.2 Hypothetical utility function curve showing relationship to percentage loss in salmonid populations.

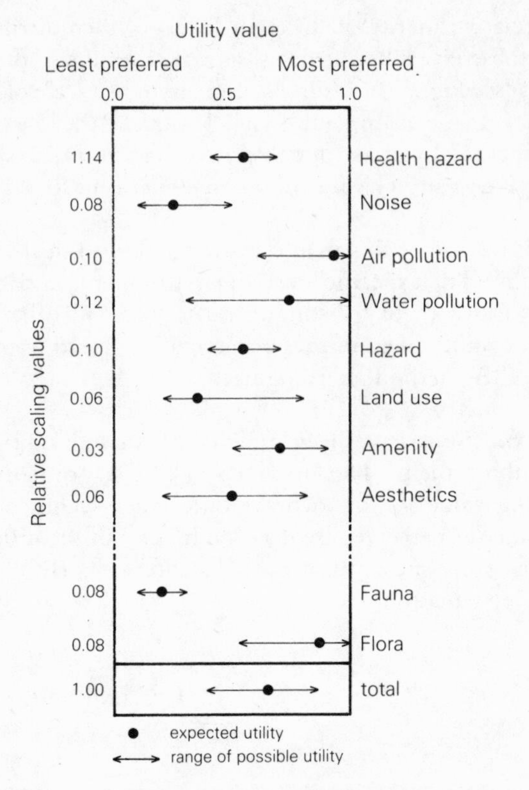

Figure 3.3 Performance profile for alternative proposals.

alternative projects or as performance profiles (Fig. 3.3). This profile shows the individual utilities of a project in relationship to ten environmental attributes. Also, it shows the composite utility over all attributes.

This method requires considerable familiarity with the theoretical basis of utility theory and is perhaps, even more complex than the numerous variations of the scaling–weighting checklists. As in the case of scaling–weighting checklists a composite index of environmental impact is obtained. It is likely that interested members of the public and decision makers will be unable to follow the steps of the process unless they are highly numerate.

The advantages of this method relate to decision making, the incorporation of probability, and sensitivity analysis. First, like all methods producing a composite index, decision making is easy as the decision is made by the method. Given the numerical utility structure, the alternative project or site with the highest utility score is the least environmentally damaging and therefore, on environmental grounds, should be chosen. Economic factors can be incorporated into the utility approach to widen the scope of the decision-making base.

The final advantage of this method resides in its ability to show changes that would occur as a result of modifying the utility functions and probability assumptions. If the calculations have been done by computer, the effects of sensitivity analyses can be seen almost immediately. Such analyses indicate the 'robustness' of the initial results and show which are the crucial variables that change outcomes.

COMMENTS ON INDEX METHODS

The main objections to these methods are well known and often repeated in the EIA literature (see, for example, Andrews 1973, Bisset 1978, and Hollick 1981). It is argued that the subjectivity involved in these computations is hidden within a spurious objectivity. Even if not hidden it is further contended that the subjective views incorporated within these methods are representative of a very restricted population, namely selected decision makers or experts. There is considerable truth in this assertion despite attempts, for example, by Sondheim (1978) and Yapijakis (1983) to increase the base of subjective inputs.

Additionally, it is argued that these methods are needlessly technocratic and complex, thereby inhibiting public participation in EIAs and review of the final results. Again, this is a valid criticism. Bisset (1978), however, has gone beyond this and argued that such methods may be devised and used for this purpose and that the aim is to inhibit wider involvement in project decision making.

One important drawback to these methods is the manner in which they compartmentalize and fragment the environment. The scaling–weighting checklist is simply a list of environmental factors, changes in which (impacts) are assessed in isolation. The same stricture applies to utility theory. Both methods focus on environmental features which can be quantified, although in no case is this an absolute requirement. However, these methods are so heavily dependent on quantification that there must be a great temptation to quantify the unquantifiable, for example, in the field of aesthetics.

The focus on single components of the environment in these methods is a major weakness. Environmental systems consist of a complex web of interrelated parts, often incorporting feedback loops. No matter how intricate and intellectually satisfying the mathematics involved, it is impossible to characterize 'system-level' impacts by considering changes in specific components in isolation and then aggregating the results.

Systems diagrams

A method which began to appear early in the EIA literature is the systems diagram. The theoretical basis for this method resides in the work of Odum in the field of ecological energetics (Odum 1971). In fact, it was Odum who first suggested the use of these diagrams in EIA (Odum 1972).

A systems diagram consists of a chart showing environmental and sometimes socioeconomic components linked together by lines indicative of the direction and sometimes the amount of energy flow between them. Systems diagrams are based on the assumption that energy flow and, therefore, different amounts of energy can be used as a common unit to measure the impacts of development.

Consequently, systems diagrams enable comparative measurements of the magnitude of different impacts to be expressed in a common unit. Activities associated with a project likely to cause impacts may be included in the systems diagram (see, for example, Fig. 3.4).

Of all the variations on systems diagrams the most comprehensive is perhaps the Activity Assessment Routine (Ecological Systems Component) developed for estimating the environmental effects of development activities on the Texas coastal region (State of Texas General Land Office 1978, Longley 1979). The ecological systems component consists of three major assessment aids, namely, ecological systems diagrams; assessment worksheets; and a series of tables and matrices to help organize the judgements of those assessing a proposal and to ensure as much standardization as possible in decision making.

The first step is to identify those actions associated with a proposal which will have environmental impacts, for example, dredging. The next stage is the identification of first-order environmental changes, known as Primary Ecological Alterations (PEAs). In the Activity Assessment Routine the Texas coast is divided into seven ecological systems. The nature of PEAs, therefore, depends on the site of a proposed project. PEAs are identified by those implementing the assessment responding to a series of screening questions. Each question is linked to the biological, physical, hydrological, chemical, or energy aspects of those ecosystems likely to be changed.

Once PEAs have been identified it is necessary to determine the direction of the expected ecological change, such as add or remove (vegetation), increase or decrease (bird species diversity). Having identified PEAs it is necessary to trace the consequences of these initial changes throughout the ecosystem. This is implemented by using the appropriate ecological systems diagram from among those specially constructed for the seven ecological systems. The components of the diagram are termed 'attributes' and changes in these attributes are called 'attribute alterations'. At this stage in the assessment various characteristics such as the direction, duration, magnitude and probability of occurrence of attribute alterations are determined. These are, in effect, secondary and higher order ecological impacts.

The direction of energy flow between components has already been discussed. Duration refers to the time period over which an impact occurs, for example, short- or long-term, or quantitatively as the specific number of months or years. The magnitude of a change is based on two factors. These are alterations in energy flow between ecosystem components caused by a project and the areas over which these changes occur. This approach to determining magnitude is perhaps the most questionable part of the method because it involves certain simplifying assumptions, for example, that a small ecological change over a large area is equivalent to a large ecological change within a small area. This part of the method depends on the local knowledge of the assessors and demands careful attention to be paid between the 'real' world and the conceptual world of energy diagrams. Finally, probability refers to the

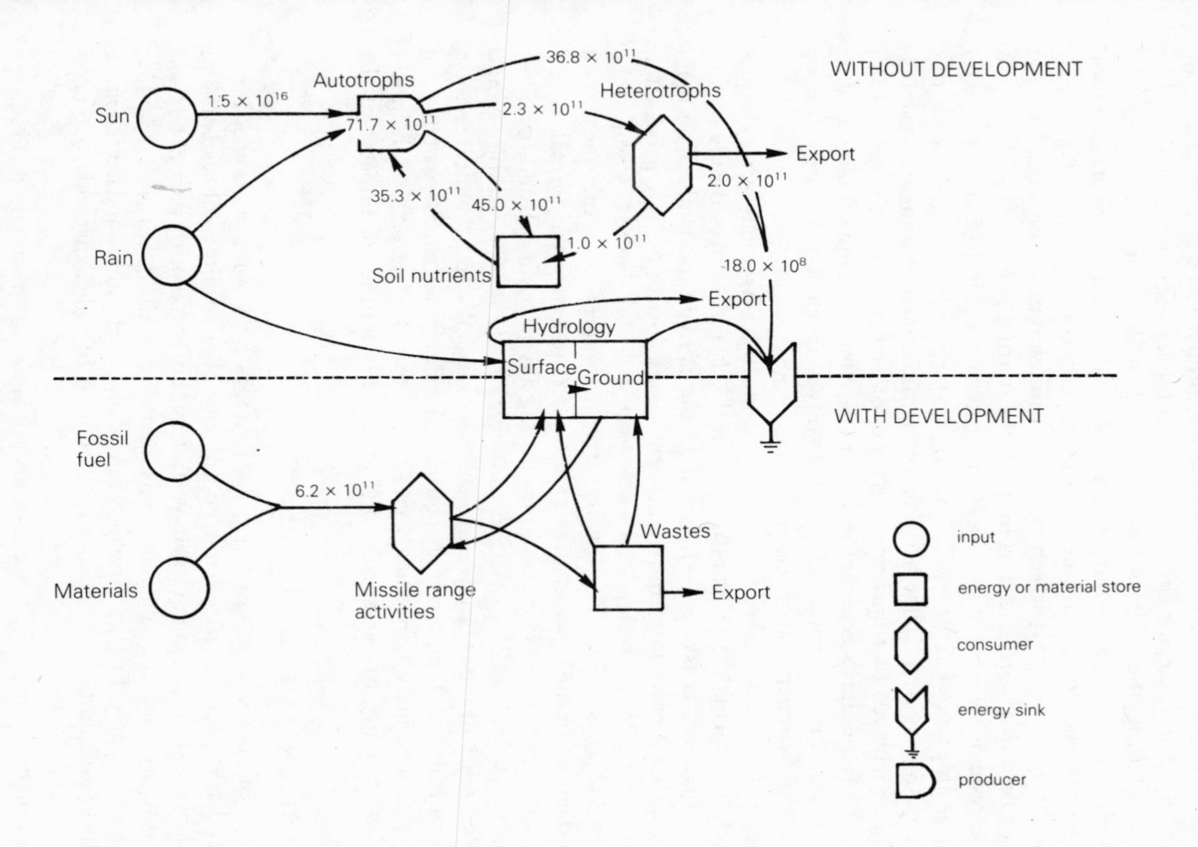

Figure 3.4 Energy flow diagram for a missile site development (after Gilliland & Risser 1977).

likelihood of an expected impact or attribute alteration and can only be expressed qualitatively by such terms as 'certain to occur' and 'may occur'.

The most important characteristics are magnitude and duration. These are used to help determine which secondary and higher order impacts should be investigated using the relationships contained in the appropriate ecological systems diagram. To save resources, it is advocated that each attribute alteration should be screened to determine whether its consequences should be traced further. As a result of this screening only a limited number of secondary and higher order impacts are traced from the initial starting point.

These systems' diagrams, like most advocated for use in EIA, focus only on ecological impacts. Lavine *et al.* (1978), however, have argued that this approach can provide a bridge between economic and environmental systems. This can be achieved by transferring all types of energy flow in the natural and man-made (economic) environments into a common unit. These can be converted into money terms by the use of known or calculated energy : money ratios for different national or regional economies.

Lavine *et al.* have described the use of systems diagrams to put money values on the environmental and economic systems likely to be affected by a 133 km two-lane highway crossing both wetlands and agricultural lands. The main impacts were air pollution from vehicle emissions, loss of organic production from alterations in the water level of the wetlands and the loss of soil to the highway and its associated borrow area. In energy terms, air pollution would cause a decrease in the capacity of the atmosphere to undertake normal energy transformations thereby decreasing its existing ability to absorb and transport gases and particles. The loss of organic production and soil regeneration would affect the ability of the environmental system to fix solar and chemical energy. Other environmental impacts were considered insignificant, in energy terms, in the analysis. The analysis indicated that $81.9m/year (1975 prices) would be lost from the economy of the region due to the environmental impacts of the highway.

COMMENTS ON SYSTEMS DIAGRAMS

This type of approach has two main advantages over index methods. First, systems diagrams acknowledge the complexity of environmental systems and that a change in one parameter can have multiple effects on other parameters and on the system as a whole. Secondly, this method uses measures of energy flow to compare impacts. Energy flow data can be obtained using standard scientific procedures and, unlike the quantitative content of index methods, are non-controversial.

Unfortunately, there are three major drawbacks to this method. First, the construction of systems diagrams for particular ecosystems with associated data on energy flow can be time-consuming and expensive. Once constructed, periodic revision may be necessary to take account not only of natural variations, but also man-made perturbations. Secondly, not all important ecological relationships can be characterized by energy flow. For example, the breeding of a rare bird may depend on the protection of a particular plant species. Removal of the plant might render the bird extinct, although no change may have occurred to the energy relationships between the bird species and the components of its environment. Thirdly, the use of systems diagrams is, at present, confined to ecological impacts. Attempts to incorporate socioeconomic impacts are still fraught with conceptual and practical problems.

Simulation modelling

The use of simulation modelling in EIA, often under the heading of Adaptive Environmental Assessment and Management (AEAM) is based upon the work of Holling and his associates at the Institute of Animal Resource Ecology at the University of British Columbia, Canada (Holling 1978). AEAM developed in response to a number of perceived weaknesses in EIA practice. First, EISs were becoming increasingly lengthy and unwieldy as a result of the volume of environmental data being included. It was considered that EIA was exhibiting a 'measure everything' syndrome to ensure that results could not be challenged for lacking comprehensiveness. It was felt, also, that EISs were deficient in impact prediction. Also, it was considered that communication between EIA personnel and those responsible for decisions on the future of projects and their management was breaking down. This resulted in a reduction of the influence of EIA on decision making.

To overcome these alleged weaknesses of EIA, AEAM uses small workshops of scientists, decision makers, and computer modelling experts to construct a simulation model of the systems likely to be affected by a development. The key component of AEAM is the workshop in which the participants have to reach consensus on the important features and relationships which characterize the systems studied. The qualitative output from the workshop is 'translated' by modelling experts into a model consisting of quantitative relationships (as far as possible) between the selected parameters. Likely broad outcomes (impacts) resulting from the introduction of exogenous factors, such as development projects or resource management strategies, can be seen quickly by operating the model under different assumptions. Constructing the model shows areas where data are deficient and allows appropriate research work to be carried out to provide the data. Periodic workshops allow the model to be refined, but not necessarily made more complex, as additional data become available.

AEAM has been applied to a variety of development and management situations, for example, an assessment of the environmental effects of the Alberta oil-sands development (Staley 1978) and an analysis of the environmental and socioeconomic consequences of different management strategies for the Nam Pong multi-purpose water project in Thailand (Mekong Secretariat 1982). The ideas of AEAM were used, also, in the assessment of the main environmental, economic and social impacts likely to arise from Salto Grande dam and reservoir on the borders between Uruguay and Argentina (Gallopin *et al.* 1980).

A review of experience in applying AEAM has been issued by Environment Canada (1982).

COMMENTS ON SIMULATION MODELLING

The philosophy behind AEAM has had considerable influence on EIA thinking and practice. The modelling approach deals explicitly with the interactions between environmental variables and permits investigation of the full ramifications of a project or management strategy. There have been attempts to produce models which incorporate social and economic concerns, but these have been limited to a few parameters such as per capita incomes and population growth.

There can be no doubt that the periodic use of multidisciplinary workshops containing EIA personnel and decision makers is a useful tool for restricting the scope of assessments to the key issues. However, the composition of these workshops is, of necessity, limited. The extent to which a wider involvement could be encouraged while still achieving the objectives of the workshop is debatable. Again, this is a problem shared with many other methods. This method is prone, also, to the common tendency to quantify relationships on the basis of uncertain data.

Some other general comments may be made about AEAM. A reading of the literature produced by proponents of this method (for example, Sonntag *et al.* 1980 and Everitt 1983) shows a tendency to play down the ability of the models to make useful predictions and instead to concentrate on the insights into environmental problems provided by workshop discussions. Also, the ability of AEAM to encourage co-operation between EIA personnel and decision makers and to be parsimonious in its use of baseline environmental data is stressed. These aspects are important and are a valuable contribution to EIA, but nevertheless EIA is, or ought to be, about predictions. Consequently, judgement on this facet of AEAM must await information on the performance of the models which are already operational. It is to be hoped that such information will be forthcoming. Indeed this process has begun with an attempt to review the Nam Pong experience (Srivardhana 1983).

The literature on the application of AEAM is biased towards the management of resources, for example, forests and economically important species such as salmon. Large-scale water projects have also been assessed, but there have been few, if any, applications of AEAM to the assessment of developments such as oil refineries, power stations and pulp mills. Until such applications are reported and its utility for such projects assessed, no judgements on the wider applicability of AEAM as a generic EIA method can be made.

The ideas propounded by Holling and his co-workers was a stimulant to renewed thinking, in the late 1970s and early 1980s, on how EIAs should be implemented to make them more cost-effective, improve their predictive abilities and make the results more accessible to non-experts. All of these concerns have resulted in an emphasis on the role of 'science' or the 'scientific method' in EIA. To some extent, there has been a desire to return to scientific basics, usually considered to be similar to, if not synonymous with, 'sound ecological principles'.

Sound ecological principles: the new approach to EIA

There can be no doubt that EIAs still need a healthy injection of scientific rigour, particularly where impact prediction is concerned. Two studies from the UK (Bisset 1984b) and Canada (Beanlands & Duinker 1982) have shown that EISs usually offer vague generalizations about possible impacts which are difficult to test in a rigorous manner. Such predictions are of little value to decision makers because of their ambiguous nature. Also, without testable predictions it is impossible to use projects as 'natural experiments' from which environmental impact information can be obtained for future use in EIA.

In the late 1970s, a considerable number of different publications appeared which attempted to show a new way forward for EIA. To the work of Holling must be added the contribution of Fritz *et al.* (1980) from the US Fish and Wildlife Service on formulating an ecological modelling approach to EIA which has, as one of its main objectives the formulation of impact predictions as testable hypotheses. Further important additions to this literature were made by Doremus *et al.* (1978), Truett (1978), Ward (1978) and Sanders *et al.* (1980). Concern over the status of the ecological component of EIAs led to a major Canadian study to formulate 'guidelines' to improve the ecological contribution to EIA (Beanlands & Duinker 1982). In the United States a major study examining ways of improving the scientific content and methodology of environmental impact analysis has been completed (Caldwell *et al.* 1982).

All of these studies have certain opinions and recommendations in common. First, EISs are considered to be scientifically inadequate. Secondly, the ecological component is poorly handled both conceptually and in terms of analytical method. Thirdly, they desire the application of the classical scientific method to EIA. This involves *inter alia* conceptualization of the problem, setting of boundaries in time and space; the formulation of study designs; establishment of control and reference monitoring stations for baseline data acquisition and testing impact predictions; relevant pilot-scale experiments to investigate possible impacts (for example, air pollution on a crop species); and the formulation of impact hypotheses and subsequent testing.

COMMENTS ON ECOLOGICALLY SOUND PRINCIPLES APPROACHES

Although this emphasis on improving the ecological aspects of EIA is necessary and laudable, it is important to remember that the scope of most EIAs extends beyond ecological considerations. It is probably true that the ecological implications of proposals can be the most important for human welfare and environmental quality. However, there are other factors which have to be considered in an EIS, which cannot be regarded as part of the remit of ecological investigations: for example, the health effects of increased noise and air pollution and the problems caused for communities and farmers by severance.

The literature concerned with the new 'scientific' thinking has a certain missionary, proselytizing tone which it is necessary to treat with a degree of circumspection. Also, there are signs of a 'take-over bid' for EIA which has to be resisted. Despite this comment, there is no doubt that the way forward for EIA rests with the application of better 'scientific' procedures and the use of projects as natural experiments.

Conclusions

The demise of index-type methods has been predicted in the past, see for example Bisset (1978). Such predictions have not been borne out by events and it is interesting to speculate on the reasons for their continuing development despite repeated criticism. Index methods fulfil a need which those in favour of other types of method often ignore. This is the desire of many decision makers to be faced with an easy decision, especially when comparing a complex variety of impacts from a number of alternatives. By scaling and weighting impacts, index methods provide a means for encapsulating impacts in total 'indices' for alternatives. As such methods contain explicit rules for selection of the 'best' alternative, its identification is easy. Also, should decision makers be sufficiently interested to test different assumptions such methods can quickly show the outcomes.

Hollick (1981) has argued that index methods can only be used in a politically stable, cohesive society. This point may be a key to the survival of such methods. Many agencies and government departments responsible for development decisions, especially in relation to site selection, are politically stable and cohesive, often with a high degree of consensus on environmental values. Other interests are not involved at this stage in decision making and index methods provide a very useful technocratic way of making decisions.

Two other factors may account for the existing and likely future popularity of these methods. First, many countries have political systems which exhibit or assume general consensus. It is likely that index methods will be popular in such decision-making contexts. Also, in many countries, EIAs are undertaken by engineers and technologically trained people who have an affinity for the use of quantitative aids in their work. This affinity is often carried over into EIA. For all of these reasons it is likely that index methods will have an assured future.

These methods are not incompatible with 'scientifically' acceptable EIAs. The main strength of index methods is the ability to amalgamate and manipulate the results of EIA to aid decision making. It is important that the results of EIA be obtained in a 'scientific' manner and that the transformation of the results into notional numbers on arbitrary scales is done in such a way that the validity of the results is not violated. However, the checklist underpinning which typifies many index methods has to be removed if such methods are to be improved.

It would be useful to bring some degree of reality into the debate on the utility of index methods. Most of the discussion is hypothetical because little information on the actual operational performance of different types of method exists. This could be done by selecting a range of projects for which different methodological approaches have been adopted and comparing not only the cost and resource requirements, but also the outcome in terms of predictive ability. Such a study should be one component of methodological research over the next few years.

4 *Uncertainty in EIA*

P. DE JONGH

Introduction

In the last decade an important discussion has taken place in the Netherlands concerning the creation of Markerwaard, a polder (an area of new land reclaimed from below sea level) in the Ijsselmeer (the former Zuider Zee). In the deliberations, a number of certainties crossed the discussion tables. The engineers who had recently worked on previous Ijsselmeer polders were sure that they would be able to create a beautiful new polder, which would give rise to economic activities, to new town development, and to the establishment of recreational resorts. Biologists were sure that a severe loss of natural beauty would result.

Apart from these certainties, a number of uncertainties crept into the lengthy discussion. For example, it was unclear how high the total cost of impoldering would be or how the growth of the Dutch population would develop. Other questions arose, for example, whether the water quality in the Ijsselmeer would be affected. Even the supposed certainties became uncertain as it transpired that it might be possible to find ways of keeping some of the threatened bird and fish populations in nature reserves within the new polder. Different interest groups asked the government to apply environmental impact analysis (EIA) to the reclamation of Markerwaard. Although at the time (the second half of the 1970s) the Dutch government had just started to develop EIA regulations, a proper EIA procedure was not followed. The main, official, reason was that the discussions about Markerwaard began at a time when EIA regulations were not foreseen. In fact comparable procedures were followed as many studies were commissioned by the government and other organizations and the decision process was structured around a physical planning procedure in which public participation was one of the main elements. In reality, an important underlying argument for not applying EIA may also have been the fear of the planners that EIA would reveal the many uncertainties associated with the project.

This example shows that uncertainties play an important role in the planning and decision-making processes for major developments. As such projects will shortly come under EIA regulations in the Netherlands, problems of uncertainty cannot be ignored when preparing for the introduction of EIA. Consequently, the Dutch government has paid explicit attention to uncertainty in several research projects as part of its EIA programme in the last three years.

The relationship between uncertainty and EIA – a first glance

At first sight, based, for example, upon EIA handbooks from the late 1970s and early 1980s, there appears to be no relationship between uncertainty and EIA. In Rau & Wooten (1980), no explicit reference is made to questions of uncertainty or risk. Thus, for example, the information given about radiation impacts implies precision. Rau & Wooten consider that 'the impact of radioactive substances on biota has been determined by the Atomic Energy Commission. A series of models . . . [is] . . . available to make useful predictions of radioactive impact upon the flora and fauna of terrestrial and aquatic systems.'

Many of the elements of EIA mentioned by Rau & Wooten might be useful for the management of uncertainty. Their main suggestion, however, that the main problems for prediction, and maybe even for the application of EIA itself, are solved by the availability of 'models' cannot be left unquestioned. In practice, this is not the case. There will be debate concerning, for example, assumptions underlying the models, the correctness of input data and the significance of the results from applying these models. All of these discussions can be transformed, but not resolved, in terms of uncertainty: uncertainty about the specific relationships in the model, uncertainty about the way input data are assembled, and uncertainty about the values given to specific impacts.

Early EIA methods were not concerned about matters of uncertainty, as can be concluded from a study of methods conducted for the Dutch government by Environmental Resources Ltd (ERL). In this study, 29 formal EIA methodologies were analysed on the basis of a number of criteria which seemed to be relevant to the application of EIA in the Netherlands (ERL 1981). Two of these criteria, namely 'analysis of sensitivity' and 'handling data shortages' can be used to give an indication of the ability of EIA methods to deal with uncertainty. ERL concluded on the subject of sensitivity analysis, that only one method 'specifically included an examination of the sensitivity of the result to differing assumptions on ranking and weighting within the methodology. A number of others could be reworked using alternative assumptions.' None of the methodologies was able to handle data shortages, although a few provided some guidance on how this could be achieved.

It was only with the discussions about the scientific content of EIAs during the early 1980s that 'uncertainty' was first mentioned as an important issue. In the report of a workshop on the scientific content of EIA, 3 of a total of 62 conclusions refer to the issue of uncertainty (Friesema 1982). These conclusions were, first, that EIAs are mainly concerned with expected events and phenomena, while the problems associated with a project are likely to come from unexpected quarters or low-probability events. Secondly, EIAs are poor at communicating uncertainty, because they often sound more certain than is justified and use unscientific ways to communicate uncertainty. Finally, conclusions should include statistical confidence limits and probability analyses.

Canter (1983), reviewing the current status and future direction of EIA, pointed to the need for the development and appropriate usage of more scientifically defensible impact prediction techniques, including those that yield a range of predictions and associated probabilities for those predictions to occur. Impact prediction techniques which address uncertainty and limitations of predictions . . . [and] . . . techniques which enable the conduction of sensitivity analyses of the influence of input data are also needed.

However, it was mainly the work of Hickling and others on the 'strategic choice approach' introduced in the Dutch Ministry of Environment in the early 1980s which enabled uncertainty to be identified as the heading under which a number of these problems could be investigated. This work was one of the elements of the government-commissioned research programme on uncertainty described above. Whereas problems of uncertainty as such have been recognized only recently in EIA, in other related fields, for example, risk assessment, the management of uncertainty is well established (Ruckelshaus 1983, Fischhoff & Stallen 1985, Otway & Peltu 1985).

Uncertainty and EIA – further exploration

In many of the methodologies developed in the early years of EIA (see, for example, Clark *et al.* 1980) consideration of uncertainty seemed to be non-existent. In later discussions the problems of uncertainty in EIA were mainly attacked by recommendations for further studies to develop 'better' models and for additional fundamental research on, for example, the functioning of ecosystems (Canter 1977). This kind of solution to the problems has, presumably, to do with a simplistic perception of the use of information in the decision-making process.

Figure 4.1 shows the 'classical' idea about the relationship between information on impacts and the decision-making process. This concept essentially refers to a one-way system involving a flow of information towards the decision maker. With this perception, the only way to handle decision-making problems is to improve information on the impacts, mainly by developing and using 'better' models or methods of prediction. Many of the common ideas about EIA are based upon this simple concept. It implies that EIA is first of all a way of making a good report on the environmental impacts of a proposed activity or project. Once the report is finalized, the work on EIA is done and it is up to the decision maker to use the report in the proper way.

Increasingly, however, it is recognized that EIA is not only a way of providing information in reports, but also a process to help the decision maker take environmental aspects into consideration. Accommodating this realization

Figure 4.1 The classical concept of the relationship between information and decision making.

requires the modification of Figure 4.1. First, an arrow back from 'decision making' to 'information' is required. This denotes that information requirements are dictated by the needs of the decision maker. Secondly, another element between 'information' and 'decision making' must be introduced. This can best be described as the 'EIA approach'. This stage involves choosing, for example, the methods to be used in assembling the information, selecting the way information will be presented and determining the alternatives that will be studied. In effect, the 'EIA approach' is a transition box where the requirements for information from the decision maker meet the information stream from the information box. With this structure (Fig. 4.2), procedural elements such as scoping should also be regarded as part of the EIA approach (De Jongh 1985a). To make the picture complete, it is important to note that after the decision is taken it must be implemented either in the initiation of a plan or by the establishment of the proposed project. It is only at this stage that the reality of impacts will become apparent and uncertainty resolved.

Analysing the different elements of Figure 4.2, it is clear that 'information on impacts' is the more scientific part of the process, whereas the 'EIA approach' covers the more subjective elements including choices concerning the selection of methods, alternatives and the types of impacts to be studied. Thus, there are at least two types of uncertainty playing a role in EIA. These are the uncertainty at the level of impact prediction and that encountered in the elements that go to make up the 'EIA approach'. Uncertainty in impact prediction is not confined to the physical environment alone, but is also manifest in the economic and social environment.

Another type of uncertainty, which fits within the scheme outlined in Figure 4.2, can be identified. This concerns related decisions. In the case of the Markerwaard proposals, for example, government decisions to change agricultural policy might be of great relevance to decisons on the new polder.

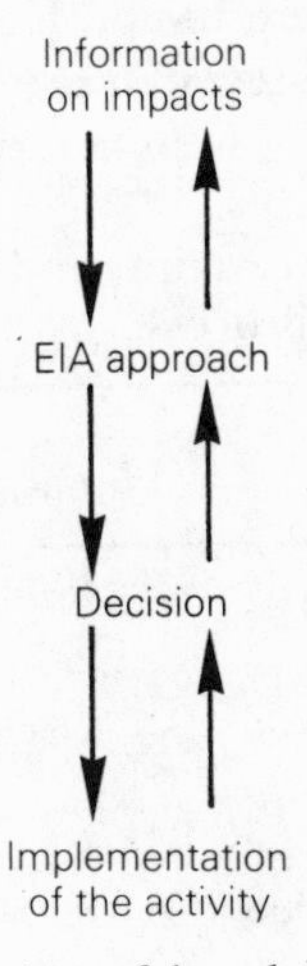

Figure 4.2 A more realistic view of the relationship between information and decision making.

Uncertainty in prediction is finally resolved by implementation. When impacts become clear, they are no longer uncertain. Of course this is a trivial statement, but in the management of uncertainty it is not without importance. Another trivial statement is that uncertainties in the 'EIA approach' are resolved by the taking of a decision. In fact, decision makers have the task of weighting different interests and, according to their value, making a decision. When, in a decision-making process, there is uncertainty in the 'EIA approach' (being related to values), further scientific research only helps decision makers by providing an alibi for delay. When decision makers are faced with a difficult political decision, the tendency to ask for more scientific research, while understandable, is not a logical way of solving the problem.

The various types of uncertainty need to be handled in different ways. Thus, uncertainty in predictions can be reduced in principle by further scientific research. Uncertainty in the 'EIA approach', on the other hand, can be reduced by means of, for example, negotiation. Uncertainty in related decisions can be reduced in principle by an agreement between the different decision-making authorities to co-ordinate their decisions. Table 4.1 contains an overview of these types of uncertainty and the principal ways of reducing them.

It is perhaps important to abandon the term 'solving the problems of uncertainty' and more appropriate to focus on the reduction of risk. This notion is significant for the management of the problem of uncertainty. This becomes evident when the differences between 'normal' scientific research and research directed at the clarification of uncertainty are described. In 'normal' scientific research, the research worker tries to resolve the environmental unknowns which he has identified as important against the standards of probability that he himself has defined. In reducing uncertainty in EIA, it is the requirements of the decision maker which define the standards and problems to be investigated, rather than the researcher. Where the problems are of great relevance to the decision at hand, very profound research may be required, perhaps amounting to basic 'fundamental' research. In other cases a simple or quick type of

Table 4.1 Different types of uncertainty and its reduction in different elements of the EIA process.

EIA process element	*Main type of uncertainty*	*Final certainty by*	*Way of reduction*
Information	Uncertainty of prediction	Implementation	Research
Approach	Uncertainty of values	Decision taking	e.g. Negotiating
Decision	Uncertainty of related decision	Evaluation	Co-ordination
Implementation			

investigation, perhaps involving no more than a telephone call, may be enough to reduce uncertainty to the level set by the decision maker.

It should be stressed that in the early years when the sole objective of EIA was the preparation of a report containing environmental information, EIA as such was meant to solve or reduce uncertainty only in prediction. Forced by law to take environmental aspects into consideration, the decision maker was placed in a situation where he lacked knowledge about the environmental consequences of his decision. Consequently, uncertainty in prediction was identified as a deficiency at an early stage in the evolution of the process. In later years, it became increasingly apparent that there were many aspects involving values in the EIA process, for example, concerning the significance of a particular impact. So the decision maker, facing uncertainties over values, often found that EIA reports were not really relevant to many of his problems. Therefore, alternative means had to be found to deal with other types of uncertainty. The scoping process, for example, is essentially a means of reducing uncertainty concerning values.

Uncertainty in prediction

Uncertainty in prediction can be reduced by research. In the framework of EIA, this means research focusing on the predictions of impacts on the environment. The Dutch government commissioned ERL to investigate uncertainty in prediction (ERL 1985) as part of a much larger study of predictive methods for EIA (ERL 1984, De Jongh 1984, 1985a).

As stated above, uncertainty is an unavoidable component of all predictions. In EIA, the objective is to provide information on the changes that will occur in the environment if a particular proposed activity is implemented. The changes that need to be identified include direct and indirect changes in the physical, chemical and biological characteristics of the environment, human health and amenity. These predictions can be made in different ways, depending on the data and expertise available, as well as on the quality of information required by the user. The extent of the uncertainty which inevitably creeps into all predictions will depend upon the data and methods used.

From the perspective of the decision maker, a number of issues concerning predictions are important. The main aspects relate to the reliability and accuracy of the information as decision makers will be concerned primarily about the extent to which the predicted result corresponds to the situation that will actually occur in the future. In addressing these concerns, two separate, but interrelated, issues can be recognized. These are the precision and accuracy of the information.

In general, the more precise the information required, the more difficult it is to obtain highly accurate information. Thus it seems to be possible to give accurate information only when the statements of likely impact are couched in very imprecise terms. For example, the prediction that if a road is built noise levels will rise, will almost certainly be confirmed upon implementation. Modelling on the other hand will give numerical values of projected noise levels, but these will only be accurate within certain statistical probability limits. In many discussions about the quality of EIAs, attention is focused on the accuracy of statements. This has led to very imprecise information being presented. A decision maker, therefore, should be aware that if more precise information is wanted, accuracy will be forfeit. Uncertainty in prediction has to do with these complexities. There are two main features of more accurate predictions. First, they usually involve more sophisticated methods, often involving the need for additional expertise and more resources. Secondly, they almost invariably require more information about the proposed activity, the local environment, and the behaviour of possible environmental contaminants.

HOW UNCERTAINTY CREEPS INTO PREDICTION

The process of prediction comprises six stages (ERL 1984). First, the scoping process involves a decision on how to describe an effect and, therefore, on what to predict. The second stage, baseline studies, requires the collection of data about the activity and the environment. Thirdly, a method for obtaining the prediction must be selected or developed. Fourthly, the method must be prepared for use, for example by calibration and validation. The fifth stage involves the application of the method to produce the required prediction. Finally, the results of the analysis must be presented to the decision maker. Uncertainty can be introduced at each of these stages.

The uncertainties in deciding how to describe an effect and, therefore, what to predict are of particular significance. It is at this stage that decisions about the required quantity of information and degree of precision are effectively made. The criteria for making these decisions include, the nature of any standards or acceptable levels of impacts against which predictions can be judged; the relative importance of an impact; the need to compare effects between alternatives; and the availability and feasibility of methods for predicting different types of impact.

Data are needed at every stage of prediction: in the development, calibration and validation of models, and in the application of the predictive method. Uncertainty may be generated during the collection of data. Here, a distinction should be made between two types of uncertainty, namely inaccuracy in measurement and sampling and the variability inherent in the data. The accuracy of a measurement is related to the precision and bias of the measuring instrument and the user (see Figs 4.3 and 4.4). Variability, however, is a characteristic feature of a natural environment. A single measurement of a particular parameter, such as temperature, could take any one of a range of values within defined statistical limits.

All predictive methods involve some model of the environment, mathematical, physical or conceptual. Uncertainties arise because these models cannot exactly reproduce what happens in reality. Three types of structural error may occur in models of the environment. First, in any environmental system many different processes may affect one variable. Models simplify this situation by

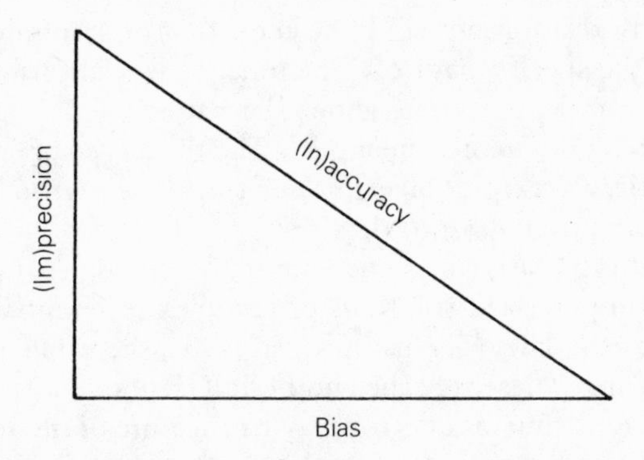

Figure 4.3 The relationship between (in)accuracy, (im)precision and bias. *Source:* ERL (1985).

assuming that only certain processses are important. These simplifications can be described as process errors. The second type of potential error in modelling is called functional error. These errors arise from uncertainty about the nature of the model with respect to a particular process. In a mathematical model, for

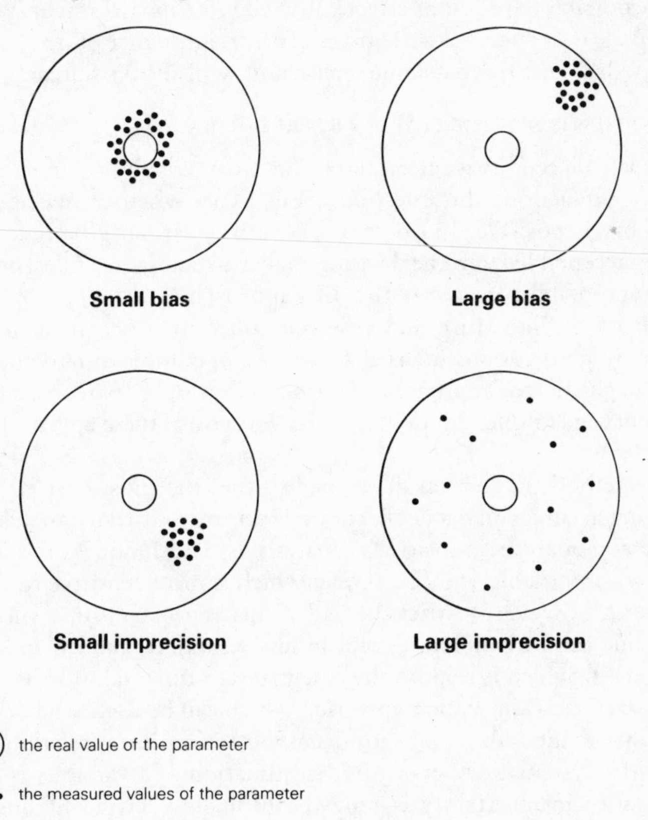

Figure 4.4 Bias and imprecision. *Source:* ERL (1985).

example, a decay relationship could be zero, first or second order, inverse or exponential. If the specific nature of the functional relationship is not known, therefore, incorrect assumptions about the nature of the decay relationship could be made in constructing the model. The third type of error in modelling can be characterized as resolution error. Such errors arise from imprecise spatial and temporal resolution of a model.

An additional type of error is encountered in certain mathematical models relying upon computerized solutions of complex differential equations. The mathematical method itself generates errors by the approximations in the numerical solutions. These are called numerical errors.

Each of these four sources of error may be a feature of model calibration and validation, whilst two causes of uncertainty can be identified in using predictive methods. The first results from using the method outside the range of circumstances for which it was developed and validated. The second source is simple human mistakes which can be a cause of major errors in prediction. Although some of these uncertainties can be removed, others such as the uncertainties associated with environmental variability are inherently irreducible. Even where uncertainities are reducible through additional research, it will never be possible to provide perfect knowledge for an assessment. While attempts can be made to reduce known sources of error, ignorance of, for example, how a system works ensures that some uncertainty will always remain.

MANAGEMENT OF UNCERTAINTY IN PREDICTION

Once it is conceded that uncertainty is an unavoidable and inherent ingredient of every prediction, the question arises as to whether management of the uncertainty is possible. In other words, can uncertainty be reduced to a level which is acceptable from the decision maker's standpoint. Clearly, defining the limit of acceptability is part of the 'EIA approach'.

Much of the literature and research concerning the management of uncertainty in prediction is focused on the use of complex mathematical models. Highly sophisticated approaches have been developed to overcome problems of uncertainty in complex modelling. An overview of these approaches is given in ERL (1985).

Two methods have been developed to find the most cost-effective way of improving accuracy in data collection. These are sensitivity analysis and Monte Carlo error analysis. Sensitivity analysis is a technique for identifying the parameter or variable within a model which is most sensitive to change. Some limitations, however, restrict the use of this approach. Only small changes in input value can be handled, except in linear models, and the inputs should be independent, which is seldom the case in environmental models. Monte Carlo error analysis is a simulation approach which can be used when there is a large uncertainty in input data and a model involves non-linear relationships.

Bias, the systematic over- or underestimation of a variable, is an important consideration in uncertainty as it may be difficult to detect and thus reduce. One method for checking bias is to use a selected method against a known standard.

If the level of the bias can be determined, measurements can be corrected accordingly. This process is usually referred to as calibration. Imprecision in data collection resulting from random differences in repeated applications of the same method is easier to assess simply by taking repeated measurements and using standard statistical tests on the results.

The available approaches for reducing uncertainty in prediction are summarized in ERL (1985). Figure 4.5 and Tables 4.2–4.4 not only give an overview of the approaches, but also indicate the way models can be developed with the aim of reducing uncertainty. Almost all approaches are based upon the assumption that mathematical models are used for prediction. Especially in EIA, however, where constraints on time and money are sometimes severe, this may not be the case. Moreover, a system is often not sufficiently understood for a reasonable model to be developed, or resources for building or improving prediction models are simply not available. In such cases, the analyst should not just ignore the possibility of a particular effect, but should indicate clearly that its consequences are not known.

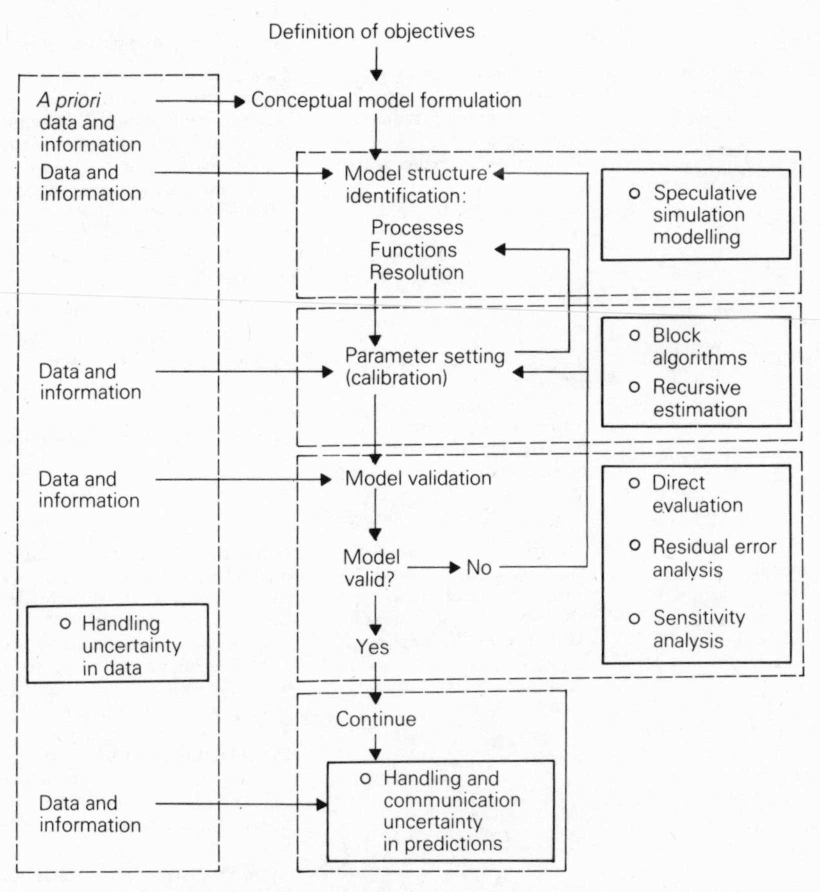

Figure 4.5 Model development. *Source:* ERL (1985).

Table 4.2 Objectives and description of techniques for handling uncertainty in model development.

Technique	*Objectives*	*Brief description*
1 Direct evaluation of forecasting ability output	To examine validity of model, i.e. how well modelled outputs fit observed data. May provide information to enable expert to propose structural changes in model.	A visual comparison is made between observed and predicted behaviour. Applicable to all types of models giving quantitative predictions. Visual comparison using graphs is particularly useful.
2 Residual error analysis	To examine validity of model by comparison of modelled and observed results; to identify non-random errors in modelled outputs. May provide information to enable expert to propose structural changes in model.	Residual errors (differences between observed and modelled values) are plotted against time, location, or input variable and the plot is examined for non-random behaviour. Non-random behaviour may suggest additional phenomena which should be included in the model. Statistical tests of 'goodness of fit' may be carried out. Applicable to all types of models.
3 Block algorithms	To provide improved estimates of parameter values and estimates of parameter error expressed as variance or a variance–covariance matrix.	Computer software based on block algorithms is used to compare observed and modelled behaviour. An iterative procedure is used to minimize the difference between model and observations by altering parameter values. The whole block of data is examined in one iteration. The algorithm also provides an estimate of parameter error. Applicable to mathematical models only.
4 Recursive algorithms	To provide improved estimates of parameter values and estimates of parameter errors as variance or variance–covariance matrix. To provide information enabling identification of necessary structural changes in model.	As for block algorithm method but only one item of data is examined in each iteration. Applicable to mathematical models only.
5 Speculative simulation modelling	To provide an indication of model structure where information and data are limited, and suggest areas for further research, enabling better formulation of model structure.	Behaviour of system is defined qualitatively. Estimates of parameter value ranges are made. Monte Carlo simulation and statistical tests are used to identify critical parameters by comparison with observed behaviour, and give an indication of important processes in model structure. Parameters are ranked in order of statistical significance to the model enabling priorities for further research into model structure to be allocated. Applicable to mathematical models only.

Table 4.3 Approaches for handling uncertainty in input data.

Technique	*Objectives*	*Brief description*
Handling uncertainty in measurement and analysis	To select measurement and analytical methods to achieve a required level of accuracy in terms of the bias and imprecision (i.e. total error) of data. These data may subsequently be used for prediction or for other purposes.	All methods of measurement and analysis are subject to bias and imprecision which together contribute to total error (inaccuracy). The selection of a method must reflect the need for accuracy; it is important for the data user to have some knowledge and understanding of the bias and imprecision in data.
Sampling programme design	To obtain information describing a system, with a required level of detail and accuracy by design of an appropriate sampling programme in terms of the size, frequency, location and randomness (or otherwise), of sampling.	The results from sampling are used to make inferences about the whole system; these are subject to uncertainty. Detail and accuracy are related to sample size and the frequency, location and randomness of sampling in relation to the system characteristics and the measurement objectives. Various statistical techniques can be used to assist in sampling design.
Sensitivity analysis	To identify those inputs which contribute most to uncertainty in prediction and so allocate priorities for further effort to improve the accuracy of predictions. Inputs can be ranked according to their priority for further research to improve the accuracy of predictions.	Inputs are ranked according to the partial derivatives of the model equation ($\partial y / \partial x$, where y is model output and x is model input) called the sensitivity coefficient. Inputs with the highest sensitivity coefficient have the greatest influence on uncertainty in predictions. The sensitivity coefficient of each input and the cost of improving the accuracy of each, are considered in assigning priorities for research effort. Sensitivity analysis is, however, subject to certain key assumptions which if violated (which they often are in environmental models) may lead to misleading results.
Monte Carlo error analysis	To identify those inputs which contribute most to uncertainty in prediction and so allocate priorities for further effort to improve the accuracy of predictions. Inputs can be ranked according to priority for further research to improve the accuracy of predictions.	Inputs are ranked according to their correlation coefficient derived from the mean of $\Delta y / \Delta x$ over many repeated probabilistic simulations (Monte Carlo simulation). Inputs with the highest correlation coefficient have the greatest influence on uncertainty in predictions. Priorities for effort can therefore be assigned as above. Monte Carlo error analysis is not subject to some of the constraining assumptions of sensitivity analysis but the mathematics/statistics are more complex.

Table 4.4 Approaches for handling uncertainty in prediction.

Technique	*Objectives*	*Brief description*
Scenario approach	To predict a range of possible outcomes (say maximum, minimum, most typical) taking into account input (variable and parameter) uncertainty.	Values of uncertain inputs (e.g. worst and best cases) are selected which give rise to the specified outcomes. These are modelled deterministically to indicate the range of outcomes.
Monte Carlo simulation	To predict the probability distribution of possible outcomes taking into account input uncertainty.	Probability density functions (PDFs) are specified for uncertain inputs (and correlations between inputs if appropriate); a large number of input sets are randomly selected and each is used to make a deterministic prediction. The results can then be plotted to give a PDF of prediction outcomes. Computation becomes very complex with correlated inputs.
Constrained parameter approach	To predict the probability distribution of possible outcomes under different scenarios, taking into account input uncertainty.	The method combines the scenario and Monte Carlo approaches. Certain uncertain inputs are set to give a number of scenarios (e.g. wet, average and dry year). The other uncertain inputs are then allowed to vary randomly within the specified PDFs using Monte Carlo simulation for each scenario.
First-order error analysis – propagation of error	To predict the uncertainty, i.e. the mean and variance of the outcome taking into account the mean and variance of uncertain inputs.	The means and variances of inputs are propagated through the model to give the mean and variance of the prediction. The method is analytical in contrast to the numerical approach of Monte Carlo simulation. The applicability of the approach is constrained by certain key assumptions, notably independence of inputs often not met in environmental models. The predictions may therefore be inaccurate and may actually be misleading if these assumptions are not met.
Generation of moments technique	To predict the uncertainty, i.e. mean, variance, skewness and kurtosis of the outcome, taking into account these characteristics of uncertain inputs.	Like first-order error analysis the statistics describing the uncertain inputs are propagated through the model. The method is again analytical. The approach suffers from fewer constraints than first-order error analysis but the mathematics may be very complicated.
Speculative simulation modelling	To predict the PDF of expected outcomes taking account uncertainty in inputs in conditions of sparse data.	The method is an extension of Monte Carlo simulation except that the population of parameter values from which random selections are made is defined by those parameter values which give rise to observed behaviour, rather than as a PDF.
Expert systems	To make predictions from a basis of lack of knowledge and data.	Models are constructed from sets of logical rules about system behaviour defined by 'experts' and used to test future responses to changes in system variables.
Probability encoding	To elicit expert opinion on the uncertainty associated with data in the form of a PDF or cumulative probability distribution.	Interactive interview techniques are available to elicit probability estimates, tools used include probability wheels. Methods for dealing with conflicting expert opinions have also been developed.

In many cases, however, a statement can be made based, for example, on the opinion of experts. These predictions can be formalized in such a way that the opinions of different experts are contrasted systematically with one another so that some indication of likely outcome is given. In the presentation of such predictions it should be stressed that these are based upon expert opinion.

Uncertainty in the 'EIA approach'

As discussed above, the 'EIA approach' is a compilation of those elements in the process which are concerned with subjectivity. However, many choices have to be made in the more scientific parts of the process, particularly in the prediction of impacts. Some of these choices have to be accommodated within the 'EIA approach' because decision makers may be interested in them for a number of reasons.

First, the decision maker will be concerned about the significance of an impact. The importance of a predicted impact in terms of, for example, a legal standard, may determine the basic need for information. Furthermore, the way the standard is defined may dictate the need to reduce uncertainty. This may require information to be presented in a particular way with a predetermined level of precision so that the impact can be compared with the standard.

Political sensitivity also plays a major role. If an issue has assumed political importance, a decision maker may demand better information simply to ensure that his decision is defensible, even though by accepted standards, the predicted effect is not significant.

Finally, confidence must also be taken into consideration. In general, the decision maker will want to ensure that he feels confident about a prediction. No general recommendations can be made as the needs of the decision maker will vary depending upon, amongst other things, his own background. Someone with a scientific background may feel confident with the type of information provided by a particular scientific group. Another decision maker may require further information or the advice of other experts in order to generate sufficient confidence concerning a prediction.

These aspects of uncertainty all have a scientific connotation. There are, however, many choices to be made in the process in which scientific knowledge plays only a minor role. These choices form the main framework of the 'EIA approach'.

CHOICES IN THE 'EIA APPROACH'

Many choices have to be made in the EIA process (ERL 1985). One of the most fundamental is the choice of impacts which should be studied and the level of analysis that is required. In many cases this choice is seen as a scientific task. Yet, as there are so many possible impacts of a proposed activity, these choices are subjective and, therefore, part of the 'EIA approach'.

A second type of choice concerns the alternatives to be investigated. Magness

(1984) considers that 'alternatives are the heart of EIA'. This in itself is an element in the management of uncertainty. As there is uncertainty about the results of predictions and, because standards are not available for many environmental parameters and impacts, the reduction of uncertainty can be found only in the comparison of alternatives.

The weighting of impacts is also an important subjective element of the assessment process. Many sophisticated methods have been developed which purport to solve this problem, see for example, Clark *et al.* (1980). In most cases, however, no real guidance can be given and *ad hoc* choices have to be made. A closely related consideration concerns the way in which alternatives are compared according to various criteria and, hence, in the selection of these criteria themselves. Here too, many methodologies have been developed. In practice, however, these problems are also resolved in an *ad hoc* manner.

An additional type of choice involves defining the limits and constraints upon particular studies. Thus, such issues as to whether long-range impacts and impacts upon future generations will be taken into account must also be resolved. Here uncertainties related to values creep into decisions as this will depend upon the perceived importance of the issues.

The remaining types of choice are of a somewhat different nature. It will be necessary to identify those who will participate in the EIA and the manner of their involvement. Clearly, this is a subjective issue which will vary depending upon a range of factors, such as the system of government as well as the cultural and traditional setting, which dictate the nature of involvement. Finally, there is the choice of the overall approach to be adopted, which is often made in only an implicit way.

Voogd (1982), in a review of multi-criteria evaluation, identified four types of uncertainty which are also relevant to this discussion of EIA. First, both involve uncertainty associated with the choice of evaluation criteria. Voogd's second category, assessment uncertainty, is comparable to uncertainty in prediction. Thirdly, priority uncertainty concerns the relative value given to different impacts by the decision maker. This type can be compared with the values in EIA mentioned above. Finally, there is so-called 'method uncertainty' which is also applicable to EIA. Each method or technique is based upon certain assumptions which are mostly arbitrary, which implies that the outcome of an evaluation is affected by uncertainty concerning choices and the consequences of assumptions.

The types of uncertainty identified by Voogd relate not only to uncertainty in prediction, but more importantly to uncertainty in value. The final category, method uncertainty, is of particular significance for EIA. The explanation that Voogd gives for this type of uncertainty shows an interesting outlook on the way uncertainty is resolved. Voogd argued that assumptions are, in effect, the solutions to problems of uncertainty. In the study of methods in EIA mentioned above, therefore, special attention was focused on the assumptions which underlie different methods (De Jongh 1983, 1984, ERL 1984). It was concluded that scientists often have great difficulties in defining the assumptions upon which their methods are based (De Jongh 1985b). Clearly scientists involved in EIA require training in the explicit handling of uncertainty.

MANAGEMENT TOOLS FOR HANDLING UNCERTAINTY IN THE 'EIA APPROACH'

Having recognized problems of uncertainty, the traditional way for solving them would be to develop improved methodologies. This would involve also the need to improve the framework for performing EIA and to handle issues of uncertainty.

One such improved method is decision analysis. ERL (1986) contains an overview of the approach in the context of handling uncertainty within environmental policy formulation and review. Many components of decision analysis also seem naturally to form elements of the EIA process. Essentially decision analysis provides a quantitative framework for making decisions when uncertainty exists. The approach rests upon two assumptions. First, there are incomplete data at the time of the decision. Secondly, in making a decision, a decision maker will be influenced by the likelihood of events occurring and the values associated with the possible consequences of events which are uncertain.

There are two key elements in decision analysis. First, it provides a consistent method for structuring and clarifying the decision problem in order to reduce the uncertainty within evaluation. Secondly, expert opinion is used to derive numerical estimates based upon subjective considerations where objective probabilities are not available. One of the main tools in decision analysis is the so-called 'decision tree', in which the different actions, along with their likely consequences, resulting from alternative decisions are shown systematically in graphical form (Fig. 4.6).

Decision analysis deals explicitly with uncertainty. Not all of the different types of uncertainty, however, can be resolved by the analyst alone. Input from the decision maker is also required. As the approach provides no mechanism for generating it, decision analysis can only be used in procedures which guarantee this input.

Improving methods for handling uncertainty is an ongoing process. In a recent article, Hobbs (1985) explored the use of an amalgamation method for EIA. Amalgamation is needed in situations where there is no clear view of the relative importance of different criteria for decision making. Sometimes this view is only prevalent amongst the analysts, but in many cases, decision makers also have no preconceived notion of the relative importance of different criteria. Hobbs (1985), reflecting the earlier views of Elliot (1981), advised against over-reliance upon this approach, citing a number of concerns. He considered that 'the precision of their numerical evaluations should not be mistaken for accuracy nor for consensus' and that 'amalgamation techniques should clarify tradeoffs and value conflicts, not hide them'. He recommended that the 'prudent course for those who use amalgamation for any purpose is to check assumptions, use more than one method, and conduct sensitivity analysis' (Hobbs 1985). These steps will help uncover uncertainties and biases and gauge their significance. The conclusions of Hobbs can be broadened to whatever

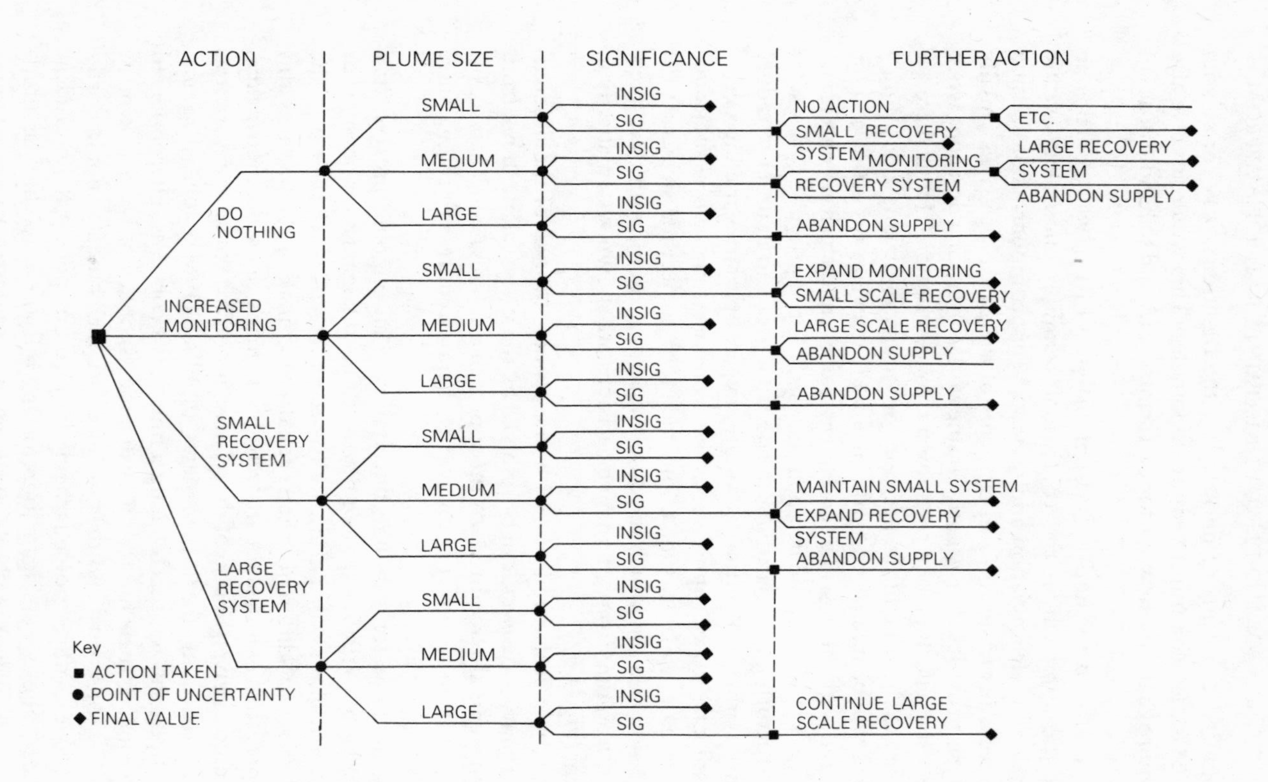

Figure 4.6 A decision tree for the problem of groundwater contamination.

improvements in methods can be made. The main problem of uncertainty related to values cannot be resolved, only clarified. This in itself is not unimportant.

In looking at solutions to the uncertainties that he identified, Voogd (1982) considered a number of approaches which could be used such as checklists, probability functions, sensitivity analysis and additional research. In his conclusions, Voogd warns against the inclination to use too sophisticated treatments of uncertainty such as probabilistic exercises, commenting that 'the theoretical elegance of those exercises is often in contradiction of their operationality'. Bisset (1980) notes that the development of complex methods involving numerical computations has been an important preoccupation in EIA methodological development. He concludes that this emphasis may be misplaced because

> as decision making procedures are made more accessible to public involvement and become more politicized, the ability of methods to provide appropriate information, for both experts and laymen, is likely to gain in importance. It is likely that the role of impact analysis in political debate surrounding contentious proposals will result in the use of methods which eschew quantitative manipulations and concentrate on impacts.
>
> (Bisset 1980)

These conclusions reinforce the view that uncertainties related to values, cannot be resolved by improving the more scientific side of EIA.

One way of resolving value-related uncertainty is to give more general and generic guidelines for the application of EIA (ERL 1981). In such guidelines the types of impact to be studied, the type of alternatives to be considered, even the methodology to be used can be prescribed. Essentially, this approach involves resolving uncertainty in specific cases by taking decisions at a higher level. Obviously, there are great advantages to this approach. There are also important disadvantages. First, it is impossible to foresee all aspects of each specific case. Secondly, strict regulations stop all creativity and prevent opportunities for protecting the environment in particular situations from being taken.

In the field of risk management, it has long been recognized that methodologies and predictions alone do not solve the management problem and that public response should also be taken into account. Communication is one of the tools which are available (Fischhoff & Stallen 1985, Otway & Peltu 1985). Thus, although the Dutch government defined standards to regulate the risk of hazards, it left a range in the standards so that the particular circumstances of a specific situation could play a role in determining the final standard that is imposed (Ministry of Housing, Physical Planning and Environment 1985). Thus, in the field of risk management it is also recognized that individual circumstances and specific values should play a role in the decision-making process.

The management tools to reduce the uncertainties described previously all fail to consider communication between the various parties involved in the process. It is clear from a consideration of experience within the Dutch Ministry of Environment, that uncertainty in values can only be managed by way of communication between not only the analysts and the decision makers, but also the different groups involved in the decision-making process.

In fact, management tools for handling uncertainty in the 'EIA approach' can be divided according to the type of relationship that exists between the decision maker and the analyst. When the analyst and the decision maker belong to different organizations or groups and their communication is more or less formal, they can be characterized as techno-scientific tools. When they belong to the same organization, or when a special forum is established in which both can meet, the tools aimed at smoothing communication can be described as socio-scientific.

Using techno-scientific tools, an analyst will be working on the uncertainties concerning values according to a brief from the decision maker. This 'information requirement stream' is canalized in structured meetings in which analyst and decision maker, or their representatives, work together. The distinction between socio- and techno-scientific tools is not strict, but is useful in understanding the different perspectives for solving problems of uncertainty. It might be worthwhile to point out a number of socio-scientific tools which have been developed in recent years. Scoping and public participation are the procedural answers to resolving uncertainty in EIA (Council on Environmental Quality 1978, 1982, ERL 1981). Negotiation, mediation and the strategic choice approach are the EIA process-oriented solutions to this type of uncertainty.

Some US and Canadian experts have concluded that the main problems in EIA are triggered by differences in interests between different parties. Susskind *et al.* (1978), for example, analysed the different interests of environmentalists and developers. They considered that environmentalists were concerned about long-term impacts and costs as well as the cumulative impacts of ecological interventions. They were averse to taking risks, fearful of the overuse of resources and likely to view the environment holistically. In contrast, developers considered projects and impacts separately, liked taking risks, had a narrow view of the environment, believed in infinite resource substitution by technological innovation and were likely to emphasize short-term benefits from investments. On the basis of these observations they concluded that, regardless of the outcome of a prediction, the weighting of values is a, and even possibly the, most important element in the EIA process.

Ultimately, litigation is one means of ensuring that the weighting of values finally will take place. If these lengthy procedures are to be avoided, however, negotiation between interested parties should be integrated into the EIA process. Negotiation which is an integral part of the EIA process, as well as being involved in the 'EIA approach', can be seen as an attempt to resolve this type of uncertainty.

Negotiation represents a more advanced stage in the evolution of EIA procedures than public participation, as it involves different interests assuming equal status. The first steps in the process of dispute resolution through negotiation, therefore, involve identifying the parties that have a stake in the outcome and ensuring that each interest group is represented (Susskind *et al.* 1978). A later development in this field is a more sophisticated approach called 'mediation' (Harter 1982, Curtis 1983, Susskind & Ozawa 1983, Susskind *et al.* 1983). According to Curtis, environmental mediation is 'a voluntary process where parties, with the assistance and guidance of a mediator, understand and compromise a dispute about a proposed project'. There are five objectives of environmental mediation. First, it is designed to ensure that all parties which have an interest are represented. This avoids the potential of new issues and organizations arising subsequently. Secondly, it should achieve a settlement to the mutual satisfaction of all parties concerned. Thirdly, continued dialogue between the parties on future related matters should be facilitated. Fourthly, a settlement in a relatively short time period and at low cost should be achieved. Finally, the community at large should view the settlement as just and fair.

US and Canadian developments in EIA mediation have helped to solve environmental disputes and in so doing have reduced levels of uncertainties to such a degree that decisions can be taken. This approach has not been used in Europe. However, in view of the promising results of comparable approaches to the development of environmental policy, it is likely that mediation will be one of the more important management tools in EIA in the future.

Susskind *et al.* (1983) evaluated the negotiation and mediation processes for resolving environmental regulatory disputes and came up with an important condition for their successful application. They concluded that 'only when all key parties are uncertain of their power to win, or of the total costs of perpetuating the conflict, or when they recognize that there are various agreements that are better for both of them than no agreement, will they start negotiating'. Moreover, they stress the importance of two further factors. First, there should be a mediator to steer the process. Secondly, there may be a need for contingent agreements to be reached during the negotiations. These permit certain activities on a temporary basis, pending the accumulation of further information concerning their environmental and social effects.

The use of mediation as a tool for the management of uncertainty in the 'EIA approach' is a choice finally to be made by the decision maker or his representative. As Wondolleck (1985) concluded, 'environmental decision makers need to understand that perhaps the most valuable impact that they can have in a given situation is not always in determining what decision or outcome is finally reached but often in how those decisions are made'.

In the Netherlands, another set of socio-scientific approaches seem to have been successful in resolving environmental disputes. These are called strategic choice approaches (Hickling 1974, 1975, Hickling *et al.* 1976). The major elements of the strategic choice approach are the involvement of a group consisting of the main stakeholders, the use of a facilitator (comparable to a mediator) and the explicit definition of uncertainty.

In the handling of uncertainty using the strategic choice approach, a number of elements play a part. The type of uncertainty, primarily in understanding how the environment functions, in values and in related decisions, is important. Major concerns, however, are the relevance of the specific uncertainties to the decision at hand and the difficulties of reducing individual uncertainties. A range of management tools is available depending upon the specific problems that are encountered. Thus, an explicit decision about the relevance of a particular uncertainty can be made. Similarly, assumptions can be made in order to remove uncertainty. Coupled with this is the level of development of a contingency plan to deal with the resulting situation if assumptions turn out to be incorrect. This element of the strategic choice approach is comparable to the continuous dialogue in the mediation process. The group can also decide to initiate research to reduce uncertainty. Consequent upon this is the decision about what will be done when the results of the research are known, in particular what actions will follow. Finally, it may be appropriate to initiate decisions at a higher level in the hierarchy. This is comparable to the need for generic guidelines described above. Application of these tools leads to a scheme for structuring the discussions around a set of decisions which have to be made. In the strategic choice approach, this is called a commitment package (Table 4.5).

The idea of back-up or contingency plans is related to another management tool for the handling of uncertainty, namely phasing the implementation of a proposal. Holling (1978), Ward (1978) and Beanlands & Duinker (1982) stress the importance of ongoing ecological research during implementation. Caldwell (1982) also suggests that the

> concept of monitoring, follow-up and feed-back would extend the environmental impact statement beyond a cautionary or action-forcing device into a continuing tool of management and evaluation. The full decision record and the feed-back loop assist an agency to assess the accuracy of its predictions, to see how mitigation measures have been working and to adopt subsequent decisions as feed-back may indicate.

Table 4.5 An example of a commitment package.

Actions to be taken now	*Research to be started now*	*Actions later*	*Back-up plan*
e.g. Phase one of the project is executed.	e.g. 1 For phase two the environmental aspects of alternatives A, B, and C will be researched. 2 Monitoring during phase one.	e.g. Execution of alternative A, B, or C of phase two according to the outcome of the research.	e.g. If monitoring during phase one shows levels of substance *X* higher than *Y* mg/l, the execution will be stopped.

In the Netherlands, EIA regulations require a monitoring programme as part of the EIA process (Ministry of Housing, Physical Planning and Environment 1985).

Concluding remarks

How to use management tools in order to reduce uncertainty is a major concern. Obviously, there is a problem of choice which is influenced, amongst other things, by the preferences of individuals and groups. Some may favour socio-scientific techniques, whereas others prefer techno-scientific approaches. It will be a challenge for the further development of EIA to combine both approaches by phasing the EIA process. This might involve the alternation of socio-oriented periods with techno- or research-oriented phases.

In the socio-scientific phases, mediation and the strategic choice approach could be used to focus not only on the subjective elements of the EIA process, but also on the definition of any uncertainties which are relevant at this stage. The research-oriented periods should be used to reduce the defined uncertainties mainly in predictions. The socio-scientific phases will continue with commitment packages, whereas the research phase will end with the submission of reports containing the results of the studies. The scoping and public participation processes could provide the raw framework for this phasing. Presumably, in many practical situations there will be a need for more than one techno- and one or two socio-scientific phases (Fig. 4.7).

The management of uncertainty is on the one hand a very old component of our lives. Daily, everyone takes decisions and, in so doing, manages uncertainty. On the other hand, as a specific field in environmental management, it is a new area of concern. Appropriate management tools seem to be available. Although at present there is only limited practical experience, socio-scientific tools have been used successfully. This is undoubtedly a development which will be explored in the years ahead.

Successful application of these techniques requires a skilled mediator to lead group activities. The process depends upon the presumption that participants have different values. These differences can be translated into the disagreements which are the basis for mediation exercises. Alternatively, these differences can be expressed as uncertainty related to values, a major element of the strategic choice approach.

With respect to techno-scientific tools, it is clear that the computerization of environmental models is a most important development and one which is still at an early stage. In some areas, for example in water management and in risk assessment, modelling of the environment and human influences upon it are relatively advanced. In these fields, quick assessments based upon existing models are sometimes possible. Undoubtedly in the future this facility will be possible in other areas of the environment, shortening the research-oriented phase of the EIA process. Moreover, this may lead to more explicit attention

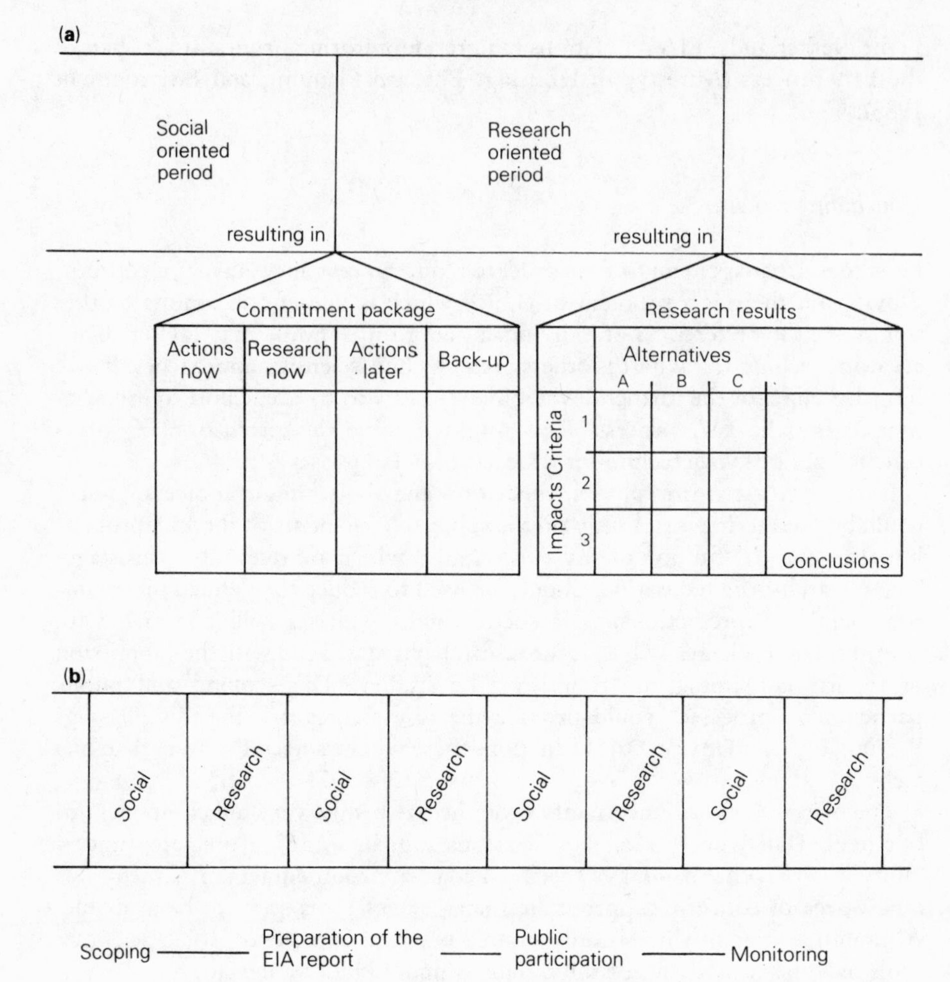

Figure 4.7 Social and research-oriented periods in EIA showing: (a) results, and (b) phasing.

being paid to the values which underlie decisions, as there will be time available for exploring the consequences of a decision using simulations. Although these developments will not eliminate uncertainty from the EIA process, they should enhance the possibility of managing it. It is likely that the management of uncertainty will be a major focus of attention in EIA and in other related fields of decision making in the years ahead. Learning from other fields of decision making could be very useful for the future development of EIA. Though most of the investigation to date has been within the area of risk assessment, experience from business management might also prove to be of great value.

5 *Environmental impact assessment and risk assessment: learning from each other*

R. N. L. ANDREWS

In concept, environmental impact assessment (EIA) and risk assessment (RA) have evolved as parallel and sometimes overlapping procedures for rational reform of policy making. With other forms of policy analysis, such as applied systems analysis, as well as cost–benefit and cost–effectiveness analysis, they share the common presumption that policy decisions can be improved by the application of explicit analysis and documentation. Both are intended to provide reasoned predictions of the possible consequences of policy decisions and, thus, to permit wiser choices among alternative courses of action.

In practice, however, EIA and RA have been nurtured by different disciplinary and professional communities in largely separate policy contexts. As a result, they have evolved differences of emphasis, both in substance and in process, which merit notice and reflection. Some of these differences reflect the varying functions of the two types of analysis, but others suggest opportunities to improve both EIA and RA by the transfer of features from one to the other.

Many of the policy decisions most in need of analysis, in fact, require some combination of both. Generally, a systematic identification of possible environmental impacts, as well as a rigorous analysis of their magnitude and probability is required. Examples include offshore hydrocarbon developments; environmental applications of pesticides; new biotechnologies; siting of potentially hazardous industrial facilities; as well as a wide range of others. Exactly what the combination of the two approaches should be, and how the breadth of impact identification should be traded off against the depth of predictive analysis for key impacts, is an important question for study.

Both EIA and RA, therefore, could probably benefit by learning from each other and, in many cases, by consolidation into a unified process. The purpose of such a process, however, is not merely to produce the most quantitatively sophisticated estimate of particular risk, nor the most comprehensive list of possible environmental impacts. It is, rather, to produce a rationale for making public policy decisions that is both well reasoned, and recognized as legitimate and acceptable by the public.

This chapter recommends two particular topics for research. The first is to develop protocols for unified environmental impact and risk assessment of proposed government actions. This process should begin with actions, such as the siting of hazardous technologies and the environmental dispersion of potentially hazardous substances, where the need for both forms of analysis is already acknowledged. These protocols should address not only the substance of the assessment, but also the accountability of the process by which the assessments are framed, executed and legitimized for use in public decision making.

The second research need is to attempt unified assessments of existing complexes of hazards for human health and ecological systems. At present, both EIA and RA are applied primarily to new proposals, such as government projects and regulations. Many of the most serious hazards, however, arise from existing situations and from the cumulative patterns of urban and industrial development rather than from new government actions. A prudent approach in setting priorities for environmental protection and risk management, therefore, is to address existing risk patterns as well as new proposals.

Professional communities

In essence, both environmental impact and risk assessments are forms of applied policy analysis, rather than purely scientific studies. That is, their purpose is to provide an acceptable basis for making public decisions, not necessarily to generate new scientific knowledge. The results of these investigations, therefore, are acknowledged to be judgements within constraints of time, money and existing knowledge. These judgements in turn are made by professional practitioners of particular forms of analysis, whose approaches are shaped both by their experience and by the norms and paradigms of their disciplines.

Environmental impact assessment has developed a large, but loose, community of professional practitioners, whose academic backgrounds are drawn primarily from ecology, natural resource management, environmental science and engineering, along with some from anthropology and sociology. The most sophisticated environmental impact assessments, such as the Trans-Alaska Pipeline System EIA, represent the results of extensive studies and interdisciplinary collaboration by teams of highly qualified experts. Most, however, are prepared by small staffs of professionals with masters-level qualifications, representing only a few key disciplines.

Risk assessment is a similarly loose label, but appears to represent some half-dozen discrete disciplinary subgroups rather than a single interdisciplinary approach. Amongst these groups, toxicologists, epidemiologists and biostatisticians, focus on health risks (mainly cancer mortality); engineers and statistical decision analysts are concerned with technological catastrophes; economists are interested in risk–benefit analyses; actuaries are concerned with probabilistic studies; while cognitive psychologists explore aspects of human perception and behaviour towards risk. Risk assessments to date appear to reflect the choice of approaches from within one of these disciplines, rather than an eclectic or interdisciplinary synthesis of several of them. Less often has this been the situation with EIA.

Both professional communities would benefit from greater interaction with each other, as each would bring a unique range of strengths and experiences. As yet, only a few papers have been written on the relationships between EIA and RA (O'Riordan 1979, Beanlands 1984a and 1984b, Giroult 1984, Vlachos in press). Only one of these has yet been published. An overview of five years' issues of the journal *Risk Analysis* turned up fewer than half a dozen mentions of EIA and no articles in which it was a central topic. Journals on EIA have perhaps paid more attention to RA as an emergent form of analysis, but have done no more to develop substantive integration.

Substance

As policy analyses, environmental impact and risk assessments should be compared in two ways. One is substantive, concerning the content of such analyses including the actions or conditions assessed; the alternative actions considered; the consequences investigated; the basis used for predicting consequences and attributing them to the action; and the treatment of uncertainty and subjective judgements. The second characterization is procedural, that is, how such assessments function as administrative processes, including their legal basis and purpose; their openness and accountability; and their role in the ensuing decisions. Substantive characteristics are explored in this section, while the following section contains a discussion of process issues.

TARGET ACTIONS

Environmental impact assessments are required for all major governmental actions that might 'significantly affect the quality of the human environment'. In practice, the majority are prepared for public works proposals such as highways, water resource developments, energy projects and public land management activities. In the United States, EIA requirements do not apply to most environmental health regulatory actions, which have been exempted by statute. In addition, many major non-governmental actions, such as hazardous industrial facilities, where the only government action is the issuing of a permit under appropriate regulations, are not included. Similarly, EIA is not enforced for legislative proposals or other broad policy actions. Elsewhere, for example in Canada and within the European Economic Community, industrial development projects are more consistently included under environmental impact assessment requirements.

Risk assessment, in contrast, is practised (albeit selectively) in both public and private sector decision-making processes. It is increasingly routine, for instance, in both the insurance and chemical industries, and is frequently used by energy production and electric utility firms. Within the public sector, risk assessments have been prepared primarily in conjunction with proposals to regulate particular substances as health hazards, and with some proposals to site energy production and industrial chemical facilities that pose risks of catastrophic accidents.

Unlike environmental impact assessments, risk assessments are not generically required by statute and, therefore, have not been produced under any common set of protocols or administrative guidelines. The primary demands for risk assessments in public decision making have arisen from three sources. These are specific laws requiring risk–benefit balancing in environmental health regulations; 'commission of inquiry' proceedings into proposals for hazardous facilities; and from more general administrative pressures for justification of proposed regulations.

By and large, both environmental impact and risk assessments have been applied only to discrete proposals for future action and to individual hazards for which government controls might be warranted. EIA and RA have been applied only rarely to existing complexes or cumulative patterns of risk to health and to the environment, such as urban areas, even though such an area might provide a more realistic unit of analysis for assessing relative risks and for setting priorities for management response. One recent exception is the Philadelphia Study conducted under the aegis of the Environmental Protection Agency's Integrated Environmental Management Division (Haemisegger *et al*. 1985).

ALTERNATIVES

The treatment of alternatives is a central issue for any form of policy analysis, for it not only affects the scope and emphasis of the analysis itself, but also determines the relationship of the analysis to the ensuing decision process. If an assessment considers only the consequences of a single action, it can perhaps produce more detailed quantitative estimates of possible consequences. However, it will also be fundamentally limited to justifying the particular proposal or, at most, identifying marginal changes which might mitigate its undesirable effects. In contrast, if the assessment is designed to compare alternative courses of action, it provides, in effect, the framework for a decision rather than mere justification of a proposal. To serve this purpose, an appraisal must be structured to emphasize differences among the consequences of alternative courses of action, rather than systematically tracing the consequences of a single course of action.

Environmental impact assessments are required to address alternatives, including the alternative of taking no action, so that the user can compare the full consequences of alternative courses of action. In practice, environmental impact assessments are often criticized for failure to consider seriously options preferred by some groups and individuals. EIA procedures, however, do allow them to introduce new and, sometimes, superior alternatives after reviewing those proposed by the agency. This occurred, for instance, in a recent assessment of alternative management plans for the Nantahala National Forest, North Carolina, USA. More than this, the requirement to consider alternatives creates a healthy pressure on the analysts themselves to focus on differences between real choices. In such circumstances, analysis is more likely to provide the basis for a final decision.

Risk assessments are more heterogeneous in their treatment of alternatives, probably because of the absence of any generic guidance on the subject. Risk assessments for health regulations, for instance, often include estimates of risk under alternative standards. This is now required in the USA for regulations that may have significant economic consequences. Risk assessments for technologies systematically identify alternative cause and effect sequences by which hazards could arise. They have also led to the identification of alternative measures to reduce risks, especially in cases such as Canvey Island, where reasonable design or operational changes could significantly mitigate risk factors (Cohen & Davies 1981). To date, however, many risk assessments have been designed more to provide quantitative estimates of the risk associated with a single proposed action, rather than comparisons of alternatives.

One promising target for future research, therefore, might be the design of comparative risk assessments to show trade-offs between alternative courses of action. A major consideration would be the need to include an evaluation of how such comparative assessments would need to differ from current approaches to RA.

TARGET EFFECTS

The selection of target effects also determines the overall scope of analytical effort in both EIA and RA. In principle, environmental impact assessments can include virtually any category of impact that might be of interest. As Munn (1979) has defined it, environmental impact assessment is:

> an activity designed to identify and predict the impact on the biogeophysical environment and on man's health and well being of legislative proposals, policies, programmes, projects, and operational procedures, and to interpret and communicate information about the impacts.

In practice, however, environmental impact assessments have emphasized possible impacts on natural ecosystems and, to some extent, human communities, but have paid little attention to health effects or other risks (Beanlands 1984a, Clark 1984a and Giroult 1984). More precisely, even for impacts whose ultimate significance might involve health, such as air or water pollution, EIA studies typically predict only the environmental fate of contaminants, rather than the effects on health itself.

Conversely, risk assessments have emphasized human health effects, especially potential mortality due to cancer or technological catastrophes. Only a few studies, for instance on offshore oil rigs, have attempted to assess other environmental hazards (see, for example, Cohen & Davies 1981, National Research Council 1982, Covello & Mumpower 1985). One important exception is a recent study by the National Science Foundation on environmental applications of biotechnology. This study recommended the use of risk assessment methods to assess their potential environmental effects (Covello & Fiksel 1985).

These differences have no intrinsic basis in the nature of the two analytical approaches; they appear to have arisen simply as artefacts of the administrative contexts and professional communities associated with each. Both environmental impact and risk assessment would be improved by eliminating such differences. This would result in health effects being incorporated into EIA and conversely risk assessment being applied to potential environmental consequences other than human mortality.

PREDICTION

Both EIA and RA are forms of applied predictive analysis. In practice, however, environmental impact assessment has much to learn from the more sophisticated approaches to prediction that have been developed in risk assessment.

Environmental impact assessments generally exhibit crude and simplistic estimates of the magnitude, likelihood, and time distribution of impacts. Prediction is typically limited to judgements that particular consequences are 'likely' or 'unlikely' (Beanlands & Duinker 1983). Exceptions exist in which, for instance, quantitative modelling of pollution dispersion is included but, for most impacts, environmental impact assessments include few rigorous predictions. For example, a study of the scientific quality of 75 environmental impact statements (EISs) produced in the United States found that over 82 per cent never used well-developed notions of probability to estimate consequences, and that none of the EISs did so systematically (Caldwell *et al.* 1982).

Risk assessment, in contrast, stresses formal quantification of probability and uncertainty. By definition, a risk assessment is a study that provides 'quantitative measures of risk levels, where risk refers to the possibility of uncertain, adverse consequences . . . most fundamentally estimates of possible health and other consequences . . . and the uncertainty in those consequences' (Covello & Mumpower 1985). A risk assessment typically includes a determination of the types of hazard posed, together with estimates of the probability of their occurrence, the population at risk of exposure and the ensuing adverse consequences (Conservation Foundation 1984). Considerable scholarship has been devoted to developing and refining methodologies for producing such estimates.

Risk assessments, of course, may be based on quite tenuous or debatable assumptions and such predictions ultimately may be unreliable, despite their apparent quantitative rigour. Hattis & Smith (1985), for instance, warn that current risk assessment practice relies on unduly narrow statistical methods for quantifying risk, at the expense of other lines of reasoning that may be more valid. Whatever its imperfections in practice, a basic virtue of RA is its normative commitment to improving the methodologies of predictive estimation. EIA may have devoted similar attention to procedures for identifying categories of possible consequences. By and large, however, it has lacked this commitment to improving methods for prediction.

UNCERTAINTY

C. S. Holling once asserted that the core issue of environmental impact assessment is how to cope with decision making under uncertainty (Holling 1978). The same is true of risk assessment. Both are intended to reduce the uncertainties associated with public policy decisions. By the same token, however, both must confront powerful temptations, common to all policy analyses, to discount issues that remain uncertain or disputed, in order to build a confident justification for a decision. The appearance of certainty and consensus is welcome to politicians. Where it is not well founded, however, it tends simply to promote cynicism about an analysis without reducing opposition to the outcome.

Even after the most thorough assessment, all public decisions ultimately must be made in the face of uncertainty about the future; about human behaviour; about stochastic events; and about ignorance of the imperfections in analysis. It is important to judge any policy analysis, therefore, not only by how much it reduces uncertainty, but also by how explicitly it acknowledges important sources of uncertainty that remain.

In environmental impact assessment there are requirements to address issues of uncertainty, but this is rarely evident in practice. Caldwell *et al*. (1982), for example, found that over 22 per cent of the EISs they reviewed never acknowledged uncertainty and that none of them did so systematically. Similarly, Reeve (1984) reported that a study of 242 draft EISs led the Council on Environmental Quality to conclude that environmental impact assessments rarely address the question of incomplete and unavailable information as required by its regulations.

In risk assessment, acknowledgement of uncertainty is similarly expected. In practice, uncertainty is often buried in arbitrary assumptions or ignored if it cannot be quantified. Despite its apparent rigour, risk assessment like EIA, is ultimately a very 'soft' process of 'artful theorizing to construct an appropriate picture of the world for informing specific choices' (Hattis & Smith 1985). An important topic for future research in both EIA and RA, therefore, is to refine methods for providing explicit and systematic treatment of uncertainty.

SUBJECTIVE INFORMATION

Subjective information refers to statements by experts or laymen reflecting concerns, value preferences, and judgements that cannot be validated objectively. Such information is unavoidably present in both EIA and RA, wherever uncertainty or disagreement exists (Otway & Thomas 1982). It is important, therefore, to identify how each treats such information.

Environmental impact assessment adopts at least three specific procedures designed to ensure explicit identification of subjective concerns and disputes. These include requirements for identifying 'controversial' impacts, whether or not the agency considers them significant on objective grounds; a process of 'scoping' in which all concerned parties may formally participate in defining the terms of reference for an assessment at the preliminary feasibility study stage;

and a formal review of the draft analysis by all relevant agencies and interested citizens – these comments must be made public and explicitly answered by the initiating agency.

Risk assessment incorporates no formal requirement for identifying subjective information or divergence amongst judgements. In practice, it frequently fails to acknowledge such information or to treat its presence as a legitimate issue. Many risk assessments, for instance, display a strong normative commitment to the concept of 'expected value', and a corresponding disdain for fears that exceed these values. Such fears are not regarded simply as differences in judgement to be acknowledged and discussed. Rather, they are considered groundless and, therefore, illegitimate, even though the concept of expected value is neither widely accepted by the general public nor legislatively approved as the basis for decision making (see, for example, Popper 1983). In addition, the United States National Research Council has identified a lengthy list of study design points in risk assessments at which assumptions must be made. There is, however, no routine procedure for ensuring explicit debate of such judgements (National Research Council 1982).

Thus, risk assessment could probably be improved by the development of explicit protocols for the treatment of subjective and disputed information. The procedures used for this purpose in environmental impact assessment may provide one set of useful models.

SUMMARY

It is clear that as substantive forms of analysis, EIA and RA, while differing in practice, are intrinsically similar in concept. An ideal example of each, in principle, could provide roughly comparable information. The output from both types of analysis is designed to clarify a decision maker's understanding of the alternative available courses of action and to present the best possible prediction of the significant consequences likely to result. Variations in EIA and RA, as currently practised, represent differences in focus and emphasis. Some of these differences reflect the respective strengths and weaknesses of the two approaches.

Each type of analysis, therefore, could benefit substantively from the adoption of some aspects of the other. Indeed, both would probably be improved by the development of a unified form of applied analysis combining their respective strengths.

Process

The most important differences between environmental impact and risk assessment, however, are differences not of substance, but of process. In practice, the two forms have functioned, not only separately, but also differently as administrative procedures. Although these differences are perhaps most pronounced in the United States, where RA and EIA have developed in distinctly separate legal contexts, they are evident elsewhere. While substantive content is important, therefore, no less important is how well each functions as a process for framing and legitimizing public decisions.

PURPOSE

In the United States, EIA originated from a statutory requirement which is not only enforceable by citizens in the courts, but also binding upon all governmental administrative decisions (with the exception of the environmental health regulations mentioned previously). The intent of the EIA requirement is not simply better analysis, but rather administrative reform. It was designed to be an 'action forcing procedure' compelling agencies to pay attention to the law's substantive purpose (Andrews 1976).

The National Environmental Policy Act directed all federal agencies to prepare a 'detailed statement' of environmental impacts, adverse effects, alternatives and other matters to accompany every recommendation, report on a legislative proposal, or other major federal action that might significantly affect the quality of the human environment (*U.S.Publ.L.91–190.42. USC 4321–4347*). The statement must also be circulated for comment to all other agencies having relevant jurisdiction or special expertise, and made available (with all comments) to the public. Similar requirements have since been adopted by over half of the US state governments and by some local authorities, as well as by many other nations and even by some transnational organizations.

EIA was explicitly conceived as an administrative reform to force government agencies to become more publicly accountable. Authors of the legislation perceived agencies not as systematic rational decision makers, but as narrow advocates of particular missions at the expense of other values and consequences. While some believed that more complete information alone would lead to better decisions, in practice EIA has drawn its primary effectiveness from the threat of public embarrassment and judicial challenge (Andrews 1976).

Like EIA, risk assessment grew out of a broad movement toward expanded use of rational techniques for analysing and justifying government decisions. Unlike EIA, however, it developed first as a management technique in the hands of experts. It was used in part to improve decision making with respect to engineering technologies and in part to justify those decisions against public fears and opposition.

Risk assessment emerged in the mid-to-late 1970s as an administrative requirement in the form of both statutes and executive orders requiring not only more extensive documentation to justify proposed risk regulations, but also the 'balancing' of risks against economic costs and benefits (Atkisson *et al.* 1985). Practice in the USA, therefore, has been limited largely to the environmental health regulatory agencies and confined to a consideration of the environmental health, particularly cancer-mortality, risks of their decisions.

While risk assessment's substantive purpose is not unlike EIA, its political uses, at least in the United States, have been rather different. Whereas EIA was adopted to increase accountability to citizen groups, RA was adopted to increase internal management control in order to foster consistency across actions and programmes (Environmental Protection Agency 1984). On the part of some advocates, it was envisaged as a means of increasing accountability not only to oversight agencies, but also to business lobbyists seeking to limit risk regulation.

ADMINISTRATIVE PROCESS

To be useful in decision making, an assessment must be not only accurate, but also legitimate. It must deal with the full range of issues in a process that is not only open to public scrutiny and debate, but also well reasoned, even-handed, and candid about unresolved uncertainties. Numerous studies have shown, for example, that those with an interest in the outcome of a public decision may hold quite different views regarding the terms of reference for an analysis, encompassing problem definition; objectives and goals hierarchies; environmental conditions; expected consequences; and alternatives (see, for example, Mason & Mitroff 1981, Kleindorfer & Yoon 1984 and Vari *et al*. 1985). Susskind (1985) summarizes a substantial body of research showing that joint negotiation of the scope and methods of analysis and, even, of the group of experts who will conduct it is a crucial step in producing legitimate analyses of controversial issues. Both EIA and recent literature on strategic planning and facility siting provide valuable insights that might enrich the practice of risk assessment.

Environmental impact assessment functions as an explicitly open analytical process, having enforceable opportunities for public involvement in designing and critiquing an analysis, which guarantees that conflicting views must be considered as a matter of record. The scoping process, the requirement that controversial impacts must be discussed explicitly, and the review and comment procedures all contribute to this openness. They serve to make the resulting analysis a reasonably thorough and publicly tested record of the issues involved in a proposed decision.

Risk assessment, in contrast, frequently functions as a more arcane expert process, couched in technical terms such as risk probability, dose–response curve and expected value that have little meaning to most laymen. In addition, the process also often lacks procedures for public involvement in the design and critique of an analysis.

A common response to this observation is that risk information is simply too technical to be understood by laymen, and that such decisions, therefore, are best left to agency experts. Indeed, it has been shown that laymen do perceive risks differently from 'experts', overestimating some and underestimating others (see, for example, Fischhoff *et al*. 1981). However, 'experts' are also prone to certain types of misjudgement. A more fundamental argument for opening the process, however, is the consideration that such decisions are not merely technical choices, but matters of public governance that happen to be framed by technical assumptions. O'Riordan (1979), for instance, warns that in many risk assessments 'scientific rationality is overwhelming political rationality'. More recently, Starr (1985) has argued that public acceptance of proposed actions depends more on public confidence in risk management than on any quantitative estimate of risk consequence, probability, or magnitude.

There are two main reasons for reforming current RA practice. First, many of the assumptions in an assessment are characterized by uncertainty which, generally, is not stated explicitly. Secondly, the public does not necessarily accept the concept of 'expected value' as a basis for risk decisions. Consequently, it is probably wiser to make decisions on risk more understandable rather than more quantitatively sophisticated. Similarly, debate should focus on options for risk minimization rather than on the refinement of risk estimates.

INFLUENCE ON DECISIONS

Despite differences in substance and process, environmental impact and risk assessment appear to have had similarly modest, but beneficial, effects on public decision making. Both have produced far more extensive documentation related to proposed decisions than was previously available; have served to deter 'extreme' proposals, involving high risk and high cost; have created incentives to identify mitigative measures to reduce risks; and have given birth to communities of professional practitioners. The gradual entry of these communities into hitherto narrower, mission-oriented, administrative agencies has probably served to broaden perspectives and moderate biases.

Thus, on considerations of process as well as on substantive issues, both EIA and RA would probably benefit from the development of a unified form of analysis that incorporates their respective strengths. Such a unified analysis, however, must incorporate not only substantive elements, but also include explicit procedural mechanisms for negotiating the terms of reference of an assessment, for openly debating its assumptions and judgements where uncertainty exists, and for developing and legitimizing a consensus on its conclusions.

Prospects for unified analysis

In only a few instances, all recent, has risk assessment been incorporated into environmental impact assessment studies. Beanlands (1984b) reports that risk assessment is now an EIA requirement in Canada, being conducted most recently as part of the assessment of oil and gas development in the Beaufort Sea produced in 1983. In Canada, RA has also been applied in EIA for a range of proposals including hydrocarbon projects, nuclear power plants, forestry projects and hazardous train derailment.

In the United States, administrative regulations require the inclusion of 'worst-case analysis', a sort of risk assessment, in an EIS in situations where the information on possible impacts necessary for an informed decision, is not available and would be too costly or impossible to obtain. To date, only a few such analyses have been prepared, but even so, three of them have been contested in the courts. The Council on Environmental Quality (CEQ) has recently proposed, amid substantial controversy, to drop this requirement (Reeve 1984, CEQ 1985).

The controversy over worst-case analysis illustrates well the conflict between substance and process in evaluating policy analyses. As substance, worst-case analysis is not the favoured approach of the professional risk assessment community, because it emphasizes speculation about the worst conceivable outcomes, rather than precise estimates of most probable ones. As process, however, it is one of the few available 'action-forcing' mechanisms by which an unwilling agency can be compelled to acknowledge risks and uncertainties that it would rather ignore.

In the case of the proposal by the Bureau of Land Management concerning the aerial spraying of chemical pesticides for forest insect control, the litigants cited published scientific studies, not mere speculation, as evidence for possible adverse effects, pointing out that the agency simply had no expertise of such health effects and had made no attempt to acquire it. If CEQ rescinds the requirement for worst-case analysis, therefore, and leaves the analysis to the agency's discretionary judgement, the result may not be better risk assessment, but no risk assessment, since there will be no legal basis on which to challenge its omission (Northwest Coalition for Alternatives to Pesticides 1985).

If worst-case analysis is not included, therefore, some alternative 'action-forcing' mechanism is necessary to provide an open and legitimate forum for debate and to compel unwilling agencies to acknowledge risk and uncertainty. One mechanism might be to retain the requirement, but to use the scoping process to define what reasonable range of worst-case scenarios should be considered. If some other, more preferable, procedure for risk assessment is incorporated into EIA its requirement should be stated in such explicit terms that the opportunity for external legal pressure to demand its inclusion in a particular case is kept open.

BENEFITS

On intellectual grounds, both EIA and RA would be improved by combining them into a unified analytical process. This would effect both substantive and procedural advantages.

Substantively, environmental impact assessment would benefit from the greater sophistication of risk assessment in the treatment of predictive analysis and probability. In any event, EIA should incorporate more explicit considera-tion of health effects. Risk assessment, in turn, should be applied to a broader range of risks than just mortality from cancer and catastrophic accidents. As a process, risk assessment has much to learn from experience with environmental impact assessment. Such areas as scoping, comparative analysis of alternatives, formal procedures for incorporating subjective values, and integration into non-regulatory decision-making processes are particularly well developed in EIA.

Practically speaking, many actions need both environmental impact and risk assessment. In these instances, a more useful analysis would be obtained by combining the two. Among the most obvious examples of such actions are decisions to site facilities for energy production, hazardous waste treatment and disposal, and other industrial facilities; environmental applications of biotechnology; and even more mundane programmes, such as, pesticide application for agricultural and forest management. While US federal guidelines may be slow to merge them, practice in Canada and Europe, as well as in some US state and local governments is already to do so, for instance, in assessing possible impacts of waste incinerators.

Substance, process and outcomes

There is good reason for optimism, therefore, about the prospects for integrating environmental impact and risk assessments into a unified analytical process. Research and experimental applications will be needed to develop such a process, but the idea is both feasible and timely.

One subtler, but fundamental, issue remains unresolved by this recommendation, however, namely the situations that warrant such analyses. Some promising target actions have been suggested, but in a broader sense the most important causes of hazards, to both human health and environmental processes, often lie in situations where, at present, there may be no specific proposal to trigger such an assessment. Examples include urban encroachment in floodplains and other areas of natural hazard, and around hazardous industrial plants, as well as some industrial and agricultural uses of toxic chemicals, where effects on ground water, human health, and other outcomes are now attracting increasing concern.

Thus, in addition to developing unified analyses for proposed actions, an important subject for research would be to apply similar analyses to existing complexes of hazards that threaten human populations and ecosystems. In this situation, the purpose of the analysis would not be to evaluate a single proposal, but rather to set priorities for hazard management. Given a highly urbanized area or an ecological region, the key issues would be to identify the important hazards that would warrant a management response, to determine priorities and to develop alternative management strategies to mitigate them. This task would require development of a somewhat different approach to assessment. Clearly, such an effort would have substantial payoffs both for advancing the methods of risk assessment and for improving the effectiveness of risk management.

Acknowledgement

An earlier version of this paper was presented at the Task Force Meeting on Risk and Policy Analysis under Conditions of Uncertainty, International Institute for Applied Systems Analysis, Laxenburg, Austria, 25–7 November 1985.

6 *EIA in plan making*

C. WOOD

Introduction

In principle, an environmental impact assessment (EIA) system could apply to all actions likely to have significant environmental impact, irrespective of their type. Thus, the potential scope of a comprehensive EIA system could encompass the approval of policies, plans, programmes and projects at all levels of government. Lee & Wood (1978) have suggested a 'tiered' system of EIA applied to a chronological sequence of category of action, as shown in Figure 6.1.

There are several advantages associated with an EIA system that is not confined to projects. In particular, higher levels of action not subject to environmental assessment may generate projects which are misspecified or which lack sufficient alternatives. Similarly, there may be time savings if environmental data are collected when a higher-order action is proposed rather than after an urgent project is suggested or if the assessment of a plan or

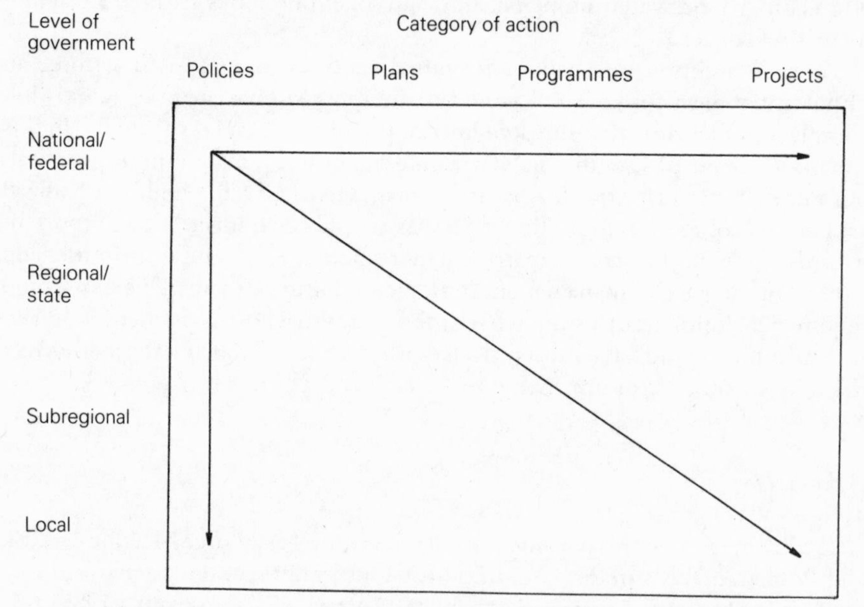

Figure 6.1 Categories of action and levels of government within a comprehensive EIA system.

programme obviates the necessity to undertake numerous small project EIAs.

In the United States, the potential benefits of extending the EIA system to higher-level actions were recognized in the National Environmental Policy Act which was phrased to require EIA for all types of federal actions significantly affecting the environment including the passing of legislation. However, the vast majority of US environmental impact statements (EISs) have related to projects. This largely reflects the complexity of non-project EIA and also poorly developed methods for undertaking EIA for higher-order actions. Recently, there has been some interest in undertaking areawide, that is plan, EIAs in the USA. Rodgers (1976) has described the advantages of plan assessment, claiming that if

> every component of the comprehensive plan – and every proposed amendment – were subjected to the rigorous evaluations required in the preparation of an EIS, the quality and usefulness of the plan should improve. Conversely, if proposed policies and projects were in conformity with a comprehensive plan which had been subjected to an environmental impact assessment, there would be no necessity to evaluate each separate proposal.

Recognition of the potential benefits of extending EIA to plans is not confined to the United States. Environmental goals are often explicit or implicit in the land-use planning process and, frequently, there is provision for considering environmental matters in plan making in many countries. As land-use plans frequently form the context for project authorization, plan making is logically the first higher level to which an EIA system might be extended. The Netherlands government has recognized this by introducing an EIA system which applies to both plans and projects. It was originally intended that the European Community (EC) EIA system would eventually apply to plans as well as projects (Commission of the European Communities 1980).

This chapter is concerned with EIA and plan making. The next section discusses the making of land-use plans. First, the nature of the plan-making process is described. Secondly, a brief review of practice in incorporating environmental goals in plans is presented. The following section is devoted to a discussion of the integration of EIA and plan making. This is concerned principally with the differences between the EIA of plans as opposed to the integration of EIA into plan making. Again, there is some discussion of current practice. The subsequent section describes some of the methods available for the EIA of plans. These are classified according to the four main stages of the planning process, namely, the formulation of goals and objectives; survey, prediction and analysis; generation and evaluation of alternative plans; and decision, implementation and monitoring.

The penultimate section contains an analysis of current practice in the EIA of plans based upon examples from the United States, the United Kingdom and the Netherlands. A concluding section draws together the main issues and puts forward some suggestions for the further development of EIA in plan making.

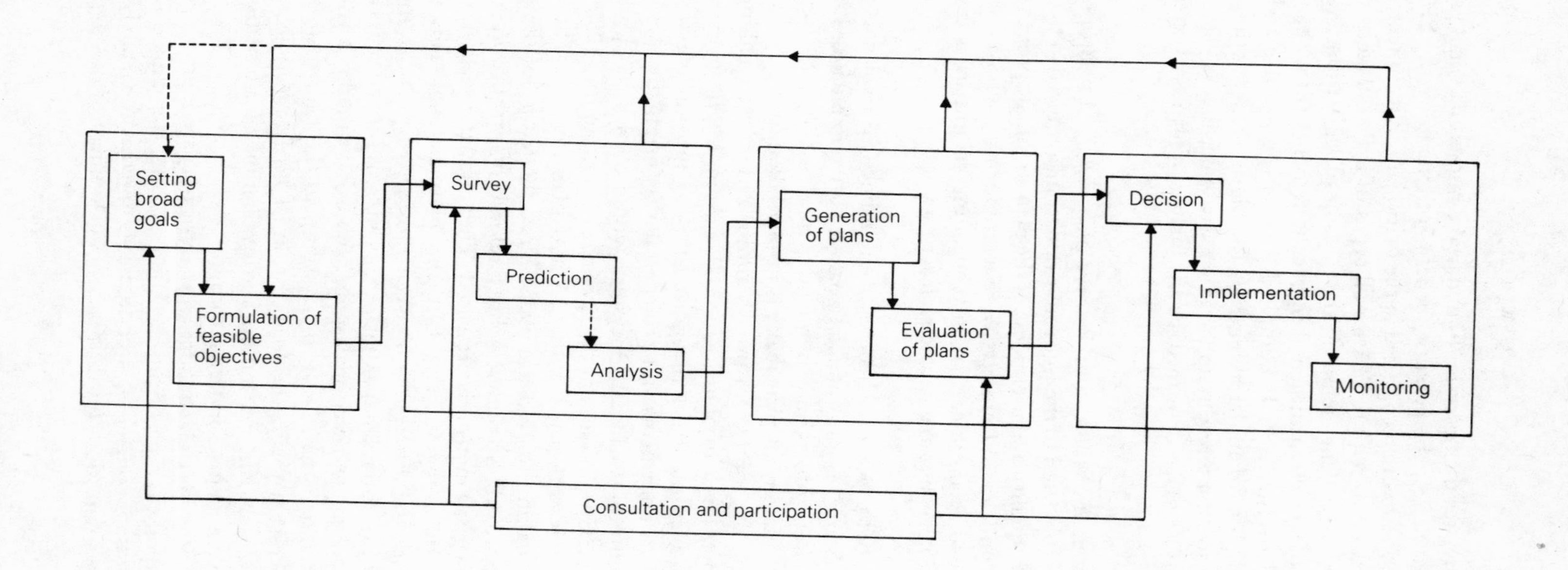

Figure 6.2 Stages in the physical planning process.

The plan-making process

Physical or land-use plans are prepared to assist in the regulation of the spatial distribution of activities and environments within a prescribed geographical area. Many different types of plan exist, covering differing sizes of area (from national to very local) with different legal enforceability (from mandatory to informal) and prepared by, or for, a variety of public authorities. They include detailed zoning or land-allocation plans for a part or the whole of a local area and land-use development plans of a more strategic nature for subregions, regions, or even nations. Table 6.1 presents some examples of different types of plan. It is apparent that, at its most detailed, a plan for the development of a very small area may be almost indistinguishable from a proposal to undertake a major project.

The plan-making process followed in practice varies considerably. While some plan makers seek to produce an enduring blueprint of the desired state of the area at some future date, especially where the area is small, others will expect the desired future state to alter over time in response to changes in external circumstances. Some plan makers rely heavily on a pragmatic approach, based upon the qualitative assessment of likely future situations and of the impact of alternative plans upon them. Others utilize an approach with greater emphaiss on quantitative planning techniques. There is, however, some consensus about the nature of plan making. Most plan makers would agree that the plan-making process falls into four main stages (Fig. 6.2 and Table 6.2). They would also accept the dynamic nature of the planning process in that insights gained at later stages in the process frequently lead to a re-examination of conclusions reached at earlier stages (see, for example, Roberts 1974). They would further concur that a degree of public participation and consultation is essential in plan making and that this should take place at each of the main stages in the process (Healy 1986). In particular, most land-use planning systems make provision for the publication of the results of the survey of conditions and of the draft plan, or alternative plans, before final decisions are reached.

Environmental goals, together with many other types of goal, are often either explicit or implicit in the plan-making process. Perhaps the most notable

Table 6.1 Types of land-use plan.

Level of government	*Type of plan*	*Example*
National/federal	National land-use plan	Netherlands national physical planning reports
Regional/state	Regional land-use plan	Oregon state land-use planning goals
Subregional	Subregional land-use plan	British county structure plans
Local	Local land-use plan	Netherlands allocation plans
Small area	Small area land-use plan	British local plans for parts of districts

Table 6.2 Characteristic features of different stages in the physical planning process.

Formulation of goals and objectives:	Establishment of broad planning goals and more detailed objectives which, at later stages in the planning process, will be refined into more specific operational form.
Survey, prediction and analysis:	Survey of existing conditions, prediction of future conditions over the planning period (without plan implementation); identification of planning problems; revision of objectives and possibly even goals in the light of these.
Generation and evaluation of alternative plans:	Development of plans likely to meet agreed goals and objectives; development and application of evaluation criteria, possible revision of objectives.
Decision, implementation and monitoring:	Plan revision in light of implementation problems and monitoring, commencing with revision of objectives (and goals).

proponent of the integration of environmental considerations into physical plans was McHarg (1969) who devised a set of indicators to guide developments to 'areas of opportunity' and away from 'areas of resistance' such as vulnerable sand dunes. McHarg's thinking has been widely disseminated and plan makers in many countries would claim that their plans were concerned with the environment. There are sometimes statutory provisions for incorporating environmental matters within plan making. Certainly, some land-use plans in Britain, for example, must contain policies for the improvement of the physical environment and many include policies relating to, *inter alia*, pollution control (Wood 1976).

There is, therefore, a number of common elements in most plan-making procedures relevant to incorporating EIA within plan-making processes. First, there is statutory recognition of environmental goals within the broad context of plan making. Secondly, there are provisions for the preparation of planning documents. Background information in the form of a survey of existing (including environmental) conditions, an indication of future prospects and problems as well as sometimes the identification and examination of alternative planning strategies are produced. In addition, the plan itself, which frequently contains policies for improving the environment, is prepared. Finally, there are provisions for consultation and public participation on the basis of these documents and for consequent revision of the plan during subsequent stages in the planning process.

Integration of EIA and plan-making processes

Environmental impact assessment is intended not only to ensure that environ-

Table 6.3 Requirements of an EIA system.

1 A draft EIA report should be prepared, containing:

(a) a description of the proposed action and its purpose;
(b) a description of the environment expected to be affected by the action;
(c) an assessment of the magnitude and significance of the probable environmental impacts of the action, including an assessment of the compliance of the action with approved environmental plans and policies and a discussion of measures for mitigating impacts;
(d) an assessment of the environmental impacts of alternatives to the action;
(e) a non-technical summary.

2 The draft EIA report should be available prior to the first significant stage in the relevant decision-taking process.

3 There should be public participation and wide consultation in the consideration of the draft report.

4 A review of the draft EIA report should be prepared (perhaps in the form of a final EIA report) which should:

(a) summarize the main comments received;
(b) include modifications to the actions originally proposed;
(c) accompany the proposed action through the remaining stages of the public authority's decision procedure.

5 Provision should be made to post-audit the EIA.

mental factors are included in decision making, but also to show how environmental factors have been included both in the consideration of alternatives and by the provision of information to the public. However, there are considerable variations in the detailed procedural requirements of the numerous EIA systems in place around the world (see, for example, O'Riordan and Sewell 1982). For example, some systems make little provision for examining alternatives. Many EIA systems are confined to projects and, amongst these, there are substantial differences in the types of project assessed. Others such as the EC EIA directive make no provision for the auditing of EIAs (Council of the European Communities 1985). Nevertheless, most systems involve many of the elements which are summarized in Table 6.3 (Lee & Wood 1978). The main tasks and activities involved in EIA can be represented in diagrammatic form (Fig. 6.3).

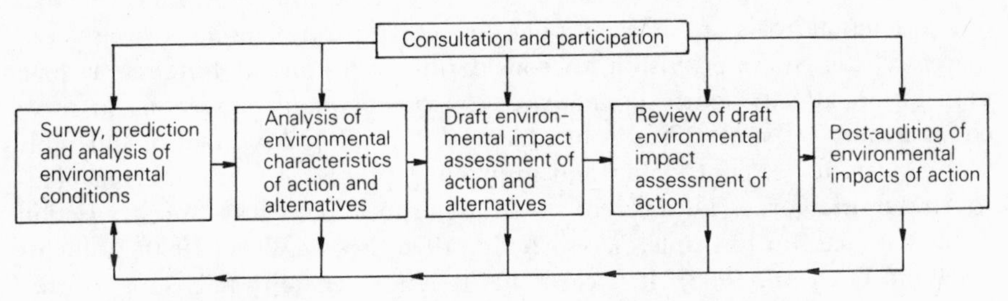

Figure 6.3 The main steps in the EIA process.

Figure 6.3 makes it clear that the activities involved in EIA should follow a logical sequence. In practice, however, they represent a series of iterative steps involving feedback as new information is generated, giving rise to an amended proposal. Holling (1978) has described this integration of the principles of EIA within the design and decision-making process in relation to projects as adaptive environmental impact assessment. It can apply equally to plans. It will be observed from Figures 6.2 and 6.3 that there are several similarities between the EIA and the plan-making processes. Figure 6.4 (adapted from Lee & Wood 1978), in which they are combined, is but one way of expressing the relationship between the plan-making and EIA processes (see, for example, Jones 1981).

Thus, it is possible to envisage two ways in which the environmental assessment of plans might be carried out. The first is to undertake an EIA of a plan, either during plan making or after it has been prepared, a procedure in which separate documentation presumably would be required. The second is to integrate EIA with the plan-making process, in a similar fashion to the adaptive assessment of projects. In this procedure, it could be argued that no separate documentation would have to be prepared, but some evidence of EIA would need to be furnished within the documents associated with the plan-making process. Foster (1983) has explained that, besides the 'EIA of plans' and 'EIA in plan making', information from a specific project EIA may be used as an input in plan preparation and that this process is sometimes erroneously referred to as 'plan EIA'. Some of this confusion undoubtedly arises from the similarities in the geographical scales of certain projects and plans. Lyddon (1983) has stated that, within the UK land-use planning sytem 'the use of the term EIA to describe planning and analysis at other [than project] levels tends to be confusing'.

The principal advantages of undertaking an EIA of a plan is precisely that confusion is removed as a separate document is produced. While integrated EIA could result in a clearly identifiable report on the EIA, there is a danger that it would not be undertaken in practice. Nevertheless, there are substantial advantages in this approach. Not least is that it would cause minimal disruption to existing procedures and should ensure that environmental matters are considered throughout the plan-making process.

It is essential to guard against the argument that there is no need to apply EIA to plan making on the grounds that it is already carried out. While it is true that environmental goals are often implicit in the land-use planning process and, therefore, that some provision for considering environmental matters in plan preparation already exists in many cases, the planning provisions in most countries fall well short of rigorous EIA procedures. It follows that considerable scope exists for strengthening and improving practice within existing plan-making frameworks. The current situation in most countries is summarized in Table 6.4 (see, for example, Wood & Lee 1978, Lee & Wood 1980, Williams 1984 and Lee *et al.* 1985). It is clear that practice generally falls short of that necessary for EIA and plan making to be genuinely integrated.

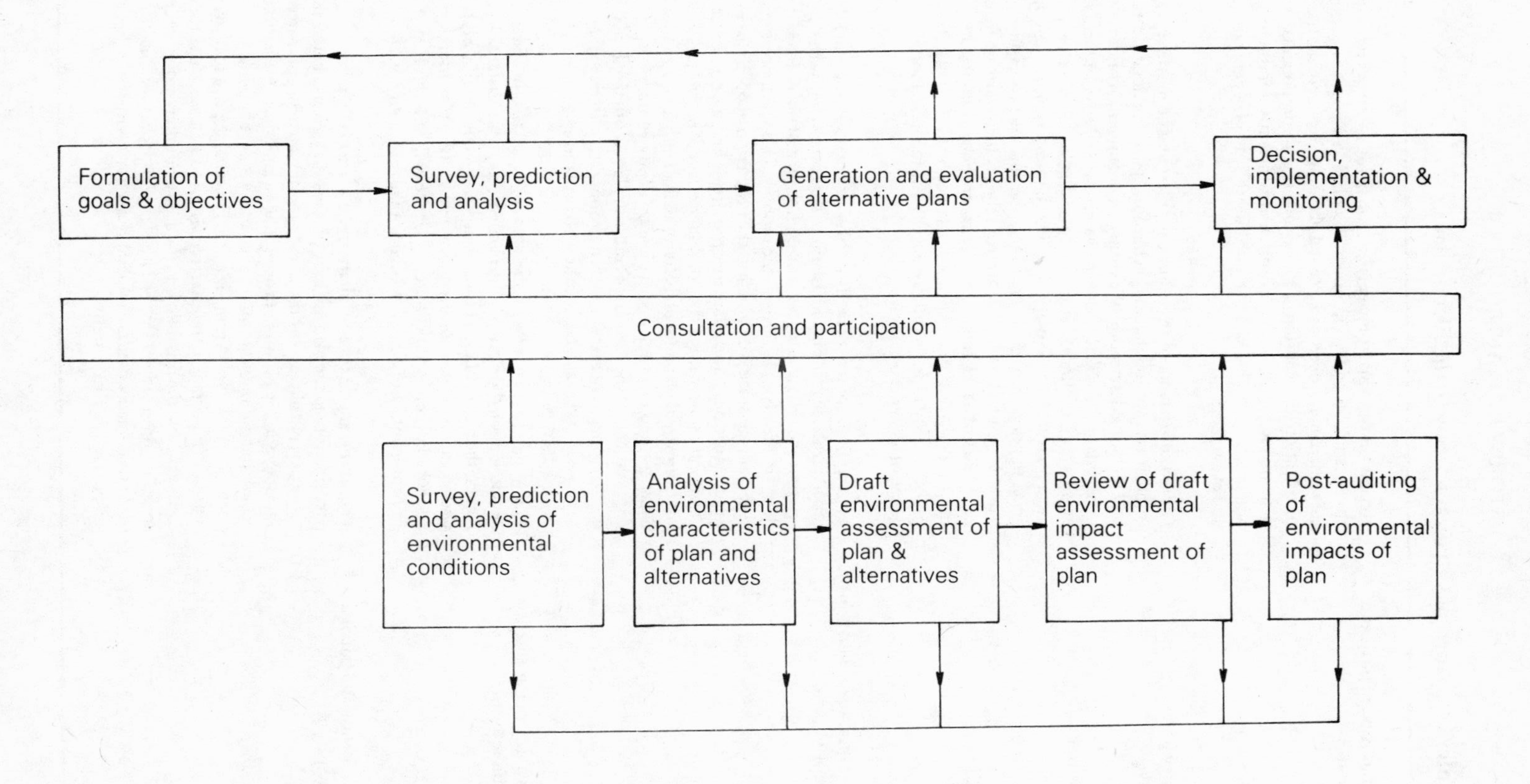

Figure 6.4 Relationship between the plan-making and EIA processes.

Table 6.4 Current situation relating to the EIA of plans.

Formulation of goals and objectives	Recent land-use planning legislation in many countries refers to the pursuit of broad environmental goals. However, although these statements of environmental planning goals provide the justification for inclusion of EIA as an integral component of the plan-making process, they are often insufficiently comprehensive and precise to enable EIA to contribute fully to plan generation and evaluation.
Survey prediction and analysis	Given the existence of environmental planning goals, there is an implicit assumption that the background planning documents will include a description of existing environmental conditions, prospects and problems, and an examination of environmental implications of alternative planning strategies. However, it is most unusual for this to be an explicit statutory requirement. In general, practice in relation to data gathering and processing varies very widely. Quite apart from the problems arising from incomplete data relating to the environmental phenomena surveyed, there are inadequacies in the sophistication of data analyses, difficulties in forecasting environmental conditions and deficiencies in presentation.
Generation and evaluation of alternative plans	The required content of land-use plans is often specified, at least in broad terms, in planning legislation. This usually includes reference to certain categories of environmental phenomena but these frequently only relate to the more traditional environmental aspects of land-use policy such as the protection of open spaces, the designation of areas of nature protection and the conservation of ancient monuments. The combined effect of imprecise environmental planning objectives, an inadequate environmental data base, and incomplete specification of environmental requirements in the content of land-use plans is often reflected in insufficiently systematic use of environmental criteria in the generation and evaluation of alternative plans.
Decision, implementation and monitoring	The weight given to environmental as opposed to other factors in deciding upon the plan to be implemented varies substantially. Again, the extent of plan monitoring varies considerably both within and between countries. However, where plan monitoring does occur, there is only limited evidence of the monitoring of the environmental impacts associated with plan implementation.
Consultation and public participation	The arrangements for consultation and particularly public participation, in the plan-making process vary greatly between countries, but also according to the type of plan being prepared. The stages in the planning process in which public consultation occurs and the amount and type of background planning documentation which is made publicly available prior to it, are as critically important in the effectiveness of an EIA system as the form of it, and the time allowed for it. In a number of cases such consultation takes place at too late a stage and/or is conducted on the basis of inadequate documentation to serve satisfactorily the purposes of an EIA system.

Methods for the EIA of plans

Notwithstanding the logic of applying EIA to plan making, there are four principal difficulties involved in the EIA of plans which do not apply to the assessment of projects. First, although a project normally relates to a single development, a plan is concerned with many potential developments in different locations. The latter type of impact assessment, therefore, is likely to involve greater analytical complexity because it entails the estimation of multi-source impacts. Secondly, the nature, scale and location of the activities comprising a plan are generally not as precisely described as for a development project. Therefore, it is often not possible to achieve the same level of precision in estimating a plan impact as a project impact. On the other hand, the more strategic a plan is, the less numerically precise impact predictions need be. Thirdly, practical experience in the comprehensive and systematic environmental assessment of plans is less than corresponding experience with projects. Similarly, there is less literature available to guide those undertaking such assessments. Finally, the nature of the investigations of existing and projected environmental conditions required for plans may be less clear-cut than for projects. On the other hand, the period available for plan preparation and approval is normally longer than for project authorization so that time constraints on gathering new environmental data by survey are less severe.

Table 6.5 classifies examples of assessment methods, according to the stages in the EIA process. In addition, methods which are appropriate for the various identification, measurement, interpretation and communication tasks involved are highlighted (Lee & Wood 1980). Many of these methods are similar to those used in project assessment, although the particular form of any method may differ between the two types of assessment. Many of those listed such as checklists and diffusion models refer to method groupings and the individual methods which each group contains will differ somewhat. Similarly, some are alternatives, and a single assessment is not expected to make use of them all. By contrast, certain items, such as checklists or matrices, may be used at more than one stage in the assessment process, though not necessarily in the same form.

There is obviously much work to be undertaken to overcome many of the weaknesses summarized in Table 6.4. A number of suggestions can be made to improve practice in the four stages of the plan-making process.

FORMULATION OF GOALS AND OBJECTIVES

In general, environmental objectives in plan making are not sufficiently comprehensive. This stems from weaknesses in the identification process which could be remedied through the preparation and use of checklists or similar instruments. Standardized checklists for this purpose do not yet seem to be in use. The lack of precision in environmental planning objectives is often a reflection of the absence of clear environmental quality standards. However, this situation should improve as more comprehensive systems of quality standards are established.

Table 6.5 Classification system for EIA methods for plans.

Tasks	*Methods (examples only)*
Stage 1 Formulation of goals and objectives	
Identify aspects of the environment to be studied, for which goals and objectives might be formulated.	Checklists.
Determine 'first-round' environmental goals and objectives.	Environmental agency consultation to determine existing and planned environmental quality standards and targets; public consultation on basis of preliminary statement of planning issues.
Stage 2 Survey, prediction and analysis	
Identify aspects of existing environmental conditions for which information is to be sought.	Checklists.
Collate existing environmental data and identify gaps in information.	Consultation with specialist agencies, use of existing maps and aerial photographs, use of data collation and retrieval systems.
Obtain additional environmental data to meet remaining deficiencies. Identify gaps that cannot be satisfactorily filled.	Review and revision of existing monitoring systems, special surveys using a variety of techniques (aerial photography, field sampling, etc.).
Predict main social and economic changes over the planning period (assuming no new plan implementation).	Population and economic forecasting techniques applied to the planning area.
Predict physical resource use and waste generation levels associated with forecast economic/social changes.	Checklists and matrices for identification use. Resource and waste coefficient analyses.
Predict magnitude of impact on environmental quality.	Environmental agency consultation, screening procedures, resource depletion, diffusion and damage analysis, landscape assessment techniques.
Analyse future environmental conditions: compliance with 'first-round' environmental goals and objectives; review of goals and objectives.	Agency and public consultation on basis of survey report; mapping and overlay methods.
Stage 3 Generation and evaluation of alternative plans	
Develop alternative plans and determine environmental evaluation criteria.	Intuitive techniques. Delphi forecasting consultation. Trade-off, cost–benefit, goals achievement matrix analyses. Lists of standards.
Describe the relevant features of each plan and assess physical changes, changes in resource use and in waste generation associated with its implementation.	Checklists; resource and waste coefficient analyses; accident and uncertainty analysis.
Predict magnitude of environmental impact associated with plan implementation on: air, water and land (including mineral resources), living receptors within the environment.	Checklists, consultation, screening procedures, resource depletion, diffusion and damage analysis, landscape assessment techniques.
Assess importance of impacts: determine compliance with environmental quality standards; investigate response of affected parties; aggregate individual environmental impacts.	Lists of standards, social surveys, public participation. Scaling and weighting systems, overlay methods.
Stage 4 Decision, implementation, monitoring	
Select plan for implementation.	Application of evaluation criteria; agency consultation and public participation; plan modification in light of these.
Monitor environmental impact arising from plan implementation; review implementation process and plan in light of this.	Environmental monitoring systems.

SURVEY, PREDICTION AND ANALYSIS

Inadequacies in the available data relating to existing environmental conditions are often due to failures in defining data requirements, to organizational problems and monitoring constraints. Problems created by neglecting significant parameters of existing environmental conditions can be overcome using checklists, environmental matrices or similar instruments which aid identification. Existing, usable data may be fragmented and inaccessible. Possible remedies range from simple systems of documentation of data sources through to the establishment of environmental data banks with retrieval systems. Some existing data may be unusable because they have not been collated with the needs of an EIA system in mind. A solution lies in a review of continuing sources of environmental monitoring data, taking into consideration other sources that such monitoring systems serve. The data deficiencies which remain can only be remedied through new survey work within the limits of time and other resources available. Careful specification of both the type and detail of new data is clearly necessary.

The prediction of future environmental states raises considerable technical difficulties. Forecasting future environmental conditions in the absence of the plan is dependent on the quality of the economic and population forecasts for an area. Although these forecasts are subject to their own limitations, they should reveal both expected industrial output and changes in the distribution of activities (see, for example, Lee & Wood 1983). Only very general indications of future resource use and waste generation can be made, because the characteristics of the expected developments can be determined only very approximately.

The diffusion of wastes from multi-point waste sources, often required in plan assessment, is a complex process. Some models for calculating likely concentrations, however, exist. Assessment of the likely magnitude of secondary environmental impacts of resource abstraction, clearly of importance in plan making, is not well developed. Finally, quantitative relationships between the magnitude of damage to receptors and pollutant concentrations are not well established. It would seem desirable to develop environmental quality standards or targets, taking into consideration available evidence on damage, and to assess expected or planned development in terms of their likely compliance.

GENERATION AND EVALUATION OF ALTERNATIVE PLANS

Plan alternatives which satisfy environmental goals and objectives are likely to be generated when two conditions are met. First, environmental planning objectives and the environmental survey must be specified satisfactorily. Secondly, adequate provisions for agency and public comment on plan alternatives must be made before they are finalized.

Evaluation criteria for comparing alternative plans are often, like planning objectives, inexactly specified and create an element of uncertainty at the plan evaluation stage (see, for example, Lichfield *et al.* 1975, Local Government Operational Research Unit 1976 and McAllister 1980). Where several are capable of satisfying the criteria, some additional means of ranking alternatives based upon a comparison of their respective advantages and disadvantages is needed.

Assessing the environmental impact of each alternative plan involves determining the likely economic, population and environmental consequences that would result from implementation. It therefore gives rise to the same technical difficulties as apply to the prediction of likely future environmental conditions.

The relative importance of different impacts has often been determined using weighting and scaling systems in plan evaluation. Goals achievement matrices, cost–benefit analysis and other ranking methods explicitly or implicitly involve weighting the magnitude of each advantage and disadvantage before aggregating and comparing them. These weighting systems are open to criticism. However, provided that their assumptions are made explicit, that they are subject to sensitivity testing to establish their suitability and that they are used consistently, such systems are a potentially useful tool when presenting a summary of significant environmental impacts. The 'testing' of weights through public consultation is clearly desirable.

DECISION, IMPLEMENTATION AND MONITORING

Provided that evaluation criteria have been carefully defined and the appropriate parameters of each plan alternative have been satisfactorily measured as well as their significance assessed, the plan decision itself should not raise any technical problems. In practice, however, difficulties do arise because of deficiencies in these items. In addition, plan implementation and monitoring are frequently inadequate. Auditing the environmental impacts of plan implementation, therefore, is frequently necessary.

CONSULTATION AND PUBLIC PARTICIPATION

Various communication techniques have been developed to make the role of public participation in the planning process more effective. It is necessary to ensure that the content of planning reports, including estimates of environmental impacts, are presented to the public in a sufficiently clear and comprehensive manner. In addition, the methods of consultation used should be appropriate for obtaining representative and reliable community assessments of the significance of probable impacts. Exclusive reliance on traditional public inquiries or hearings may well be insufficient for this purpose and should be supplemented by other methods.

There is some agreement that direct public participation operates least satisfactorily where the public concerned is spread over a very large area and where likely impacts can only be identified in general and approximate terms. This may mean not only that different methods of public participation may be needed in the case of, for example, regional plans and local plans, but also that the general effectiveness of direct public participation may be lower in the former case. There is a clear need for further development of the various public participation and communication techniques presently available.

Case studies of EIA in plan making

There is very little experience of the use of EIA in plan making. This is particularly true in Europe, where none of the EC member states, apart from the Netherlands, has made any provision for EIA in plan making. The EC directive on environmental assessment adopted in 1985 makes no reference to the EIA of plans. Nevertheless, it is quite likely that, as more experience of project EIA is gained, there will be pressure to extend its use to plan making. Experience in the Netherlands, the UK and the USA will now be considered by means of some case studies.

THE NETHERLANDS

Jones (1981) has described the EIA of an industrial estate plan in Brabant. The area concerned, 500 hectares, was small enough to be similar to that for a project, but there were considerable uncertainties associated with the types of activities to be carried out on the estate. Four scenarios involving a no-development option, a nationally representative industrial mix, a large-scale industrial mix and a metal and construction industries mix were considered. Sufficient information was provided to predict quantities of air emissions, aqueous effluents, solid wastes and noise emissions. The scenarios were then converted into various zoning allocations to anticipate where impacts would occur and four alternatives evaluated. Models were used to calculate environmental quality and the results were compared to environmental norms and standards to ensure that these would be met or to identify where changes in the industrial mix would be necessary. The provisional EIS was made public, but no choice of a particular alternative was made. Although problems were encountered, especially in communication with the public and in arranging meaningful public participation, it was concluded that EIA should be incorporated into the plan-making process. It was also recommended that an 'environmental management plan' should be developed to implement the mitigation measures arising from the EIA (Jones 1981).

A national plan and a plan for East Gelderland were investigated to determine how far the various stages of the plan-making process and the associated procedures could be considered to fulfil the requirements of an EIA system (in't Anker & Burggraaff 1979). They concluded, on the basis of a thorough analysis, that the 'present physical planning process in the Netherlands has characteristics that enable environmental impact assessment to be integrated into this process without a great deal of change being necessary'.

They also questioned whether giving such weight to environmental factors in plans was justified given the importance of social and economic factors. The data presented, however, are insufficient to judge whether change in the Dutch plan-making system is necessary for EIA to be genuinely integrated into it. From a consideration of the case presented by Jones (1981), it appears that there will be more difficulties in implementing EIA at the plan-making stage than the authors indicated.

UNITED KINGDOM

In the UK there have been few attempts to apply EIA to plans though Foster (1983) describes the use of 'strategic EIA' to identify sites for development, especially in Scotland (see also Lyddon 1983). Collins (1986) has described the integration of EIA into the plan-making process for the Mersey Marshes area in Cheshire. He records the efforts of Cheshire County Council to assess the effects of possible industrial expansion on two large (250 hectares) sites close to an existing petrochemical complex. In the process of preparing the plan, various environmental baseline surveys were undertaken and widespread public participation and consultation took place. The key issues which emerged were air pollution, hazard, further alternative sites, and the future of existing villages. Extensive use was made of existing British guidance on EIA and various environmental safeguards were written into the plan. It was originally intended that the environmental impacts of alternative scenarios would be assessed, but it proved in practice to be impossible to gain sufficient co-operation to do this. The inspector appointed to hear the public inquiry on the plan disagreed with many of the land-use proposals it contained. Ironically, he suggested that a policy requiring that detailed development proposals should be subject to an EIA, as a means of determining the most appropriate safeguards, should be strengthened to help determine appropriate land uses. This policy now reads:

> where development is proposed of such nature or scale that in the opinion of the local planning authority it could substantially affect the local environment by reason of atmospheric or water pollution, noise levels, hazard risk, visual impact or impacts on transport systems, the authority, before determining the application will require the applicant to provide a statement of the anticipated environmental effects of the proposed development and of the measures to be incorporated to minimise the effects (Cheshire County Council 1986).

While this example did not involve a 'formal EIA' sufficient consideration of environmental issues was intended to be included in the four stages of the planning process for it to be regarded as an example of the integration of EIA in the plan-making process. As in the Dutch example, the geographical area covered by the plan was quite small. Certainly, Collins (1986) felt that the experience gained held out enough hope that EIA might offer a better way to deal with complex environmental issues 'to encourage us to pursue the approach further'. There is, however, no separate chapter on EIA in the Mersey Marshes Local Plan and the precise extent to which EIA was actually integrated into the plan-making process fell below the original intentions.

UNITED STATES

Foster (1983) has reported that only ten areawide (plan) EISs had been prepared in the United States by the beginning of 1983, but EISs for some other types, such as forest plans and coastal management plans, had also been undertaken.

Though a methodology for assessing comprehensive plans has been suggested (Hall 1977), the most complete guidance has been prepared on behalf of the US Department of Housing and Urban Development (HUD 1981). This guidance makes it clear that areawide EIA differs from project EIA in the USA in several respects (see also McElligott 1978 and Merrill 1981). First, areawide alternatives are not single-action choices, but define an overall pattern of development. Secondly, areawide alternatives combine numerous individual public and private actions. Finally, areawide EIA can compare the cumulative effects of many individual developments and reveal issues that are easily overlooked at the project level. An additional advantage of areawide EIA, according to HUD (1981), is that it can save time because it often obviates the need to undertake numerous project EIAs.

One of the ten plans listed by Foster (1983) is the New Castle County, Delaware, Areawide EIS (HUD 1978). This concerned residential development planned on about 5000 hectares in a rural area on the fringe of Wilmington following the opening of a new interstate highway. Socioeconomic, physical and environmental issues relating to the location and timing of growth as forecast in the New Castle General Comprehensive Plan were assessed in some detail. The EIA identified geographical areas and issues where there was conflict with existing requirements. The main areas of interest were the generation of pollutants, damage to forest and agricultural land, and possible loss of ground water supplies. The Comprehensive Plan was modified in the light of the mitigatory measures identified by the EIA.

This is the most formal example of EIA in plan making reviewed here. It represents an instance of the application of EIA after the plan has been prepared, rather than the integration of EIA into the plan-making process. There is no evidence of the assessment of alternatives, though modifications to the plan were made as a result of the EIA process. Further, the EIA is concerned only with HUD's area of interest and excludes consideration of, for example, industrial development. Despite its partial nature, however, this is an example of an EIA of a plan which could be extended and developed to provide a model for plan assessment, especially if used in conjunction with the HUD guidance.

Conclusions

There appear to be profound advantages in extending EIA from projects to plans. The physical planning system in most places bears strong similarities to the EIA process and the two procedures could be integrated in most countries, certainly in Europe, without radical upheaval or major legislative change. The methods involved in EIA in plan making are broadly similar to those employed in project EIA, but with a number of significant differences. Generally, however, these methods are less well developed for plan, than for project, EIA and greater difficulties and imprecision arise in their use.

Though the logic of undertaking EIA in plan making is irrefutable, there is little evidence that, in practice, it is taking place to any substantial extent. The reasons for this have to do with institutional resistance to change, resource restrictions and methodological difficulties. Nevertheless, some examples of EIA in plan making do exist and those involved in them have indicated the benefits which have accrued. The difficulty of applying EIA in plan making is shown by the fact that most of the examples of such EIAs tend to relate to plans for small areas, rather than larger areas.

The future development of EIA in plan making probably lies in the integration of EIA in the plan-making process rather than in the somewhat artificial EIA of plans. However, it will be essential to specify the minimum requirements of an EIA and to demand that evidence of compliance with them is presented formally. Otherwise, there is a danger of insubstantive 'lip-service' adherence to EIA.

Clearly, it would be desirable for environmental aspects to be included within each stage of the plan-making process in much the same way that EIA is sometimes integrated into the various stages of the project appraisal process. As with the environmental management of projects, the considerable benefits of integrating EIA into plan making will need to be clearly demonstrated and a number of further steps will have to be taken before its use becomes common.

There are several obvious requirements for the further development of EIA in plan making. One is the preparation of EIA case studies relating to plan assessment for use as models by other practitioners. Manuals for the practical application of EIA in plan making of the kind developed by HUD (1981) are also required. Similarly, training programmes in the use of EIA in plan making should be initiated. In addition, the application of EIA in plan making should stimulate further attention to such matters as the review of existing environmental monitoring systems and methods of data assembly and dissemination. Development of communication techniques to allow more effective public involvement is also needed. There is also a requirement for research into several of these areas and into techniques for forecasting environmental conditions and the impact of alternative plans upon them. Finally, legislative initiatives of the type being taken in the Netherlands will be needed. If this proves to be successful the Commission of the European Communities might be emboldened to resurrect its own proposal for the use of EIA in plan making.

Part III

EFFICACY OF EIA

7 *Monitoring and auditing of impacts*

R. BISSET and P. TOMLINSON

Introduction

A systematic examination of the literature on pollution has revealed almost no mention of monitoring prior to the Stockholm conference in 1972 (Harvey 1981). Since then, there has been a significant increase in the attention paid to monitoring. In the environmental science literature, however, 'considerable confusion has resulted from the contradictory way in which terminology relating to the monitoring concept has been used' (Harvey 1981). For the purposes of this chapter, therefore, monitoring is defined as an activity undertaken to provide specific information on the characteristics and functioning of environmental and social variables in space and time.

With this definition, it is clear that monitoring fulfils a number of important functions in EIA. The main uses of monitoring data are in impact monitoring and in 'audit studies'. Again, the term 'audit' does not have, as yet, an agreed meaning in environmental science literature. Increasingly the term is used to describe the process of comparing the impacts predicted in an EIA with those which actually occur after implementation in order to assess whether the impact prediction process performs satisfactorily (see, for example, Andrews *et al.* 1974, Bisset 1984). In this chapter the significance of impact monitoring and audits in EIA are considered in detail.

Impact monitoring schemes

The aim of impact monitoring is to detect an impact if it has occurred and to estimate its magnitude. An essential part of the process is to establish that the perceived change is a consequence of the project and not the function of some other cause. The changes might result, for example, from natural variations in the parameter monitored or may be the result of some other development in the vicinity and, thus, not related to the project under consideration. This is not an easy task and great care and attention has to be paid to experimental design in order to achieve this objective.

The only way of ensuring that an impact can be assigned correctly to a project is to use 'reference' monitoring locations, comparable to the controls in classical

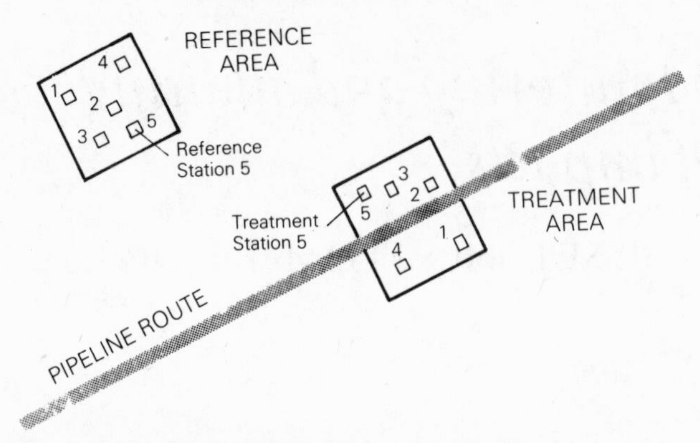

Figure 7.1 The use of reference and treatment areas in audit.

scientific experimentation. For impact monitoring, it is necessary to use pairs of 'reference' and 'treatment' locations. The treatment locations are situated in an area where the impact is expected to occur. For each treatment location a reference location must be selected which is similar in all important respects, except that it is situated in an area where the impacts are not expected to occur. Several stations for sampling must be established within each area, because the differences between the locations can only be assessed when the variability within each is known (Fig. 7.1).

Skalski & McKenzie (1982) have described the features of an impact monitoring system for aquatic environments, but the principles are applicable to other situations. Monitoring must begin in the pre-operational (baseline) period with the establishment of the paired treatment–reference locations. Sampling should begin as soon as possible and continue into the operational period of the project. There are two main reasons for this. First, pre-operational sampling can be used to assess the success of the pairing scheme, for example, in establishing that the variable behaves in the same way in both treatment and reference locations. Secondly, pre-operational monitoring establishes the relationship between the parameter, for example the population of a shellfish, in both the treatment and reference locations. This would give, for example, the proportional abundance which can be compared later with data collected during the operational stage. An impact can then be defined as a statistically significant change in the proportional abundance of organisms at reference and treatment locations between pre-operational and operational periods.

Skalski & McKenzie (1982) stress that 'the objectives of a monitoring program must be established explicitly before the field design for the monitoring program can be properly determined'. Monitoring objectives are of two types. First, the variables to be monitored must be selected. Secondly, the magnitude of a change which is ecologically significant or important to detect must be defined. In addition, it is necessary to set probability levels to prevent a significant impact being identified when in fact there has been none, or no impact being identified when one has occurred. Traditionally, ecologists work to the 5% probability level.

Once these criteria have been incorporated into the monitoring scheme it is possible, using standard statistical techniques, to determine the sampling effort required. This will determine the frequency of sampling, the number of stations required and the period over which samples must be collected. For example, Skalski & McKenzie calculate that, for some aquatic ecological impacts, monitoring need only be undertaken for two years in the pre-operational period and for two years in the operational phase. This type of programme would fit well within the timescale of many EIA studies and does not imply extensive commitments of resources and money to impact monitoring.

Impact monitoring requires the formulation of hypotheses for successful application. In impact monitoring the most frequently used statistical analysis will involve tests of significance, that is, whether a change in a measured parameter (a suspected impact) is statistically significant. Such tests only have meaning when they are made against an *a priori* null hypothesis, H_0, which can be falsified. However, H_0 must be chosen to be falsifiable not just on theoretical considerations, but also with respect to practical considerations of time, cost and technical constraints.

This discussion of impact monitoring and hypothesis setting is most applicable to biological impacts on individual organisms or populations. These impacts, however, are often the most difficult to identify and describe. Fortunately, expensive and time-consuming monitoring programmes may not be required for all variables. It is important that the monitoring objective for each impact variable be stated clearly before a programme is devised. In addition, it is essential that both ecologists and statisticians be involved in the formulation of the monitoring objectives before a scheme is started. Experience in the past has shown that impact monitoring data have been of very little use in actually assigning impacts to a particular project. Most recent studies of the success of impact monitoring schemes cite lack of clear objectives and inattention to sound statistical principles as the main failing of attempts to monitor impacts (Green 1979, Bisset 1981).

Experience of project performance and impact monitoring has shown that attention to the detail of data collection is no substitute for beginning monitoring exercises with clear conceptual objectives and a practical and workable framework (Carley 1982). The conceptual framework establishes the objectives of the monitoring and helps to ensure that the monitoring programme can supply the requisite data. The aims of monitoring should be formulated clearly and set out in explicit guidelines to ensure that no deviation from the required monitoring programme occurs, because changes in sampling procedures may invalidate comparisons of monitoring data. Once the objectives and guidelines have been established it is essential to design an appropriate institutional structure which can implement the monitoring and collect, interpret and publish (if possible) the results.

Uses of monitoring data

Impact monitoring has a number of benefits. First, monitoring environmental and social variables can identify harmful trends before it is too late to ameliorate or prevent them. This is a very important benefit for decision makers. Impact monitoring, therefore, provides an 'early warning' device which alerts those managing the project or the environment to possible harmful impacts before the full potential for damage is realized. This, in itself, is sufficient reason to monitor potential impacts.

Secondly, such monitoring can be used to improve knowledge about the impacts of various projects on specific environments. Written accounts of the actual impact of past projects are sadly lacking in both EIA and general environmental literature. Such experience does exist, but it is very limited in relation to the number of major projects implemented throughout the world in the last decade. It is this lack of knowledge which, in part, makes impact prediction such a difficult and uncertain task. Increased impact monitoring in different localities for a variety of projects would be a considerable step in increasing knowledge of the impacts of development. Such knowledge can be used to improve future EIA studies and would make a contribution to reducing the time and effort currently expended on individual EIAs. As more and more data become available, the need for impact monitoring will decline with corresponding savings in time and money.

At present, decisions on projects are made on the basis of considerable uncertainty. All EIAs involve predictions of future events and, as such, exhibit varying degrees of uncertainty. Part of this uncertainty is inherent in the problems of prediction *per se*, but some of the uncertainty arises because of the lack of knowledge concerning the actual impacts of projects on particular environments. It is precisely because of this prevailing degree of uncertainty that many mitigating and monitoring schemes are implemented. They are undertaken to try and avoid possible mistakes and harmful impacts. In many cases they are expensive and involve considerable commitments of staff and other resources. It is possible that monitoring might reveal that certain expected impacts did not occur and that future decisions on similar projects need not consider mitigating measures or monitoring schemes which have been implemented previously (Johnson & Bratton 1978).

Additionally, information on the utility, accuracy and comprehensiveness of methods and predictive techniques would be beneficial. At present, there is little knowledge of this topic which means that considerable resources are expended in attempting to devise 'optimal' EIA methods and techniques.

There is a strong argument for 'random' monitoring schemes to be instituted for selected variables. These could be environmental or social components which might be directly affected, but which are considered not to be seriously at risk in a particular case. Alternatively, they might be indicators which integrate multiple stresses on particular ecosystems. Obviously not everything can be monitored but, in addition to monitoring those impacts expected to occur, there is room for the occasional monitoring of other variables at major projects in order to see what can be learned. This topic has been considered little in the literature, but merits attention.

Audit studies

A number of audit studies are examined in this section. Additional examples can be found in Tomlinson (1987). The Wisconsin Power Plant Study examined the effects of Phase 1 (500 MW) of the Columbia electricity generating station in Wisconsin (Institute for Environmental Studies 1977, Environmental Protection Agency 1980). The study examined a large variety of impacts and instituted extensive monitoring schemes to detect the impacts of the station. The analysis of ambient sulphur dioxide (SO_2) changes and their impacts is typical of one of the individual studies. Monitoring data were collected for two years before the plant commenced operation in 1975. Using these data, in conjunction with impact monitoring data collected once the plant was operational, the performance of the Gaussian plume model with respect to SO_2 levels used in the EIA was investigated. The study concluded that the model was successful in predicting annual average concentrations, but inadequate for simulating hourly averages. It was found that the model tended to slightly under-predict for certain atmospheric conditions and to overestimate for others.

Within the United States, a study of the 'off-road' motor cycle race across some 250 km of desert terrain in California and Nevada is comparable in terms of the number and range of impacts investigated (Bureau of Land Management 1975). The aim of the study was to determine the extent and the nature of the impacts and decide whether they were serious enough to prevent future races. Long-term effects, including recoverability of damaged sites and wildlife numbers could not be determined. There were sufficient data, however, for impacts on air quality, archaeological sites, soil characteristics and land surface effects to be determined. It was found, for example, that certain air quality predictions included in the EIA were confirmed, while others were not. Similarly, the area affected by the race was 31% greater than predicted.

The study exemplifies many of the problems encountered when an attempt is made to analyse the effects of development and compare them with pre-operational predictions. The audit report contains admissions concerning the unsatisfactory nature of some of the available data. Monitoring of wildlife disturbance was implemented in a time-constrained fashion without control sites. The 'before' and 'after' comparisons were not made at sufficient sites to give a useful indication of disturbance and damage to wildlife numbers. Monitoring studies were only directed at determining immediate consequences.

A number of studies, known as Environmental Technical Specifications (Tech Specs), were initiated by the US Nuclear Regulatory Commission to determine some of the impacts of operational nuclear power plants. Twelve plants were examined and the main foci of attention were aquatic ecological effects, the characteristics of thermal plumes and the impact of chemical discharges.

The EISs predicted that there would be effects on all biotic groups, but that these would be minimal and unimportant. Analysis of the monitoring data showed that there was no evidence of significant effects of plant operation on these groups. It is stressed by Murarka (1976) that analysis of the monitoring data did not prove anything apart from the fact that no impacts were detected within the frame of the available monitoring data and statistical analyses carried out.

It was realized that long-term impacts could be occurring, but were undetected because of the lack of appropriate monitoring data. For the three plants investigated, problems involving lack of pre-operational data, changes in monitoring programmes and sampling procedures made it difficult, if not impossible, to detect and separate man-made impacts and natural biological fluctuations. Despite the interpretive limitations which must be placed on the conclusions, Murarka was satisfied that EIS predictions about the aquatic ecological effects of power stations were accurate.

Similar results were obtained from analyses carried out on other nuclear power stations as part of the Tech Specs programme. For example, Gore *et al.* (1979) could not attribute any changes directly to plant operations and this finding agreed with the predictions. However, they point out that several changes were qualitatively identified, but not statistically validated. Also, in a few cases, where statistically significant impacts did occur, the influence of factors other than power plants could not be ruled out. Despite this, it was thought that significant impacts would have been detected had they occurred.

Audits were carried out on predictions related to thermal plume behaviour at three of the nuclear power stations (Marmer & Policastro undated). An analytical model was used to predict the size, shape and temperature characteristics at the Zion and Kewaunee power stations, whereas a hydraulic model was used at Quad Cities. All predictions indicated a wide range of plume characteristics due to variability at the sites. Analysis of monitoring data covering two operational years indicated that the predictions from the hydraulic model were generally accurate, whereas the predictions of the analytical model were not. Marmer & Policastro concluded that pre-operational predictions using the analytical model did not take account of localized environmental conditions. From this limited comparison it appears that hydraulic modelling is to be preferred to analytical modelling, but more cases would have to be examined to enable this tentative conclusion to be confirmed.

The studies discussed above have, in the main, concentrated on environmental impact. There have been other studies which have focused on socioeconomic impacts, especially for power stations. In fact, the analysis of actual socioeconomic impacts and the testing of techniques used to predict the nature and scale of these impacts is more advanced than for environmental impacts. Major studies have been undertaken in the USA by the Denver Research Institute (Gilmore *et al.* 1980), Leistritz & Maki (1981) and Murdock *et al.* (1982) and at Oxford

Polytechnic in the UK by the Power Station Impacts Research Team (1979).

The objectives of the study on the Coal Creek power station in North Dakota (Leistritz & Maki 1981) were to evaluate the accuracy of impact assessments, as well as the strategies implemented to mitigate impacts and manage induced growth. In this case, it was found that the impacts on population, public services and the fiscal characteristics of the local communities were generally consistent with the predicted impacts. The projection for housing requirements for construction workers failed, to some extent, because many workers from outside the local areas preferred temporary accommodation (motels, boarding houses and recreation vehicles such as campers) to the more 'permanent' type of mobile home. At a more general level, a tendency was found for predictions to be more correct for large areas when figures were aggregated than for the individual constituent communities. This is an important observation as it compares with the findings from the Columbia Power Station Study, that the air pollution model predicted more accurately at the aggregate level than at a small scale.

Gilmore *et al*. (1980) reported that the timing and magnitude of construction employment differed substantially from the estimates made prior to the commencement of the project. In most cases, there were construction delays while the magnitude of construction employment (both peak and total man-years) usually exceeded the initial projections by a wide margin. Similarly, the geographical extent of the impact area was greater than had been estimated in the pre-impact studies. Workers tended to commute on a daily or weekly basis from a much larger area than had been anticipated. The studies also showed that the local service-to-construction worker ratio did not exceed 0.2 in most cases. Most projections examined during this research project used much higher ratios and, therefore, tended to overestimate secondary impacts. Finally, most projections were based upon peak construction levels which exist for only a few months. The case studies indicate that there is usually little service response to this peak, further aggravating the problem of overestimating the scale of the employment multiplier.

The CEMP audit study

A detailed study of four major developments in the UK, based upon an analysis of EIA reports and other comparable project documents, has been undertaken by the Centre for Environmental Management and Planning (CEMP), University of Aberdeen (Bisset 1984). The projects considered were the Sullom Voe and Flotta oil terminals, Cow Green reservoir and the Redcar steelworks. The reports were analysed to identify all environmental impact predictions. In addition, information on subsequent impacts irrespective of whether they had been predicted was also collected. Abstraction of impact predictions was a complex and time-consuming task. Table 7.1 shows that there were considerable differences in the number of impact predictions identified in the four studies

Table 7.1 Number of impact predictions for four major developments in the UK.

	Sullom Voe	*Flotta*	*Redcar*	*Cow Green*
Inappropriate form of prediction	11	5	3	0
Design changes	0	6	160	18
Conditions not satisfied	6	395	0	6
No appropriate monitoring data	8	36	36	7
Total not auditable	25	442	199	31
Total number of impact predictions	52	459	220	60

with the EIAs for Flotta and Redcar the most comprehensive. The EIAs for Redcar and Flotta included techniques for predicting noise and air pollution impacts and oil spill behaviour and generated a large number of individual predictions.

Frequently, predictions were expressed in vague, imprecise and even 'woolly' language. Whenever possible, the predictions were interpreted 'objectively' to enable them to be audited. In some cases, however, this could not be done and these predictions, consequently, could not be audited. Design modifications after the EIA had been prepared also prevented some predictions from being audited. An additional process of elimination was necessary. A number of predictions were contingent upon certain assumptions concerning environmental conditions. For example, many oil spill predictions for the Flotta terminal were dependent upon assumptions, such as tide and weather conditions at the time of spillage, which had not occurred prior to the audit. The validity of these predictions also could not be assessed. The final restriction on ability to audit predictions was the lack, or unsuitability, of monitoring data.

The major conclusion from Table 7.1 is that a large number of the predicted impacts could not be audited. The numbers finally audited constitute only 3.7% and 9.5% of the total impacts predicted for Flotta and Redcar respectively. The proportions for Sullom Voe and Cow Green were significantly greater. Paradoxically, fewer impacts had to be tested at Sullom Voe and Cow Green than for the Flotta and Redcar projects, yet more intensive and varied monitoring programmes were established at the former sites.

In some instances, it is impossible to come to firm conclusions regarding the accuracy of predictions, because the available monitoring data give only a general indication of accuracy. One particular problem which was faced when assigning accuracy to a prediction should be highlighted. Many predictions do not contain any reference to the time period within which an impact might be expected. Although monitoring data might show that such impacts had not occurred, the possibility remains that they might occur at some time in the future. Some predictions, therefore, were classified as accurate or inaccurate,

Table 7.2 Accuracy of the predictions of impacts from four major developments in the UK.

	Sullom Voe	*Flotta*	*Redcar*	*Cow Green*
Accurate (definitive)	18	4	3	14
Accurate (time-dependent)	0	3	2	0
Inaccurate (definitive)	9	1	2	11
Inaccurate (time-dependent)	0	4	4	2
No conclusion	0	5	10	2
Total auditable	27	17	21	29
Total number of impact predictions	52	459	220	60

but only within the time constraint of the audit. These impacts are described as 'time-dependent' in Table 7.2

The number of accurate and inaccurate predictions is shown in Table 7.2. Only 12% of all predictions could be audited, but with these it was possible to come to firm conclusions on their accuracy in most cases (82%). In the Redcar, Flotta and Cow Green cases approximately 50% of the predictions were accurate, as were 66% of the Sullom Voe predictions. There was a tendency for inaccurate predictions to indicate impacts which subsequently did not occur. Although an attempt was made to determine the extent to which inaccurate predictions under- or overestimated actual impacts, it is impossible to come to any conclusion concerning this factor. For example, the results from the Redcar and Cow Green case studies show opposite situations. In the Redcar case study, 5 predictions overestimated actual impacts and 1 underestimated, whereas for Cow Green the converse was true with 7 predictions underestimating and 2 overestimating impacts. No conclusion could be drawn from the results for Flotta and Sullom Voe.

No explicit EIA method was used in preparing any of the EISs. Therefore, the research focused on whether the *ad hoc* approaches used were comprehensive and covered all impacts. It was found that a number of impacts occurred which were not included in EISs. Most were 'secondary' impacts occurring at distant locations. A few direct impacts, however, were omitted from the EISs. It could be argued that a formal, but simple EIA method such as a two-dimensional interaction matrix would have aided the assessment process. No conclusions could be made concerning individual prediction techniques.

Implications of results from EIA audit studies

The results of the audit studies discussed above show that the ability to predict impacts accurately is not widespread. For some impacts and for certain projects, such as annual SO_2 levels at the Columbia power station and most socio-economic impacts of the Coal Creek, predictions seem to have been accurate.

An important conclusion, but as yet based upon only tentative evidence, is that a number of predictive techniques may be better at predictng impacts at an aggregate, rather than a more specific, level. This was true for all socioeconomic impacts of the Coal Creek station and for SO_2 levels at the Columbia power station. A major failure of these studies, however, is that little attention seems to have been focused on the coverage of impacts, that is, whether all subsequent impacts were predicted prior to project authorization. Only the CEMP study considered this aspect and concluded that EIAs had failed to predict the entire range of impacts known to have occurred.

The main conclusion of the CEMP study raises a rather more fundamental issue. The nature of most EIAs makes it very difficult, in some cases impossible, to audit the predicted impacts of a development. Impact predictions are not phased in a way which allows auditing and they can become obsolete very easily, for example, as a result of design changes. In addition, existing monitoring programmes are often not very useful in providing data to allow predictions to be tested in a scientifically acceptable manner.

The US studies encountered similar problems. Often assessment of actual impacts had to be based upon partly subjective judgements of environmental change. Assessments based upon monitoring data should have a firmer foundation, but a number of the US experiences also showed that monitoring data were often not appropriate for the detection of impacts, nor for determining whether they were the result of the project rather than another source of environmental disturbance.

Conclusion

The need for an examination of the accuracy of impact predictions can be illustrated by a review of the objectives of EIA. EIA is intended to provide decision makers with an understanding of the environmental consequences of a proposed action or project. This objective is achieved by the use of environmental information which is often characterized by scarcity and uncertainty, predictive techniques with unknown error margins and evaluation methods which assess and present information to decision makers in a variety of ways.

A major failing of EIA practice has been the common use of EIA to obtain a development permit, rather than as a tool to achieve sound environmental management either within the project objectives or on a broader regional and national basis. Presently, the emphasis is directed towards the approval procedure with little attention being given to the post-approval stage. This situation prevents an evaluation of the performance of various EIA activities and, therefore, inhibits the process of using and refining the existing procedures to achieve their maximum utility. As a consequence, there is a need for a feedback mechanism in EIA which involves the transfer of knowledge from the actual environmental effects of a project or action to future EIAs (see Fig. 7.2). This can only be achieved through audits.

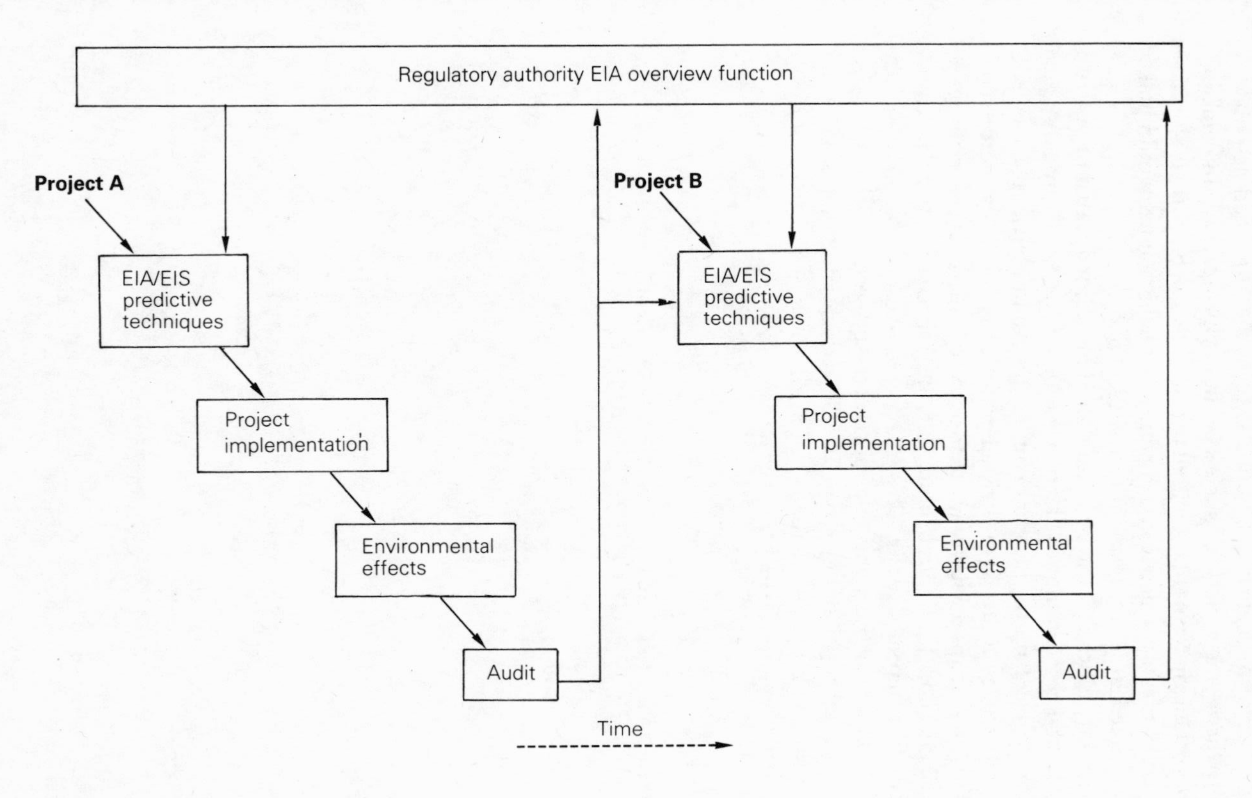

Figure 7.2 Schematic illustration of the role of audits.

Auditing can be used to test mitigating and monitoring schemes. Such schemes are undertaken to achieve certain objectives, for example, to prevent expected impacts or to enable actual impacts to be identified and described. In many cases, there is a variety of such schemes which can be implemented. Auditing of schemes in terms of the extent to which they achieve their objectives would be useful for future decisions. Again, knowledge gained would make a contribution to resource saving and future success of these schemes.

One of the difficulties that can be encountered in auditing is that of mistrust, since decision makers seem not to like their rationale questioned by subsequent investigations which have the benefit of hindsight. Auditing could be viewed as a tool for criticizing the decision-making process and hence be perceived as a threat. This is perhaps a viewpoint that is to be expected, especially when decision-makers know little about auditing. Issues related to confidentiality and access to data files may also occur. This situation is likely to arise in those cases where environmental assessments and monitoring programmes have not been designed with auditing as an objective. The objective of auditing is, however, not to examine the rationale of the decision maker, but rather to study the quality of the information available to the decision-making process.

8 *The evaluation of assessment: post-EIS research and process development*

B. SADLER

Introduction

During the past fifteen years, environmental impact assessment (EIA) has been adopted in various parts of the world in order to analyse and mitigate the effects of development proposals. The legislative and institutional frameworks for applying this approach vary considerably among countries and even within federal states, such as the United States and Canada (O'Riordan & Sewell 1981). As a formal procedure, EIA is distinguished by certain characteristics which are common to most, if not all, systems. It is, above all, a predictive exercise directed at the identification and evaluation of the significance of potential changes induced by programmes, projects and activities (see Munn 1979). The emphasis understandably and, perhaps, inevitably is on pre-decision analysis leading to the preparation of an environmental impact statement (EIS) or similar document which establishes terms and conditions for project approval.

The paradox of impact assessment, as conventionally practised, is that relatively little attention is paid to the environmental and social effects which actually occur from development or to the effectiveness of the mitigation and management measures which are adopted. A lack of follow-up after a project has been approved is a major constraint on the advancement of EIA practice (Sadler 1986a). It means that there is no opportunity to learn from and utilize the results of case experience. Yet, learning by trial and error is essential for coping with the uncertainties which characterize complex ecological and social systems (Holling 1978).

An investment in retrospective research can repay major dividends in designing and implementing the adaptive approach to environmental assessment and management, the main currency of international theory and practice (Clark & Munn 1985). The opportunities inherent in this relationship are becoming widely realized. In Canada, for example, EIA audit and evaluation research is under way on several fronts. The directions taken and the findings of selected studies are outlined in this chapter in order to highlight elements of general interest to administrators and practitioners. There are three main objectives. First, the role, scope and contribution of post-decision analysis and evaluation to improve the EIA process are reviewed. Secondly, a strategy for applying process review is proposed. Finally, present trends in research and development in this field are discussed with reference to work being undertaken in Canada.

Organizing perspectives

At the outset, it is important to place retrospective (often referred to as 'postdictive' – the converse of predictive) evaluation of EIA in a comparative context. In the present discussion this is undertaken at two levels. First, reference is made to the emerging trends in thought and practice which provide not only the context but also the impetus for the development of post-EIS analysis. Secondly, the main components of such analyses are briefly distinguished and delineated.

A continued expansion in the role and scope of EIA has taken place since the passage of the National Environmental Policy Act of 1969 in the USA. The main stages in the evolution of the field in Canada for example, can be organized into three distinct phases (Boothroyd & Rees 1984). The initial focus of interest was on the methodology of impact prediction. This broadened to include the administrative procedures for EIA and now encompasses its relationship to the larger framework of resource management and development planning.

A new paradigm of EIA is emerging. One of its fundamental premises is that the impact assessment process requires two supporting provisions to work effectively (Cornford *et al.* 1985). First, a policy-planning context, sufficient to permit an evaluation of the significance of potential impacts, is needed. Secondly, an implementation-management system for monitoring, controlling and evaluating the effects of development is required.

The importance of developing the proper context for EIA is a major theme in the literature (Clark 1981, Wiebe *et al.* 1984). A similar emphasis is now being placed on follow-up activities, including research to measure performance against prediction and practice against intention in order to improve analysis and administration. This area of investigation is concerned with research on, as opposed to in, the EIA process.

A brief clarification of the relationships between monitoring, audit and evaluation is necessary, since these terms are used differently and often inconsistently within the literature. The definitions adopted here are drawn from previous work by Bisset (1980), Sadler (1985) and Munro *et al.* (1986).

Figure 8.1 is a schematic summary of the relationship of post-assessment follow-up activities within the overall process. Evaluation refers to the generic process of analysis and interpretation, and incorporates monitoring, surveillance and audit programmes. The process of evaluation involves making subjective, policy-oriented judgements about the effectiveness of EIA procedures and results. The concept of audit, in contrast, implies an independent and objective examination of whether practice complies with expected standards.

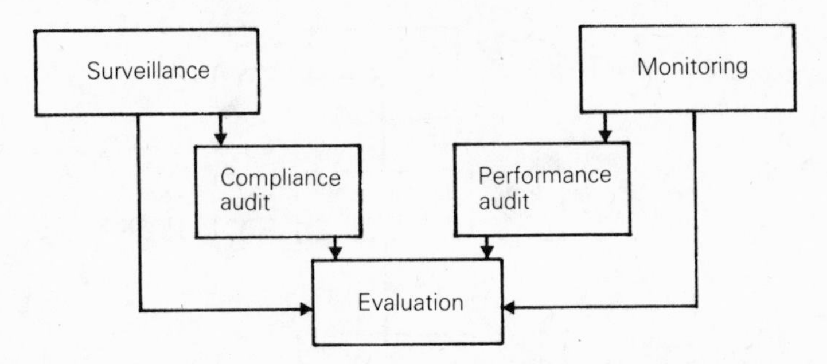

Figure 8.1 Generalized model of post-assessment activities based upon emerging practice in North America.

It involves a search, or a verification, of a system of records derived from surveillance and monitoring programmes (see Bisset 1980 and Chapter 7 for further discussion). Audit, thus, can be considered a distinct and restricted form of evaluation concerned only with establishing conformance and characterized by certain data preconditions. Where these are absent or inadequate for the purposes of an EIA evaluation, research can be supplemented by, or based upon, other methods such as survey, consultation and observation.

The role and scope of post-EIS evaluation

Without some form of feedback, impact assessment is a static linear exercise rather than a dynamic iterative process (Larminie 1984). At present, the mechanisms designed to achieve this do not seem a well-developed part of the institutional systems established for EIA in various countries (see O'Riordan & Sewell 1981). This is certainly the case in Canada (Sadler 1987). The result is the well-known tendency of every EIS to duplicate information and undertake unnecessary analysis, because the results of previous experience were not monitored or evaluated. This constant tendency to 'reinvent the wheel' represents a hidden tax upon impact assessments.

The addition of an evaluation suffix to the EIA process would build continuity into both the project approval and development cycles. Figure 8.2 illustrates this function and the linkages by which practitioners and administrators can learn from case experience and apply the results to future actions. It underlines the important part which evaluation plays in establishing a systematic follow-through in EIA and decision making. As a formal procedure, this approach may be considered an integral element of development control or a separate end phase of audit and review. The notion of evaluation as a continuing process is advanced here. Depending upon purpose, it may be deployed at various stages in the implementation phase which follows completion of the EIS and the granting of project approval.

During the implementation phase, evaluation is designed to meet three basic objectives. First, in project regulation, evaluation should ensure that activities

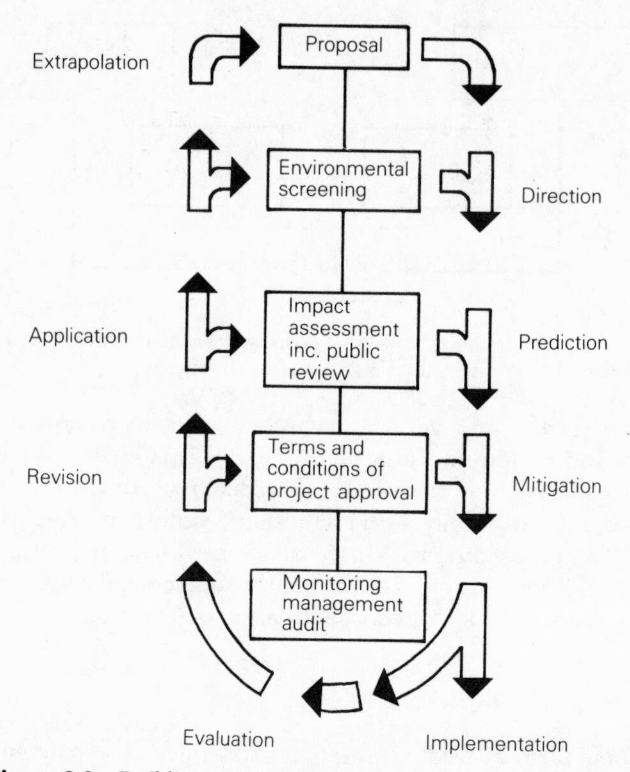

Figure 8.2 Building continuity into environmental assessment and project development.

conform with operating conditions previously established on the basis of an EIS. Secondly, it should facilitate impact management by providing an opportunity to manage the unanticipated effects through modifications to mitigating measures and project design. Finally, evaluation should aid field development through improving the practice and procedures of environmental assessment and its supporting processes. The relationship of these factors to the monitoring and research activities identified in Figure 8.1 are summarized in Table 8.1.

An important distinction in the streams of analysis identified here relates to the scope and immediacy of use of information. In general terms, surveillance and monitoring programmes for project regulation and impact management are user-oriented and action-specific, geared to development control. Field development research, in contrast, is process-related and, consequently, has broader applications and longer time horizons. This approach, amongst other things, may build upon and extend, through comparative analyses, the results of compliance and performance audits. Such reviews can prove useful for refining similar projects, as well as helping to advance the general state of the art of environmental assessment and management. The emphasis in this chapter is on the latter.

Table 8.1 Applications of EIA evaluations.

Objectives	*Components*		
	Surveillance	*Monitoring*	*Audit*
1 Project Regulation	★ (primary)	★ (tertiary)	★ (tertiary)
2 Impact Management	★ (secondary)	★ (primary)	★ (secondary)
3 Field Development	★ (tertiary)	★ (primary)	★ (primary)

Application
★ Primary
★ Secondary
★ Tertiary

Notes:
a While monitoring has primary application in impact management and field development the same programme will not necessarily fulfil both objectives.
b Field development connotes general long-term improvements in the theory and practice of EIA.

EIA process evaluation can take a number of forms. A typology of research is developed in Table 8.2 and shows that EIA practice may be analysed from three standpoints. First, evaluation can be used to assess the capability of impact prediction and mitigation methods. This could establish whether, for example, the important project-related impacts and mitigation responses were identified accurately in an EIS or that the cumulative and secondary impacts were traced properly. Secondly, the effectiveness of administrative procedures used for an assessment and review can be gauged; for example, whether the guidelines for an EIS were clear and coherent and whether the opportunities for public review were fair and credible. Finally, the utility of the process for decision making can be assessed by showing for example, whether the process yielded sound, relevant and focused information concerning project effects and their implications and whether there was a clear indication of the level of confidence and significance placed on key functions.

Table 8.2 A typology of EIA process evaluation.

Type of research	*Elements of analysis*	*Canadian example*
1 Technical/scientific	Adequacy of baseline studies and pre-project monitoring	Kiell *et al.* (1985)
	Accuracy of impact predictions	Zallen *et al.* (1985)
	Suitability of mitigation measures	Spencer (1985)
2 Procedural/administrative	Efficiency of guidelines for EIA	Erickson (1985)
	Fairness of public involvement measures	Case *et al.* (1983)
	Degree of co-ordination of roles and responsibilities	Janes & Ross (1985)
3 Structural/decision-making	Utility of process for decision making	Sadler (1984)
	Implications for development	Munro *et al.* (1986)

The last category, determining the utility of the process for decision making, lends itself to a macro framework for EIA process evaluation. It suggests a policy-focused review of the relevance and value of EIA as an instrument for planning and management (Sadler 1984). While this will be difficult to determine unambiguously, there seems to be considerable value in cross-linking the performance of the scientific and administrative aspects of EIA. A recent, large-scale study of ways to improve the methodology of impact analysis found, for example, that the most promising route to reform were institutional rather than technical (Caldwell *et al.* 1982). Moving in this direction in process evaluation means a commensurate increase in the scope and generality of value judgements and, eventually, contact with the more generalized literature on the strengths and weaknesses of EIA. The main difference is that the forms of evaluation discussed here are functional elements of the EIA process, built in to provide feedback, facilitating both immediate and longer-term improvements in practice and procedure (Sadler 1986a).

A strategy for the application of research

The next step is to consider the way evaluation can be most productively incorporated and utilized within EIA processes. A meta-strategy for research and development is proposed in this section. The term 'meta-strategy' implies the combination of aim, means and decision criteria which impart direction to, and can be accommodated within, the various legal and administrative arrangements for EIA. It is axiomatic that post-EIS evaluation must be integrated within systems of decision making in a cost-effective way. As a general rule, this means that the analysis and subsequent feedback of information should be roughly consistent with the order of the problem examined.

The key to the application of these principles is a decision protocol which relates the degree of confidence in an EIS and similar support documents to the terms and conditions of project approval (Sadler 1986a). It is based, amongst other things, on an explicit attempt to categorize the extent of the uncertainties associated with project-induced change (Cornford *et al.* 1985). As Table 8.3 indicates, this is done through establishing the reliability of data and attaching qualifications to scientific understanding of the ecological process likely to be affected. This systematization of the level of understanding becomes the vector for further improvements in EIA via correlation with the decision categories outlined in Table 8.3. In general, the lower the degree of confidence in impact predictions, the more stringent the terms and conditions attached to project approval and the greater the requirement for post-EIS analysis and evaluation will be. Depending upon the correlations established, some or all of the following activities will take place during and after project implementation: routine surveillance of compliance with the terms and conditions set; monitoring of effects to allow operational changes to mitigation and contingency plans to be made; additional monitoring and research to test cause–effect hypotheses; and audit and evaluation to maximize the use of the lessons learned.

Table 8.3 A decision protocol to promote learning from experience.

	Information gradients in impact analysis			*Decision categories for project approval*		
Confidence level	*Data set ratings*	*Process knowledge*	*Approach permitted*	*Type of approval*	*Colour code*	*Terms and conditions of implementation*
1 High	Good	Proven cause–effect relationship	Statistical prediction	Unqualified	Green	Normal standards and regulations apply.
2 Fairly high	Sufficient	Evidence for hypothesis	Quantitative simulation	Qualified	Yellow	Subject to special requirements, e.g. monitoring programme or post-project audit.
3 Fairly low	Incomplete	Postulated linkages	Conceptual modelling	Conditional	Orange	Must conform to stringent environmental controls; additional information to be filed prior to construction, research management experiment imposed.
4 Low	Poor	Speculation	Professional opinion	Deferral	Red	Abandon project; or moratorium on development until major redesign or special studies completed, e.g. contingency planning or pilot project for demonstration and evaluation purposes.

Large-scale developments which are particularly controversial or precedent-setting would be appropriate projects for experimental research and management designed to lead to better functional knowledge of the relationships within and between ecosystems and institutions. In certain cases, when technology is unproven and the operating environment is hazardous and sensitive, phased development may be implemented through a series of small-scale pilot projects and studies. This is essentially the approach recommended by, for example, the Beaufort Sea Environmental Assessment Panel (1984) for long-term commercial production of offshore oil and gas in the ice-infested seas of the Canadian Arctic.

By these decision conventions, post-EIS evaluation becomes a catalyst for translating the principles of adaptive assessment and management into practice. The basic concept is to make the trial and error approach an integral part of the assessment process. Large-scale projects are considered experiments for 'learning by doing' with particular emphasis on both the capabilities of administrative procedures and management practices, as well as the strengths and weaknesses of tools and techniques of analysis.

A systematic feedback loop is thereby created which leads to increased effectiveness of EIA processes through improved tracking of projects (Fig. 8.3). It should lead also to long-term efficiencies in project design, analysis and implementation. To achieve these aims, however, presupposes other conditions which are not dealt with here – notably the requirement for a computerized data bank for clearing information to line agencies and other institutions. This kind of development could help end the constant duplication of data which currently plagues EIA.

Canadian trends in research and development

EIA is one of the most formalized and visible processes for development planning and resource management in Canada. The federal Environmental

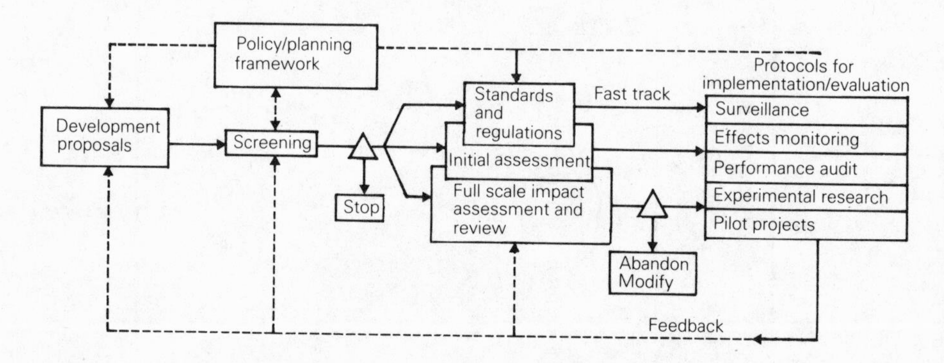

Figure 8.3 Learning from experience: an evaluation strategy for adaptive EIA.

Assessment and Review Process (EARP), for example, has been in operation for more than a decade. During that time, thousands of projects and activities have undergone screening and preliminary assessment, perhaps several hundred have been subject to an Initial Environmental Evaluation and a score of major development proposals have been formally referred for public review. Most of the provincial governments in Canada have a similar record of application.

Until recently, relatively little attention was paid to analysing the results of environmental assessments. This oversight is an understandable and perhaps inevitable consequence of the pressures of decision making and time and resource constraints. It is now widely recognized, however, that more attention should be paid to follow-up activities in order to develop more cost-effective practices. From an initial focus on impact monitoring, the scope of interest has broadened to encompass compliance and performance audits as well as related evaluations of administrative and management procedures. While none of these activities is yet firmly institutionalized in Canada, there are several important research initiatives under way. Three are of particular note. First, Environment Canada and the Federal Environmental Assessment Review Office (FEARO) recently commissioned a series of audits and evaluations of EISs. Secondly, Munro *et al.* (1986) have carried out a state-of-the-art review of these audit studies under the auspices of the Canadian Environmental Assessment Research Council. Finally, an international conference on EIA audit and evaluation was held in October 1985 at the Banff Centre at which the above studies, augmented by additional papers, were discussed (Sadler 1987). The more important findings emerging from this work are given here to provide a preliminary indication of the performance of EIA in Canada.

IMPACT PREDICTION AND MITIGATION CAPABILITY

A major review of the scientific quality of impact assessment in Canada was undertaken by Beanlands & Duinker (1983). From an ecological perspective, it revealed a dismal record. The EISs examined were: descriptive rather than predictive; largely lacking in a rigorous approach to analysis and interpretation of data; and provided results of questionable value either for decision making or subsequent testing and replication. Mitigation measures proposed to ameliorate or offset impacts were largely based on generalized principles and not grounded in specific findings. Finally, the lack of monitoring to test predictions and to facilitate impact management, appeared to be the exception rather than the rule.

Subsequent environmental audits were often constrained by the absence of formal monitoring programmes and pre-project baseline data (see, for example, Ruggles 1985, Zallen *et al.* 1985). Some of the studies also underlined the more fundamental limitations imposed by the present stage of development of biological sampling procedures and statistical techniques (Everitt & Sonntag 1985). This conclusion is supported by other research which suggests that existing models of environmental change are too coarse-grained for specific prediction (for example, Hecky *et al.* 1984). Although this generalization usually holds, certain types of project-induced changes are better understood than others. The 'reservoir paradigm' for hydroelectric development is a case in point (see Rosenberg *et al.* 1981).

The track record of the impact predictions audited in the Environment Canada studies was mixed. While the nature and direction of major effects were correctly identified, more often than not, statements of the magnitude of change were typically in error (Munro *et al.* 1986). Environmental impact statements still tend to be generalized, tentative and qualitative. At a recent workshop, the auditors themselves generally concluded that prediction capabilities were reasonably well developed in the physical and chemical sciences, moderate to fair in the ecological sciences, and relatively low in the social sciences. In other words, the level of understanding and accuracy in prediction varies with the directness of effect. Cumulative impacts from multiple sources represent a particularly taxing issue of impact assessment that is beginning to receive serious attention (Beanlands *et al.* 1986).

The fact that many project-induced effects cannot be accurately predicted in complex environments has led to a renewed interest in impact mitigation and management in Canada. Mitigation practice, by and large, tends to involve no more than the application of sound construction techniques (Jakimchuk *et al.* 1985, Moncrieff *et al.* 1985, Spencer 1985). There are, however, examples of relatively sophisticated and successful mitigation based on, for example, simulation modelling (Ruggles 1985). The routine and pragmatic character of mitigation means that there is often incomplete information on which to evaluate its effectiveness. Munro *et al.* (1986) also note that mitigation actions make it difficult to measure the accuracy of predictions, even when the effect of intervention is specifically incorporated into forecasts. This is not yet common practice.

These findings have several methodological and procedural implications for the conduct of post-EIS evaluation. It was evident from discussions at the Banff conference that the design and implementation of various kinds of monitoring, particularly for regulatory compliance, impact management and scientific validation, are reasonably well understood and quite widely agreed to. So, in general terms, is the role and scope of EIA audit. The methodology and techniques for carrying out this activity, however, are embryonic and individualistic, in part because of the difficulties of tracking and analysing non-quantitative predictions. Building an audit 'trial' needs to become an integral part of the initial design of project EIA.

PROCEDURAL EFFECTIVENESS

The recent series of audit and evaluation studies also dealt specifically with the efficiency and effectiveness of EIA procedures. The recurrent theme, which transcended differences in institutional arrangements, was the discontinuity between impact assessment and the subsequent phase of implementation and review. A lack of follow-through from EIA and authorizing decisions is evident even at the rudimentary level of surveillance to ensure compliance with the terms and conditions established for a project. This is partly the result of ambiguous environmental conditions stipulated for construction projects (Rowsell & Seidl 1985) and partly a consequence of unclear definitions of regulatory rules, responsibilities and the procedures for reporting to the appropriate authorities. At the federal level, for example, these problems stem from jurisdictional fragmentation, ambiguous mandates and insufficient resources for follow-up (McCallum 1985). Typically, considerable latitude is left to project proponents to decide just how much time and effort will be devoted to monitoring, mitigation and management of large-scale and non-routine development (Kiell *et al.* 1985).

All of this tends to run counter to the purpose of undertaking EIA in the first place. It means environmental protection, in the final analysis, tends to be *ad hoc*, uncertain or incomplete (Jakimchuk *et al.* 1985). Not only is the cost-effectiveness of EIA called into question, the proponents who voluntarily undertake environmentally sound implementation practices may be placed at a comparative financial disadvantage. From a regulatory standpoint, the lack of continuity between assessment and implementation raises issues of accountability. This question takes on considerable significance in connection with the formal review phase of EIA, into which interest groups and individual citizens, typically, put considerable time and resources. At the very least, they have the right to expect that this effort will result in agreed recommendations being acted upon.

On a wider front, public involvement was a focal element for analysis of the procedural effectiveness of EIA. This is a perennial subject of debate in Canada and positions are determined by the role played by the participant (Sadler 1980). The major source of concern regarding the cost-effectiveness of public review of EIAs is the time taken to complete them, and the delays occasioned to projects. One of the reasons, according to Everitt & Sonntag (1985), is that review agencies tend to consider that they are dealing with a new issue, often discounting previous work undertaken to resolve similar problems. The adversarial character of public reviews was identified by participants at the Banff conference as the main factor prolonging the EIA process. According to interest groups, this occurs precisely because project proponents often ignore local concerns and knowledge, even though industry has made a concerted effort at community consultation in recent years (Fee Yee Consulting Ltd. 1985).

The process of decision making at base, is one of bargaining and negotiation. Dorcey & Martin (1985) in a comparative analysis of the administrative and analytical procedures utilized in two major projects in British Columbia, argue that substantial innovation and success has characterized impact assessment, monitoring and management. As a result of continuing public controversy, however, the innovations have not been well recognized. The authors conclude by highlighting the interdependence of participant skills and procedural developments that are crucial to determining success. Since the former aspect is not well documented in the literature on EIA, it is worth stressing that the process is enormously dependent on the ability of the key actors to integrate complex information, to recognize clearly the implication of their interests and values, and to be able to make trade-offs in both a disciplined and a democratic manner (Sadler 1986b).

On the positive side, the difficulties which stem from process discontinuity are becoming well known and a number of procedural adjustments are being made to deal with them. Under the federal Environmental Assessment and Review Process (EARP), for example, co-ordinating structures have been developed to link the public inquiry with the implementation phase of development (Janes & Ross 1985). These exemplify certain important human and organizational aspects referred to in a number of studies. The gradual build-up of a climate of trust and confidence fostered by working exchanges between proponents and regulators over a period, ranging from eighteen months up to three or four years, is a crucial enabling condition of environmentally effective project implementation. It is a moot point whether this leads to a dangerous compromise which is against the public interest. What is clear is that wide distrust is a major handicap to achieving a more effective EIA process (Sadler 1983).

CONTRIBUTION TO DECISION MAKING

The audit and evaluation studies described above did not explicitly consider the contribution of EIA to decision making. It does seem generally agreed, however, that this activity has resulted in better decisions and more environmentally sound development than otherwise would have been the case. Still at issue is the cost-effectiveness of EIA, that is, whether the results are commensurate with the time and resources expended on the activity. This sort of determination, of course, is difficult to make, even for a particular sequence of project decision making. Environmental assessment and review processes for major developments, in particular, are not only complex and fluid, but also moulded by the interaction among numerous actors with different roles, views and abilities to press them (O'Riordan 1976, Sadler 1981). It follows that the views of success are relative, often dependent on the affiliation of the participant.

A study of the assessment and decision-making process which unfolded in response to an application to develop a new port, designed to support offshore oil and gas exploration in the Canadian Arctic, illustrates this problem (Fenge *et al.* 1985). This analysis demonstrated that it was difficult to secure agreement on the facts of the matter between the key parties and that much of the information required to satisfactorily explain the progression of events was unobtainable or became rationalized after the fact. While the study was structural rather than evaluative, it did conclude with a framework for analysing the contribution of the assessment to decision making. This was subsequently modified and applied to interpret the conflict over the siting of the port (Sadler 1984).

For present purposes, three basic questions can be identified as being important. First, was the final decision correct in the light of this retrospective review of the information generated during the assessment process? Secondly, was the assessment process undertaken in a timely and efficient manner? Finally, was the assessment process reasonably equitable in its treatment of all parties?

The short responses to these questions in the Arctic port case are respectively 'yes', 'no' and 'partly'. Others, however, would certainly judge the first issue differently. There are extenuating circumstances with respect to the second issue and the third response is partly dependent upon the second. As attempts to increase the efficiency of EIA invariably constrain the consideration of some issues, efficiency (e^1) and equity (e^2) tend to be inversely related. In the analysis of effectiveness (E), therefore, $E = e^1/e^2$ (Sadler 1983). This trade-off, more than the difficulties associated with determining the utility of EIA for decision making, should be carefully borne in mind when searching for improvements in practice and procedures.

Retrospect and prospect

A much greater commitment is now being made in Canada and other countries to promoting and undertaking post-EIS research. This work complements and draws upon the efforts being made by the competent authorities to establish more systematic implementation procedures that will provide continuity to assessment. Monitoring, impact management and environmental audit, important activities in their own right, can pay extra dividends when integral elements of an overall evaluation and review process. A decision protocol, which links implementation and evaluation requirements to levels of confidence in impact predictions, is advanced as a means of achieving disciplined and cost-effective feedback. It offers a long-term strategy for implementing the concept of adaptive environmental assessment and management.

In the interim, present trends in research and development of EIA audit and evaluation augur well for improving the state of the art. Recent papers and discussions at the Banff international conference established future directions for research (Sadler 1987). Several consolidated recommendations have wide applicability and are outlined by way of conclusion.

There is a need to standardize methodologies for EIA audit and evaluation. Pilot studies of selected projects should be undertaken jointly by industry, government and environmental interest groups to demonstrate and test procedures.

For the purposes of audit and evaluation, it is clear that impact predictions should be stated as testable hypotheses. In addition, monitoring and mitigation programmes must be recognized as sources of information for testing these hypotheses as well as important elements in the management of the proposal under consideration.

Public participation represents an important focus for evaluation because it drives many innovations in EIA practice. Clearly, socioeconomic monitoring will be required to provide the data necessary for such analyses.

Finally, as the major purpose of retrospective evaluation is to improve EIA practice, it is important that the results are fed back efficiently into the process.

Thus, mechanisms should be developed for national and international dissemination of the results of pilot studies and other audits and evaluations in order to improve ongoing project EIA and promote developments in this field.

Acknowledgements

The research for this paper was undertaken as part of work contracted by the Federal Environmental Assessment Review Office, Ottawa. While this support is gratefully acknowledged, the views expressed are those of the author and do not necessarily reflect those held by the Office.

9 *Training requirements for environmental impact assessment*

N. LEE

The case for training

The last fifteen years have seen a remarkable expansion in the provisions made, or envisaged, for the environmental impact assessment of environmentally sensitive projects (see, for example, Lee *et al*. 1985). Through a variety of laws, application decrees and non-mandatory provisions, EIA procedures have been inserted into the planning and decision-making arrangements for such projects in many countries and international organizations.

This is a considerable achievement, particularly given the difficult economic conditions which prevail. Having accomplished it, there is a natural tendency to assume that 'the main job has been done' and that the smooth and efficient implementation of environmental impact assessment will follow automatically. However, this is unlikely to be the case unless adequate prior provision is made to raise the knowledge, understanding and technical capabilities of those likely to be engaged in the EIA process. Thorough, objective evaluations of practice, in the early years following EIA implementation in particular countries, are few and far between. However, one such study has revealed that, though the potential benefits from using EIA are considerable, the extent to which these are realized in the early years may be limited by 'teething problems' traceable to those engaged in particular activities and tasks within the EIA process having insufficient experience and expertise (Council on Environmental Quality 1976).

The potential benefits from EIA implementation are considerable. They include more effective compliance with environmental standards; improvements in the design and siting of plant; savings in capital and operating costs; speedier approval of development applications; and the avoidance of costly adaptations to plants once in operation (Cook 1979, Dean 1979, Canter 1983). However, these benefits may not be fully realized for a variety of reasons which include incomplete understanding of environmental relationships and gaps in basic environmental data; delays resulting from weaknesses in the management of the EIA process; the production of overlong and poorly organized EIA study reports; inadequate organization and use of the consultation process; and unsatisfactory handling of the EIA within the decision-making process.

Although these deficiencies may arise from a variety of causes, one of the most important is believed to be a lack of sufficient, appropriately trained and experienced personnel in the day-to-day operation of the EIA process. It is for this reason that preparatory arrangements for the introduction of EIA procedures should be accompanied by a careful review of EIA training needs to ensure that steps are taken, at a sufficiently early stage, to remedy any training deficiencies that are identified.

In fact, remedying such deficiencies should not be a major or costly exercise, provided training needs are satisfactorily researched and new training initiatives are implemented as part of a coherent EIA training strategy. Developing such a strategy is likely to involve considering the following factors, namely, the existing and projected EIA training needs in the country or organization concerned; the existing provisions for EIA training and the main deficiencies in that provision; and the improvements to existing training facilities which should be made and how these should be implemented. Each of these issues is examined more fully below and related, by way of illustration, to the situation prevailing in a number of the member states of the European Economic Community (EC), based upon a major study for the Commission of the European Communities, reported in Lee (1984). However, many of the conclusions reached are believed to be of broader applicability.

Identifying training needs

EIA training needs may be identified by exploring two more specific issues, namely, determining who needs EIA training and the content of EIA training which is required. In order to answer the first question it is necessary to identify the structure of the EIA process, and the different kinds of personnel who participate in this process. In answering the second question, the roles performed by personnel within the EIA process and the tasks inherent in each of these roles must be considered. This is illustrated in the form of a personnel–training needs matrix (Table 9.1).

Table 9.1 Personnel–training needs matrix.

	EIA tasks to be performed			
Personnel types	*Task 1*	*Task 2*	*Task 3*	*Task 4*
EIA manager	√	√		√
Technical specialist		√	√	

A tick indicates the existence of a training need in the task indicated for the personnel shown. In practice the quantity, depth and nature of the need should also be specified as these vary.

WHO NEEDS EIA TRAINING?

In order to provide an answer to this question, it is necessary to determine, in some detail, both the arrangements already made or planned for EIA in the country or organization concerned and the types of personnel participating within the EIA process. EIA arrangements differ, in detail, between different countries (Lee *et al.* 1985). Notwithstanding, there are many similarities in the basic features of the EIA process in most countries (Table 9.2) and it is these which form the basis for the following analysis of training needs.

It is only after the specific structure of the EIA process in a particular country has been detailed that it becomes possible to identify satisfactorily the likely types and numbers of organizations and personnel involved in its operation

Table 9.2 Basic features of an EIA system.

EIA is mainly limited to projects, drawn from both private and public sectors, likely to have significant environmental impacts, although the longer-term trend is to extend the same kind of approach to other types of actions (for example, plans and programmes).

The *environmental impact studies* (sometimes called environmental impact statements) which are prepared for individual projects will normally be expected to cover the following types of items:

- description of the main characteristics of the project;
- estimation of residues and wastes that it is likely to create;
- analysis of the aspects of the environment likely to be significantly affected by the project;
- analysis of the likely significant effects of the proposed project on the environment;
- description of the measures envisaged to reduce harmful effects (this may be extended to include a consideration of alternatives to the proposed project and the reasons why they were rejected);
- assessment of compatibility of the project with environmental regulations and land-use plans;
- non-technical summary of the total assessment.

The main *procedural* elements of the EIA process will normally include the following:

- the developer (often with assistance from consultants, regulatory bodies and other organizations) prepares an environmental impact study which is submitted, along with his application for project authorization, to the competent authority;
- the study is published (possibly after checking its adequacy) and is used as a basis for consultation involving both statutory authorities, possessing relevant environmental responsibilities, and the general public;
- the findings of the consultation process are presented to the competent authority;
- the assessment study and consultation findings accompany the proposed project through the remainder of the competent authority's authorization procedure.

In a number of cases, these basic features are further elaborated, for example, by making arrangements for the preliminary *screening* of projects, for *scoping* the coverage of studies, for *independent panels* to vet the studies made for major projects, and for *monitoring* the environmental consequences arising from the implementation of the project.

in that country, including those involved in subsidiary or inherent roles. However, though the details will differ from one country to another five main types of organizations are likely to participate. First, proponents of actions subject to EIA will be involved. These may be individual developers or development organizations in both the private and public sectors. Secondly, competent authorities will be empowered to authorize actions subject to EIA. These may be local, provincial, regional or central authorities, depending upon the nature of the authorization procedures in force. Thirdly, other authorities may be engaged in advisory and review activities relating to EIA, including pollution control, nature conservation or planning authorities, special EIA advisory or review panels and the chairmen of public inquiries. A number of organizations such as consultancy firms and research institutes may assist the proponent or the competent authority in their roles within the process. Finally, various interest groups may be involved in advisory, review and public participation activities.

Within these various organizations a number of major groups of people are likely to be concerned with EIA-related activities. Elected representatives and business leaders will be involved in making decisions on projects for which EIAs have been prepared. Senior administrators in all types of organizations such as developers, competent authorities and pollution control agencies will be actively engaged in the EIA process, for example, in advising decision makers and in selecting EIA project managers. EIA project managers in developer, consultancy and competent authority organizations will be involved. Technical specialists in developer, consultancy, competent authority, environmental protection agency, research institute and other organizations will contribute to the preparation of EIA studies or respond to subsequent consultations. Members of EIA review bodies, review panels, appeal boards and public inquiries will be expected to participate. Finally, members and representatives

Table 9.3 Estimates of number of personnel likely to be involved in EIA activities.

		Assumptions	
Type of personnel	*10 EIAs/year*	*50 EIAs/year*	*100 EIAs/year*
Elected representatives, business leaders	40	175	300
Senior administrators	40	175	300
EIA project managers	20	90	150
Technical specialists involved in preparing studies	80	350	600
Technical specialists involved in commenting on studies	40	175	300
Members of review bodies, etc.	15	70	120
Members of environmental interest groups	40	175	300
Total (approx.)	275	1210	2070

of environmental interest groups will be involved in the EIA public participation process.

A large number of factors influence the total number of personnel likely to be involved in EIA-related activities but the estimates contained in Table 9.3 are believed to provide useful 'order of magnitude' guidance for determining the likely scale of EIA training needs. In calculating the likely number of EIAs that will be undertaken each year in any particular country, it is desirable to include those which may be undertaken on a non-mandatory basis as well as those that are statutorily required.

The most likely number of mandatory EIAs undertaken each year in the majority of member states within the EC may lie between 10 and 100, but to this should be added the simplified project assessments which will continue to be

Table 9.4 Major activities and types of personnel involved in the EIA process.

Major activities	*Types of personnel involved (examples only)*
1 Deciding if a study is necessary	Senior officers in competent authority, project leader in developer organization and certain of their support staff.
2 Scoping of study	Project leaders in competent authority and developer organization (with support staff), environmental control agency specialists, representatives of interest groups.
3 Managing the study	Developer/consultancy/competent authority project leaders.
4 Preparing specialist technical contributions to the study	Technical specialists employed by developer/consultancy/competent authority/other environmental control agencies (some of the work may be subcontracted to a number of different consulting groups).
5 Preparing the study report	Developer/consultancy project leaders (with support assistance, possibly including professional writers).
6 Reviewing the study report	Competent authority project leader (with technical support assistance), members of review panel (if in existence).
7 Organizing the consultation and public participation process	Competent authority project leader (with support assistance), chairman of public inquiry.
8 Participating in the consultation process	Environmental control agency specialists, interest group representatives, general public.
9 Synthesizing the findings of the consultation process	Competent authority project leader, chairman of public inquiry.
10 Using the findings of the study and consultation process to reach a decision on the proposed action	Competent authority senior officers and elected representatives.
11 Monitoring and post-auditing the environmental consequences arising from project implementation	Technical specialists employed by developer, competent authority and other environmental control agencies; competent authority senior officers.

required under existing legislation. Thus, it would appear that there are, potentially, numerous people who will be employed in EIA-related activities. However, it is important to recognize that, for the most part, this will only be one, for many only a small, element of their total responsibilities. A major challenge in formulating an EIA training strategy will be in meeting the training needs of such a relatively large, but diverse, number of people who only have a part-time commitment to undertake EIA-related work.

During the initial stage of the implementation of a formal EIA system, it is likely that virtually all involved in the EIA process could benefit from some EIA training. The estimate of training needs for the first year in most member states is thus probably in the region of 300–2000 depending upon population. Subsequently, annual training of perhaps 25–50 per cent of these numbers for 4–5 years might be necessary, after which the percentage might fall to 10–20 per cent, to cover the training of replacement staff and periodic updating of existing staff.

WHAT TRAINING CONTENT IS REQUIRED?

The training that different types of personnel require depends upon the activities in which they engage within the EIA process and on their background knowledge and experience. The main types of activities and the categories of personnel who are likely to undertake them are illustrated in Table 9.4.

However, in order to define the content of the EIA training needs more precisely, it is necessary to identify the different tasks involved in each of the activities in Table 9.4 and to identify the methods which are helpful in carrying them out. The total number of EIA tasks and methods for which some training may be required is very large (Lee & Wood 1979). Those relating specifically to the preparation of an EIA study (which is only part of the total EIA process) are illustrated in Table 9.5. The assessment methods involved are of different kinds and may be grouped as shown in Table 9.6.

Thus, the training needs of different types of personnel may be quite diverse. At the same time, it has been found from studies of individual countries (Lee *et al.* 1985) that EIA training needs can be usefully grouped into three broad categories. First, general awareness is required to convey to political and business leaders, senior administrators, consultants and technical specialists a good basic understanding of the main procedures and assessment methods involved in EIA to enable them to achieve a better understanding of its role and general significance in environmental management and decision making. Secondly, specialized technical training is necessary to provide technical specialists with a command of the methods appropriate for undertaking specific assessment tasks such as predicting the diffusion patterns of particular air or water pollutants. Finally, EIA project management training could provide potential EIA project managers with the broad knowledge and project management skills which are required to organize and co-ordinate not only the preparation of EIA studies, but also the associated consultation and review activities. These three types of training need are used in the remainder of this

Table 9.5 A classification of EIA tasks and associated methods.

Tasks	*Methods (examples only)*
1 *Description of proposed development*	
(a) Identify aspects of project for which information is to be sought distinguishing, where necessary, between different stages in the proposed development (e.g. construction and operating phases) and between different levels of screening.	Checklists, consultations with developer.
(b) Determine resources to be used in construction and initial operating phase, wastes to be created, physical form of the development.	Data sheets, engineering drawings, etc., prepared by developer; mass balance analysis; accident and uncertainty analysis (this continues through a number of assessment stages).
(c) Forecast future resource use, waste generation, etc., over the expected life of the development.	The same methods apply as in (b), but methods of production and technological forecasting are also relevant.
2 *Description of existing and projected environmental conditions*	
(a) Identify aspects of existing and projected environmental conditions for which information is to be sought.	Checklists, consultations with environmental agencies; alternatively, may be linked with 1(a) through the use of matrices or through more elaborate representations of relationships such as networks.
(b) Collate existing environmental data and identify gaps in information.	Consultation with environmental agencies and voluntary organizations; use of data bank and retrieval systems.
(c) Obtain additional environmental data to meet remaining deficiencies.	Review of existing monitoring systems; special surveys using a variety of techniques (aerial photography, field sampling, etc.).
(d) Predict future environmental conditions (without the proposed development).	Variety of available methods, ranging from simple forms of extrapolation to complex modelling studies; consultation with environmental agencies.
(e) Summary and presentation of environmental data.	Mapping, overlay methods, summary sheets.
3 *Assessment of probable impact of development*	
(a) Assess magnitude of impact (in present and future conditions) on:	
(i) air, water and land;	Diffusion and resource utilization models, physical intrusion assessment;
(ii) receptors within the environment.	ecological modelling, damage functions.
(b) Assess importance of impact by:	
(i) investigating response of affected parties;	Social surveys, agency consultation and public participation;
(ii) aggregating individual environmental impacts.	scaling and weighting systems, overlay methods, use of panel of experts.
4 *Compliance with other environmental plans, policies and controls*	
Assess likely compliance of development with existing and proposed controls.	Agency consultation, checking of extant plans.
5 *Review of alternatives to the proposed development*	
(a) Identify alternatives to be considered.	Checklist (of types of alternatives to be reviewed), consultation and survey methods.
(b) Describe project alternatives and assess their impacts.	Same methods as in 1, 2, 3 above, combined with screening methods.
6 *Preparation of non-technical summary of the assessment*	
Determine salient features of assessment and most effective means of presentation.	Communication methods.

Table 9.6 A classification of assessment methods.

Type	*Use*
Identification methods	To assist in identifying the project alternatives, project characteristics and environmental parameters to be investigated in the assessment.
Data assembly methods	To assist in describing the characteristics of the development and of the environment that may be affected.
Predictive methods	To predict the magnitude of the impacts which the development is likely to have on the environment.
Evaluation methods	To assess the significance of the impacts which the developer will have on the environment.
Communication methods	To assist in consultation and public participation, and in expressing the findings of the study in a form suitable for the decision-making process.
Management methods	To assist in managing the scoping of the study, the preparation of the impact study, the efficient conduct of the consultation process and related activities.
Decision-making methods	To assist decision makers in assessing and understanding the significance of environmental impacts relative to other factors relevant to a decision on the proposed development.

chapter as a basis for the evaluation of existing EIA training provision and the identification of training deficiencies and in making suggestions for its improvement.

Existing training provision and its deficiencies

A review of existing training should aim to record the nature and extent of facilities currently available to each category of personnel involved in the EIA process, indicating the type of EIA task for which these provisions are appropriate. This approach is illustrated in the personnel–training provision matrix (Table 9.7), which uses the same classification system as the personnel–training needs matrix (Table 9.1). Comparison between these completed matrices may be used to identify particular training deficiencies.

Table 9.7 Personnel–training provision matrix.

	Training provision for EIA tasks			
Personnel types	*Task 1*	*Task 2*	*Task 3*	*Task 4*
EIA manager	√	√		√
Technical specialist		√		√

A tick indicates the existence of a training provision, in the task indicated, for the personnel shown. In practice the quantity, depth and nature of the provision should be specified as these vary.

In analysing existing training provisions the following factors should be considered. First, the institutions such as universities, government agencies, consultancies and large industrial enterprises which undertake EIA training should be identified and the manner in which courses are financed determined. The length and context of EIA training, for example, the stage of the training and whether it is a self-contained course or part of a wider educational and training programme should be considered. The content of EIA training such as the range and level of detail of EIA tasks and skills on which training is given should also be determined. Finally, the relative use made of traditional methods of instruction such as lectures, learner-active methods including problem solving, role playing and case studies, and 'on-the-job' training should be assessed.

In addressing these issues, it is helpful to group training provisions into four categories, namely first degree level courses, higher degree level courses, short courses (at various levels) and 'on-the-job' training. As there are differences in the structure of higher and post-experience training between countries, these groupings are not necessarily the most useful in all member states of the EC. However, they do enable broad comparisons to be made between them. Each of the four types of provision mainly fulfils the training functions indicated in Table 9.8, although there may be some important exceptions. The following review relates to the situation in 1985.

Table 9.8 Main types of EIA training provision.

Type of provision	*Training function*
First degree level training	General awareness training and basic training in specialist skills mainly intended for those with little or no previous work experience.
Higher degree level training	More advanced training in specialist skills and, to a limited extent, EIA project management training; may be used both by those with little or no previous work experience and as a form of continuous training.
Short course training	Performs very diverse and often multiple functions within continuing education (general awareness, specialist technical and EIA project management training); it may be remedial, updating or integral to career development.
'On-the-job' training	In practice, often unsystematized ('on-the-job' experience rather than 'on-the-job' training) nevertheless highly valued at the practical level; may serve any of the three broad training needs and be used both by newly qualified staff and by those with previous work experience in other areas.

FIRST DEGREE LEVEL TRAINING

This type of training is undertaken through university and other forms of full-time higher education and is financed and regulated as an integral part of the higher education system in all member states. Material relevant to EIA training, although mostly not directed consciously to EIA training, is to be found in a number of specialist degree courses. Generally, these are in disciplines most closely connected to environmental activity and its regulation, the biological sciences, civil and sanitary engineering, geology and, to a much lesser extent, physics and chemistry. It is also to be found in a number of more broadly based interdisciplinary degrees, notably in environmental science, planning and architecture courses. In any one country, however, EIA training provisions can be very uneven; many courses in a particular discipline make no provision, and amongst others the extent is very variable.

Single-discipline degrees if making any provision, mainly concentrate on providing specialized technical training in some aspect of environmental assessment although some provide more broadly based 'general awareness' training. This latter form of training is more common in interdisciplinary degrees such as planning and environmental science. The extent of provision varies considerably between member states, being comparatively less in Italy, Greece, Belgium and Ireland than in France, the Netherlands and the United Kingdom (Lee *et al.* 1985). Overall, the level of first degree course provision directly relevant to EIA is low.

HIGHER DEGREE LEVEL TRAINING

This type of training is also undertaken through universities and related forms of higher education and is largely financed and regulated as an integral part of that sector. In some member states it is linked to the post-experience and continuing education systems through which it may have industrial and professional links. Also, in some countries a significant element of higher education training is carried out on a part-time basis.

Some limited provision of more advanced specialist environmental training is made in some countries, for example, in the biological sciences, but this is not usually closely targeted upon EIA training objectives. To a lesser extent still there are few postgraduate interdisciplinary courses which contain some material relevant to environmental impact assessment such as postgraduate courses in planning and in environmental technology. The treatment of topics directly related to EIA is fairly brief; much of this is at a level of 'general awareness' training, but in some cases it also aspires to undertake a limited amount of EIA management training as well. In summary, advanced level EIA training, through higher degree courses, is in very limited supply in all member states.

SHORT COURSE TRAINING

To date, France has made the most extensive provision of all member states for EIA-related short courses. This reflects both the nature of its formal EIA system and the fairly strong central government support for EIA training provision. Continuing or professional training in EIA in France is provided through short courses by a number of university departments, industrial and other associations and by training divisions of government departments. In most of the other member states there has been, at most, a handful of short courses which have been directly targeted on EIA, supplemented by the occasional conference or seminar on the subject. In most member states there are also some more specialized courses relating to, for example, the evaluation of noise nuisance, landscape impacts, forms of public consultation and public inquiries which help to meet some of the more specialized EIA training needs. The EIA-related short courses which have been provided, however, have often been criticized as being insufficiently targeted on EIA training objectives. This reflects similar experience in the United States (Wood 1985).

'ON-THE-JOB' TRAINING

EIA practitioners in all member states attach great importance to the role of practical experience in EIA work, but few organizations provide any systematic, in-house practical training for the purpose. However, properly supervised 'on-the-job' training is only practicable in fairly large industrial enterprises and government departments which have already acquired considerable EIA experience; these constitute a relatively small proportion of the total training market. In the absence of satisfactorily supervised 'on-the-job' training, various aids such as explanatory regulations and advisory guidelines, manuals and case studies can be valuable in guiding and channelling such experience.

TRAINING AIDS

Guidelines, case studies and role-playing exercises may also serve as partial substitutes for direct practical experience and so provide useful aids on the three types of 'external' training course described above. EIA manuals and guidelines only exist widely in France and, to a much more limited extent, in the Federal Republic of Germany, the United Kingdom and Ireland (Wood & Gazidellis 1985). They are also in the process of being prepared in the Netherlands.

Small numbers of EIA case studies in a form directly usable for training purposes have been prepared in the Federal Republic of Germany, France, the Netherlands and the United Kingdom. All member states are known to possess EIA-type studies which could be converted into studies for training use, but so far only limited efforts have been made to do so.

Again very few EIA role-playing exercises appear to exist and are used for training purposes. Only in France and Italy have examples been found of audio-visual aids developed specifically for EIA training purposes. In summary, in most member states there has been relatively little development to date of effective EIA training aids.

TRAINING DEFICIENCIES

EIA training deficiencies are identified by comparing both existing and projected needs and provision in 1985. It is helpful, when doing this, to distinguish between quantitative deficiencies (an insufficient amount of training) and qualitative deficiencies (weaknesses in, for example, the appropriateness and content of the training).

Quantitative deficiencies The provision for basic EIA training within first degree and related courses in the relevant discipline areas is extremely uneven in different member states. While in France there are significant, but not substantial, numbers of courses that include an EIA unit, in the United Kingdom and the Federal Republic of Germany there are very few while in others this type of training provision is almost unknown.

Advanced university (higher degree level) EIA training courses are in limited supply in a few member states, such as France and the United Kingdom, whereas in other cases, for example, Italy and Greece they are practically non-existent. This is partly due to the limited public funds available for postgraduate training in general and partly because the need for more advanced EIA training is not sufficiently recognized.

Deficiencies in numbers of short courses on EIA are somewhat different. All member states provide some specialized technical short course training of a remedial, updating nature, although the extent of provision varies considerably between member states. In general, however, the supply of EIA-related short course programmes is very limited, particularly for purposes of general awareness and EIA management training.

Systematic 'on-the-job' training is only available in a small number of organizations in one or two member states, for example, France and the Federal Republic of Germany. In most others, the development of skills in environmental impact assessment is taking place mainly through relatively unstructured 'on-the-job' experience.

In summary, significant quantitative deficiencies in the provision of EIA training facilities appear to exist in the 'run-up' period to the implementation of the EC directive, in each of the four main forms of training. These deficiencies occur in most, if not all, member states.

Qualitative deficiencies These kinds of deficiencies in EIA training are at least as important as quantitative deficiencies. They are of two interrelated kinds, namely a mismatching between existing course content and training need, and the provision of courses which are frequently regarded as 'too theoretical' and insufficiently related to practical experience.

The problem of mismatching is believed to be fairly extensive and exists in many forms. There may be an over-concentration upon procedural elements of the EIA system to the neglect of training in EIA methods. This problem is likely to be encountered especially in general awareness and EIA management courses. Specialized training courses may fail to provide sufficient understanding of the EIA context in which specialist technical skills will be used. There may be a failure to identify sufficiently clearly the course content required for EIA management training and to distinguish this in level, content and training method from general awareness training. Finally, insufficient training in particular EIA skill areas, for example, scoping, choice of predictive tools, evaluation and assessment of impact significance, communication and study management, may be provided.

The provision of courses which are 'too theoretical' partly stems from this mismatching problem. In addition, many course lecturers may have insufficient practical experience in the preparation and use of EIA studies in decision making. Good case-study material and role-playing exercises, based upon real-world examples, which can be drawn upon for training purposes may be lacking. Finally, there may be a considerable gulf between sophisticated EIA 'methodologies' which are emphasized in some training programmes, and the more modest EIA methods which have often been found to be of greater practical relevance in field situations.

One or more of these qualitative deficiencies appear to exist in many of the EIA-related courses currently provided within the member states. This is one of considerable relevance to the development of an EIA training strategy because a significant improvement in EIA training, in fact, may be achieved by improving the quality of existing training facilities, without substantially increasing total training costs. Also, increases in the quantity of training facilities will have a greater impact on the efficiency of an EIA system if accompanied by an improvement in their quality.

Improving training provision

Ideally, proposals to improve EIA training should be formulated within the framework of a coherent, overall training strategy with clearly focused objectives. The formulation of such a strategy is a matter for the individual country or organization concerned. However, in most cases, this is likely to entail better provision, both quantitative and qualitative for the main types of training need previously identified, namely general awareness training; specialized technical training; and EIA project management training.

QUANTITATIVE IMPROVEMENTS

The main quantitative priorities are likely to be increased provision of general awareness and EIA project management courses. In some countries, however, training facilities in the specialized technical assessment of particular environmental impacts will also need to be strengthened.

Over the long term, preliminary EIA training should be progressively incorporated into appropriate first degree level courses, to be supplemented later in the trainee's career by other forms of training. Therefore, steps should be taken to reduce the existing unevenness of EIA training provision in such key disciplines as engineering, biological sciences, environmental sciences, landscape architecture, town planning and chemistry, for example, by introducing short EIA-related course units where none currently exist. Brief course units on general awareness training should be introduced into law, management and business studies courses, from which it is largely absent at the present.

There should also be some increase in the provision of more advanced training at the higher degree level both in the technical assessment of particular environmental impacts and, more especailly, in EIA project management skills. This might be concentrated in a small number of university or polytechnic centres. In addition, there is a need to strengthen general awareness training as a component within existing higher degree courses in such subject areas as business management, engineering, landscape architecture, town planning, pollution studies and nature conservation.

In the more immediate future, quantitative deficiencies in training provision will have to be met mainly through a programme of post-experience short courses. Such a programme should include general awareness, project management and some specialized technical short courses. In larger countries, the potential number of trainees for general awareness courses should be sufficient to justify providing these on a regional basis for which attendance costs will be relatively low. Specialized technical short courses should be viable at the national level and, in certain circumstances, at the regional level. EIA project management courses, for which the potential market is smaller, can probably only be sustained at the national level in most countries.

Steps should also be taken to increase 'on-the-job' training by reinforcing the present practice of 'learning by doing' through the development of self-teaching modules which enable the trainee to pursue a training course, with carefully prescribed objectives, in his or her own workplace or home. These forms of distance-learning could perform an important training function for those who are unable (because of pressure of work or geographical remoteness) or are otherwise unwilling to attend externally organized courses. In general, however, it is preferable to regard 'on-the-job' training as complementary to, rather than as a substitute for, short courses.

QUALITATIVE IMPROVEMENTS

Qualitative improvements in EIA training are at least as important as increased course provision. These improvements may be realized by encouraging the adoption of clearer training objectives for individual courses and by stimulating changes in the content of, and training methods employed in, individual courses to conform more closely to their training objectives.

There are a number of important ways in which training courses might be improved to reduce mismatching between the content of training provision and training needs. Training in methods of environmental impact assessment should be strengthened and the heavy emphasis upon procedural elements of the EIA system should be reduced, especially in general awareness and EIA project management courses. Strengthening training in specialized technical training courses to provide a basic understanding of the overall EIA context in which specialist technical skills will be used may be required. The content of EIA project management courses would be improved by focusing upon the specific skills which project managers will have to practise, and sufficiently distinguishing this from the content appropriate to general awareness training. The training content in particular skill areas, notably scoping, choice of predictive tools, methods for determining the significance of particular impacts, treatment of uncertainty, communication and consultation skills should be augmented.

It is also desirable to adopt training styles and methods which match course objectives more closely. In this respect a major objective should be to encourage the wider use of learner-active methods of training and of 'real-world' examples on EIA courses. This could be encouraged through stimulating the preparation and use of more 'real-world' case studies, audio-visual aids, simulation exercises, self-teaching modules and field study exercises.

In addition, it is necessary to confront the problem of the limited practical experience in EIA work of many instructors and teachers on EIA courses. This could be partly remedied by making more realistic training aids available (see, for example, Wood & Gazidellis 1985, Wood & Lee 1987) and through greater use of EIA practitioners as course contributors. However, some support should also be provided for 'training the trainers'. An EIA training guide has recently been prepared for this purpose (Lee 1987).

IMPLEMENTATION

The implementation of training programmes of different kinds is frequently handicapped by problems of organizational inertia, lack of response, or a lack of commitment to, and understanding of, the training programmes which are being proposed. There is no reason to believe that an EIA training programme would be exempt from these kinds of difficulties. It is important, therefore, to try to ensure the practicality of any training improvements which are proposed and to take specific steps to promote their implementation.

One important feature of the kind of EIA training strategy proposed in this chapter is that it includes decision makers, senior managers and administrators within the target groups. Raising their awareness and basic understanding of EIA and their appreciation of the value of good training in its application is essential to the successful implementation of the EIA training strategy as a whole. A second important feature is that the cost of the type of training programme being advocated is relatively modest and could be justified in terms of the improved quality and cost-effectiveness of environmental management and decision making that would result.

As environmental policy matures it tends to place greater emphasis on anticipating environmental problems and on taking corrective action at the planning and design stages of new projects. There is also an increasing concern to achieve environmental goals, wherever possible, in a cost-effective manner. EIA is of central importance in applying this anticipatory principle and a well-grounded training programme is needed to ensure that EIA is implemented to this end in an efficient way.

Acknowledgements

I gratefully acknowledge the assistance of the Directorate General for the Environment, Consumer Protection and Nuclear Safety of the Commission of the European Communities for both providing general guidance and funding the work on which this study is based, and of my colleagues Christopher Wood and Vicki Gazidellis who assisted me with this project. A shorter and earlier version of this chapter appeared in Lee & Wood (1985).

Part IV

APPLICATION OF EIA

10 *The co-evolution of politics and policy: elections, entrepreneurship and EIA in the United States*

G. WANDESFORDE-SMITH
J. KERBAVAZ

Introduction: EIA scholarship transformed

The publication of *Making Bureaucracies Think* (Taylor 1984) marks a major turning point in the scholarship on environmental impact assessment (EIA). Apart from the early work of Anderson (1973), Andrews (1976), and Liroff (1976), Lynton Caldwell has been almost alone for a decade in arguing persistently that the environmental impact statement process created by the National Environmental Policy Act (NEPA) of 1969 makes a significant difference in federal agency decision making. As Caldwell is well aware, his arguments are bound to be suspect because he was a principal architect of NEPA. The ambitious, collaborative research programme on the impact of impact assessment, previewed in Caldwell (1982) and subsequently completed with support from the National Science Foundation (Caldwell *et al*. 1982), was an impressive empirical demonstration of NEPA's influence. Now there is Taylor who, despite a studied and not altogether charitable indifference to Caldwell's contributions, provides independent confirmation of the essential facts.

Taylor's work, however, is part of a broader literature on the nature and consequences of environmental policy instruments employed in the 1960s and 1970s (see, for example, Hawkins 1984, Latin 1985, Ackerman & Stewart 1985, Sproul 1986). From this literature, it is clear that policy instruments can be differentiated according to who is given the opportunity in theory to bargain in the making and implementation of policy. A more important consideration, however, in understanding the way different instruments work in practice is how a combination of resources, political circumstances, and skill in relating the two empowers some to bargain more effectively than others, and thus to shape the evolution of policy. That this combination, which we identify as the product of entrepreneurship, has been the principal factor in the evolution of EIA at the federal level in the United States and in California is the focus of this chapter.

At the US federal level, impact assessment works. We know how it works to influence project selection and design and to mitigate environmental impacts. It facilitates bargaining in the shadow of the law (Mnookin & Kornhauser 1979; Taylor 1984: 208). We also know why it works, although the essential interplay of factors internal and external to the implementing agencies is analysed somewhat differently by different analysts.

These differences, for example, in the importance assigned to public interest group litigation in ensuring the effectiveness of EIA, are reflected in recommendations for improving performance. Caldwell (1982) has always emphasized the necessity of interpreting and implementing the environmental impact statement (EIS) section of NEPA in the context of the statute as a whole, and has repeatedly looked to presidential commitment and, if necessary, intervention as the ultimate key to enforcement. Sax (1973), and more recently Fairfax (1978), and Mazmanian & Nienaber (1979) have despaired of the prospect of increasing the effectiveness of NEPA in reshaping the entrenched behaviour patterns of federal bureaucracies, without either changes in the specificity of statutory language or levels of public interest litigation and judicial intervention that seem highly improbable. There has been and remains, then, a range of views about why EIA works and what is the key variable to focus on in pushing for improvement.

Overall, however, the tendency has been to see the essential dynamic in the history of EIA in the United States as arising out of a tension and conflict between recalcitrant bureaucratic insiders and reform-minded external intervenors. Most analyses have concluded that in the early years the courts were the key actors, egged on by environmental public interest law groups and a general pro-environment climate of opinion. In more recent years, as bureaucrats have learned to play the game by the courts' new rules and as the courts themselves have learned the limits to their intervention, the reform potential of EIA has been devalued, even dismissed. It is as if students of environmental law and policy, having found that EIA has become routine and even welcomed by the bureaucrats it was imposed upon, are now prepared to abandon EIA and move on to new strategies of administrative reform.

It is a measure of Taylor's achievement that he has immediately transformed the debate on the evolutionary dynamics of EIA, which was threatening to become very dull, and has moved EIA scholarship in new and exciting directions. His ability to transform the discussion on the dynamics of EIA adoption and implementation stems from his use of a redundancy hypothesis, important precursors of which can be found in Landau (1969, 1973), to explain how and why EIA works in the USA. The redundancy hypothesis transcends all the earlier explanations of EIA that searched for a single dominant variable such as a sympathetic judiciary and argues (Taylor 1984: 252) that the successful institutionalization of precarious environmental values in the federal bureaucracy through the use of EIA involves an interplay of both internal and external factors. The hypothesis further states that the particular success of EIA is dependent on the redundant structure of outside support for probing and disclosing (in EISs) the technical premises of agency decisions that only insiders can reveal.

The real beauty of this thesis, however, lies less in its ability to integrate earlier alternative explanations emphasizing internal and external variables, than in its argument that the evolutionary dynamic of EIA (at least in the USA) is self-sustaining and self-regulating. In particular, Taylor asserts that the redundant structure of outside support for EIA, essential for making insiders influential and the whole process successful, is sustained and renewed by informal rules and expectations shared by all of the actors involved. The process works without any formal requirement that someone must make it work.

By this conception, Taylor has defused what was rapidly becoming a sterile and frustrating debate over the issue of who should bear most of the burden of keeping EIA alive and well – the President, the courts, or the public interest groups. Simultaneously, though not very successfully, he has shifted attention to the conditions that sustain or erode the ability of all actors to contribute to the marvellous mechanism of mutual adjustment that EIA has become.

Beyond these significant intellectual accomplishments, however, there is more. Taylor moves EIA scholarship in new and exciting directions, first by taking a comparative approach to the analysis of the impact statement strategy of administrative reform. Secondly, Taylor sees the study of EIA as part of a more widespread search in policy analysis for the answers to some very basic theoretical questions about policy change, and about the role of scientific and technical information in shaping such change.

Thus, Taylor is concerned initially with the question of whether, why, and under what conditions the impact statement strategy of administrative reform has advantages over other policy instruments, such as administrative reorganization, command and control regulation, and economic incentives. He asks how the relative advantage of EIA can be affected by the characteristics of the political system in which EIA reform is being contemplated or implemented. His comparative focus on alternative instruments of policy, therefore, complements the more conventional and more limited comparisons of EIA arrangements in various countries that now dominate the literature.

The second level at which Taylor breaks important new ground is to put his own and related work on EIA squarely in the middle of the current preoccupation with policy analysis and social learning (see, for example, Kaufman *et al.* 1986). He asks whether EIA can develop into a mechanism for steady improvement in the outcomes of decision making. Can EIA, in other words, constitute a forum for social learning about how to balance competing social values? His framing of this question and his attempt to provide a positive answer are the most provocative aspects of his work, and the most challenging for future EIA scholarship.

The entrepreneurial origins of policy change

The transformation of EIA scholarship signalled by the appearance of Taylor's book still leaves a number of important issues in need of attention. We are particularly concerned with Taylor's portrayal of the marvellously effective yet informal mechanism at the heart of the evolution of EIA. Taylor (1984: ch. 16) tries to distinguish his structural redundancy from the invisible hand that guides the market, from the unforced consensus at the root of scientific progress, and from the fragmented, disjointed, and incremental co-ordination of partisan mutual adjustment. To make his redundancy functional, however, it appears that Taylor ultimately relies upon the same automatic and inevitable logic of co-ordination George (1972) found so unsatisfactory in the work of Charles Lindblom more than a decade ago.

We can certainly imagine how the informal rules and expectations of an EIA process might become a powerful source of co-ordinated behaviour, as Taylor argues they do. A similar theme, for example, underpins comparable reforming work on contracts (Macaulay 1963) and on US administrative law (Stewart 1975, Ackerman & Stewart 1985), as well as on pollution control (Hawkins 1984). In the many interactions that occur among the individuals who populate Taylor's redundant structure of interests and groups caught up in the EIA process, there is persuasive evidence that informal relations exert a powerful pull. We can even agree with Taylor that informal incentives for co-ordination, if they receive institutionalized support and reinforcement, may prove to be a more powerful impetus for co-ordination than formal rules and responsibilities.

What we cannot imagine, however, is how either formal or informal incentives for effective behaviour in the context of an EIA process produce results unless somebody does a lot of work. Moreover, we see one vital aspect of this work as the making of judgements about whether formal or informal rules and expectations are going to be relied upon to produce a desired result. This is a process about which we need to know much more than Taylor tells us. If the logic of co-ordination in an EIA process is no more automatic and inevitable than it is in the national security advising process (George 1972), for example, then someone has to try to manage both their own behavioural responses to incentives and those of others.

At one level we are saying that we want to know who these persons are, what choices they make, and what cicumstances, especially political factors, influence their judgements. At a more basic theoretical level we are also saying that the behavioural effects of incentives cannot be understood independently of specific information about who responds to them and how.

The cause of theoretical parsimony is, of course, served by accounts of the history of EIA in the United States or elsewhere excluding information of the kind we suggest. Thus, for example, Taylor explains the origins and form of the 1978 NEPA regulations (40 C.F.R. § 1500 et seq.) without reference to the name and background of the first chairman of the Council on Environmental Quality in the Carter administration, and his general counsel, and without considering the intense political campaign waged by opponents of NEPA in the run up to and after the election taking President Carter to the White House.

Taylor simply sees EIA as a variant of Lindblomian partisan mutual adjustment (Taylor 1984: 306). He thinks the dynamics of the adaptation of policy produced by the application of EIA over time is somehow influenced by politics and by perceptions and evaluations of political change. That is why he says that politics and policy evolve together, reciprocally (Taylor 1984: 328). Notwithstanding, Taylor is remarkably silent about how particular people at particular points in time have perceived and evaluated political changes, such as the emergence of new legal rules of the NEPA process and alterations in the fortunes of environmental groups. He also does not explain how these same people have translated a changed appraisal of political circumstances into policy changes, except to say that the response of all actors to new political developments is mediated by the same structure of behavioural incentives. Moreover, he does not explain how, in the absence of human agency, this structure endures for long periods and ultimately becomes more determinative of policy change than either political phenomena or political actors.

For Taylor, the ultimate secret of the success of EIA in the USA lies not in the fact that sensitive and intelligent people have worked hard to make it succeed, but in the enduring and informally articulated structure of behavioural incentives that arises from multiple and overlapping centres of power. By this argument, the origins of policy change must be largely independent of individual effort, particularly in the long run, and not significantly affected by things like the outcome of an election and political appointments. This perspective on policy change undoubtedly explains why, in Taylor's account of US experience, the advent of the Reagan administration is a much less plausible reason for the present condition of EIA than the inexorable logic of constitutional redundancy that seems to shape EIA policy and practice even as successive presidents come and go.

By contrast, it seems to us important to know who or what makes the redundant structure of external critics an effective enforcer of EIA norms. Who, for example, forges the long-term relationships between actors in the EIA process that, according to Taylor, are a more powerful force co-ordinating the redundant elements of an EIA arrangement than the formal strictures of the courts (Taylor 1984: 270; after Axelrod 1981)? Similarly, who mends and re-creates the structure of redundancy when it falls apart, or is destabilized by external events?

Our answers to these questions identify two major influences on the co-evolution of the politics and policy of EIA that Taylor and all previous investigators have neglected. These are political entrepreneurs and elections.

It seems to us that without the enthusiasm, initiative, and drive of entrepreneurial actors in the political system there is no engine to turn the wheels of analytical competition that Taylor identifies as the primary cause of the development of EIA (Taylor 1984: 16). In a market, competition does not occur and cannot be sustained unless there are entrepreneurs willing to do the work of organizing firms and willing to take the risks involved in accumulating capital, acquiring inputs, and managing the production and distribution of output. Entrepreneurship is valued in the private sector, and often rewarded very handsomely. We think it is a valuable, but much under-appreciated, asset in the public sector. Indeed, unless political entrepreneurs come forward to do the political work involved in developing, implementing and changing policy it is hard to understand how and why the redundancy so much admired by Taylor is turned into an asset rather than a liability. As Diver (1982) asked 'if you have to hustle to be successful in business, why shouldn't the same be true in government?'

By the same token, explanations of the co-evolution of politics and policy that find no role for entrepreneurship fall into the Georgian trap (George 1972: 761). They analyse and evaluate policy choices as if they were the inevitable and acceptable outcome of political arrangements that are largely fortuitous, inherently stable, and immutable by the actors caught up in them. We think the history of EIA sustains a different view, an entrepreneurial thesis, by which individual drive and imagination make an impact.

The other place to look for improved understanding of the course EIA has followed in the United States is to elections, and more specifically to the political regime changes that may follow. What we mean to say here is very simply that elections, albeit crudely, take periodic soundings of the 'general social values' (Taylor 1984: 327) by which people expect to be governed, at least until the next election. Furthermore, elections change the composition and character of political regimes.

The changes can have both personal and party characteristics. One person replaces another as President or governor. One party loses its legislative majority to another, bringing in its wake, for example, changes in legislative committee chairmanships and staff appointments. Through the power of appointment, post-electoral change comes very quickly to government departments and less quickly, but just as surely, to independent boards and commissions and, eventually, to the judiciary.

In the USA, particularly at the state level where so many statewide offices are filled by direct election rather than by election of a governor who subsequently shapes a regime by appointments, the impact of elections on the composition and character of a regime can be complex and difficult to gauge. There can be no doubt, however, that theories of policy change and evaluation must take account of elections, of the regime changes that follow elections, and of the policy changes that are set in motion when one regime succeeds another. Belsky (1984) has brilliantly shown the potential for such analysis by tracing changes in federal environmental policies to the replacement of 'the Carter regime' by 'the Reagan regime' after the 1980 presidential election.

What matters most for this discussion, however, is that elections and subsequent regime changes affect the conditions necessary for the initiation and success of political entrepreneurship. In the case of EIA, the causal connection between elections and entrepreneurship and the influence that this has on the co-evolution of politics and policy can be difficult to show because EIA is unlikely to be a distinct and salient item in the electoral campaign agenda offered to voters. We think the link is not so tenuous as to make it impossible to test the entrepreneurial thesis, however, and we suggest here how that might be done, at least qualitatively.

This analysis of recent developments in the politics and policy of EIA in the United States, therefore, is set in the context of the profound transformation of EIA scholarship marked by the publication of Taylor's (1984) book. After an initial consideration of the federal situation, developments occurring in California since the publication of an earlier paper on the evolution of EIA (Wandesforde-Smith 1981) are the main focus of the discussion. There are two main reasons for this emphasis on California. First, the history of NEPA is the subject of existing accounts (see, for example, Murchison 1984). Secondly, the still undervalued role of elections and entrepreneurship can be readily appreciated in the context of California. Although one purpose is to bring the earlier analysis up to date, the main concern is to engage the conceptual and theoretical issues raised by Taylor. No serious student of EIA policy and practice can ignore Taylor's thesis and the analysis presented here shows that it can be developed into a broader framework for analysing the co-evolution of EIA politics and policy than Taylor imagined.

Our aim is heuristic rather than probative. Therefore, we shall not attempt to show that elections and entrepreneurship explain more of what has happened to EIA than other factors. We are content to argue that these variables need to be brought into any evaluation of hypotheses about the evolution of EIA, and that consideration of them appreciably enriches our understanding of the California case.

In the next section, we address the apparent limitations in Taylor's redundancy thesis as it applies to NEPA. They are attributable to his omission of political phenomena which are causally more important to the process he wants to explain than the structure of relations that he considers to be significant. This is followed by a section in which we interpret the recent history of EIA in California, using simple facts about gubernatorial and legislative elections, and information about entrepreneurs acquired from interviews and secondary sources. A final section offers some conclusions about the value added to explanations of policy change by including individual- and regime-level variables, more simply called entrepreneurs and elections.

EIA and the reformation of continuing relations: the Taylor thesis on co-evolution in the USA

EIA scholarship, particularly in the United States but also elsewhere, seems to have lost much of its former excitement. Taylor's (1984) book, however, seems to hold the promise of revitalizing interest in EIA research. Whether interest is revived and is guided in useful directions will depend in part on why people think EIA may have lost some of its attraction as an instrument of reform, and in part on how they evaluate explanations of why this might have occurred, Taylor's being chief among them for the US case.

An obvious place to start in explaining the US case is the election of President Reagan in 1980 and the subsequent course of his administration, which was and remains the most overtly hostile to environmental values since the early 1960s. It is also the only administration since NEPA was enacted to have at its head a President not committed to reasonable implementation of the statute. If there is no understanding or support for EIA at the top of the executive branch and hostility between the agency and bureau chiefs, surely, enthusiasm for using EIA in the way NEPA intended will dwindle very quickly at the operational level, where impact statements are prepared and projects either dropped or mitigated.

In some ways, this is an attractive theory to interpret the relative decline of EIA in the USA since the mid-1970s, particularly combined with what is known about top executive branch attitudes to NEPA, the continuing reluctance of the federal courts to read the Act expansively and the continuing indifference to both executive and judicial behaviour *vis-à-vis* NEPA in Congress. This essentially institutional and environmental explanation of NEPA's history is bound to seem especially appealing to people looking at events from a distance. Clearly, institutional manifestations of the changing political environment within which NEPA is being implemented and enforced are the most visible. Moreover, even close observers of EIA in the United States, including Taylor, would agree that institutional variables are important.

One problem is that other variables in addition to how Congress, the President, and the courts might be disposed towards NEPA and its goals affect the value placed on EIA as a policy instrument, and influence perceptions of how satisfactorily it performs. The availability of resources like money and manpower is one that comes readily to mind, for an EIA process starved of resources is unlikely to be much good. However, this is not inconsistent with a 'top-down' institutional explanation because an administration hostile to EIA is likely to starve the process of money and competent people.

In fact, as this simple example shows, the problem of accounting for the history of NEPA and the present state of EIA is not only in determining how many sets of variables are at work, but also, and much more fundamentally, in sorting out their relative importance. Thus, blaming the present rather dismal state of EIA in the USA (if such it is) on institutional factors either has to be preceded by a strong theoretical case for the primacy, not just the mere existence, of such factors, or it has to be abandoned.

In the case of Taylor (1984), whose theory of the co-evolution of EIA politics and policy is bound to have a central place in any future discussion of EIA experience, three sets of variables, or levels of analysis are recognized. The first, represents the 'top-down', institutional view. This is a perspective from which, in effect, Congress, the courts, and the executive branch are treated as units of analysis and discussed as if they were actors in the political system, capable of devising and pursuing strategies for the attainment of their goals.

A second level is the 'bottom-up' view. This represents the perspective of the analyst–advocates for environmental values who work inside the agencies, as well as people who are variously described by Taylor as: project managers; agency managers; agency leaders; interest group and commenting agency representatives; and lawyers both in and out of government. Clearly, the unit of analysis here is the individual, and strategic motives are again used to account for much observed behaviour.

Finally, at an intermediate level, there is the structure of the EIA process itself. It consists of both a set of rules to articulate and regulate relationships between the people inside and outside the agencies who participate in the EIA process and a forum for the airing of analytical disputes and the setting of norms for their resolution. The forum most used during the history of NEPA was the federal courts. Taylor's use of the concept of an oversight forum to air and negotiate analytical disputes is comparable to that of Yngvesson (1985) in its preoccupation with the courts. Yngvesson much more clearly recognizes that people exercise a range of choices between formal forums, such as the courts, and informal alternatives. However, neither comments on the characteristically American habit of pursuing analytical disputes in several forums simultaneously.

This framework, with its three levels of analysis, is used by Taylor to argue that NEPA has very successfully shaken up, restructured and ultimately restored to a new equilibrium the relationships among those involved in agency decision making. Thus, Taylor is not a pessimist about EIA as it has evolved under NEPA nor, more generally, about the impact statement strategy of reform. On the contrary, it is his thesis that, by improving the quality of the analytical competition that occurs within and between public agencies in the process of writing EISs and, therefore, searching for solutions to complex and difficult problems, the impact statement strategy substantially improves the likelihood that the political process of impact assessment will also be a social learning process.

The lessons are learned each and every day, on each and every project subjected to assessment. Over time, some of the lessons accumulate and are codified and made relevant to policy in the analytical norms through which the courts, or other forums, try to provide the contending parties with incentives to settle disputes, preferably among themselves and without appeal. Thus, and this is very important to an evaluation of the Taylor thesis, the ultimate test of whether EIA politics and policy co-evolve usefully and productively together over time is whether informal accommodation and stability are restored to the continuing relationships of agencies and other actors that were disturbed when the EIA requirement was originally mandated.

For Taylor, therefore, time is an important factor in weighing the significance of the first of his three levels of analysis. His look back over more than a decade of US experience is certainly sensitive to the fact that EIA practice has evolved in a changing political and economic climate. For example, public opinion on environmental issues has waxed and waned and executives and legislatures more or less sympathetic to environmental values have come and gone. This appraisal also clearly recognizes that the structure of analytical competition and the rules of the game for airing disputes in the EIA process have been altered over time.

Taylor's theory alerts the analyst to the contingent influence environmental and structural factors, particularly when unstable, can have on the performance of an EIA mandate. The principal thesis, however, is that, in the long term, it is the continuing relationship between individuals that counts.

According to Taylor, people in public agencies have always tried to survive and prosper in an uncertain world by building and sustaining networks of informal relationships. Such relationships provide information and political support and they act, above all, as avenues of accommodation through which those involved buy the time needed to negotiate solutions to the problems being addressed.

The crucial importance of continuing relations is best seen, perhaps, in the American pluralist setting (Taylor 1984: 300–7, after Lindblom 1965). In such a context, agencies typically face unclear goals, uncertain means and a multiplicity of factors to be weighed in decision making. It is the quintessential setting for Lindblomian partisan mutual adjustment and incrementalism.

We see no reason, however, to suppose that the continuing relations hypothesis will be useful only in the US context. The need to discover solutions to environmental problems by a largely informal political process of bargaining and accommodation seems to us to be present everywhere. The literature on EIA in other countries, however, has been too preoccupied with descriptions of institutional forms and elaborations of impact assessment techniques to reveal much about how the process actually works elsewhere (Rosenberg *et al.* 1981, Lee 1982).

At the heart of Taylor's vision, then, are informal relationships of mutual respect and trust, out of which agency reputations for expertise and balanced judgement are forged, and through which agency claims to exercise discretion on the basis of reputation are continually challenged and renegotiated. From this perspective, litigation and other types of formal confrontation are likely to be much less effective in persuading agencies to change their behaviour than threats to reputation, unless they bring indirect pressure to bear on the informal definition of agency expertise and, therefore, on an agency's claim to discretion and legitimacy. Taylor claims that this was the principal effect of such formal assaults under NEPA.

From this perspective, the history of NEPA is one of a single challenge and a decade-long response. It is the story of how environmentalists, beginning in the late 1960s and early 1970s, successfully used the courts as a forum to place agency reputations in formal jeopardy, and how those reputations were subsequently informally restored in the steady, often unspectacular, and always uncertain day-to-day work of impact assessment. Although a new equilibrium in continuing relations between agencies and other actors might have resulted eventually from an unstructured process of partisan mutual adjustment in the wake of the early environmentalist challenges, NEPA made the transition quicker and more effective. The key to understanding the significance of EIA in the United States, therefore, rests on an appreciation of the capacity for evolutionary policy change that resides in a properly structured politics of EIA which harnesses the redundant and overlapping pressures in a pluralist system for using analysis to gain advantage.

It follows from this analysis that, unless a new challenge is profound enough to realign the structure of formal and informal relationships evolved under NEPA, impact assessment will remain a productive way of enhancing organizational intelligence. The advent of the Reagan administration is judged not to be a threat to EIA in this sense. Hence, Taylor's conclusions are generally optimistic, which sets him apart from most other recent observers of the US scene.

We hope this brief sketch of Taylor's interpretation of the past emphasizes the influence which he believes is exerted by continuing relations on agency performance under NEPA. It is an influence contingent on the structure of the EIA process brought into existence by NEPA. It is an influence likely to be affected by short-term changes and instability in the political environment of implementation.

In the longer-term perspective of a decade or more, however, environmental influences fade in the face of the harsh reality that difficult and complex problems rarely have quick and radical solutions. Nor can they be solved by reference to predetermined standards or criteria. In the case of NEPA, Taylor argues that congressional and executive indifference to the outcomes of the EIA process was a way of managing the national environmental policy-making agenda. Furthermore, it was essential to the development of EIA as a learning process, because it gave the agencies, the courts and the environmentalists time to work out their differences and to approach difficult and complex problems incrementally (Taylor 1984: app. F).

We also hope this interpretive look at NEPA's history through Taylor's eyes will begin to raise questions about the continuing relations hypothesis. We hope it is clear, for example, that Taylor's thesis is much more complicated and demanding than previous attempts to explain what makes EIA work. Indeed, it demands insights into organizational behaviour that much of the literature on EIA has barely recognized as relevant. However, he may have made continuing relations too determinative of decision making and understated environmental influences.

We are particularly struck, for example, by the implications of the continuing relations hypothesis for the way people caught up in the EIA process spend their time. Clearly, from Taylor's account, a good deal of time is invested in worrying about reputation, and in avoiding situations of conflict and confrontation that jeopardize it. An essential aspect of this concern for reputation is making shrewd strategic assessments where one stands in relation to various theatres of external judgement. Essentially, this involves anticipating the reactions of potential critics and opponents to plans, projects and proposals, and either avoiding conflict or at least managing the forum in which it is likely to be aired.

Thus, one of the most fascinating insights to be gained from Taylor's analysis of the EIA process under NEPA comes from his depiction of the care and sensitivity with which agency lawyers and attorneys representing environmental groups monitor court opinions on NEPA cases (Taylor 1984: ch. 11). Moreover, they clearly make judgements, which change over time, of the likely prognosis for disputes that go to court and become adept at selecting an appropriate forum within the court system.

Although attorneys naturally spend more time appraising the relationships between agency and court behaviour than other actors, this is also a consideration for others, for example, in internal negotiations about the extent and quality of the environmental analysis to be included in an impact statement (Taylor 1984: ch. 5). Similar considerations enter into strategies devised by commenting agencies and interest groups for scrutinizing impact statements (Taylor 1984: ch. 7).

Given Taylor's view of the central role played by the courts and the indifference he attributes to Congress and the White House, it is understandable that the actors appear preoccupied with the courts to the exclusion of other forums. While these analytical choices on Taylor's part simplify his task they also oversimplify and distort reality, particularly by de-emphasizing the extent to which environmental politics in the decade under review was marked by conflict. Moreover, by overestimating the extent to which political stability and the passage of time are responsible for the learning that has occurred under NEPA, the Taylor thesis may undervalue the causal significance of individual strategizing and environmental change.

If in fact the people caught up in the EIA process also appraise other forums, the evolution of EIA reflects the balancing of these appraisals rather than an assessment of courts alone. If this is true, two things follow. First, empirically we need to know more than Taylor reveals about what actors think of alternative forums, and how they monitor, appraise, and choose strategically between them. The second, a theoretical issue, concerns the significance of political conflict and instability for policy learning.

From Taylor's perspective, the NEPA litigation of the 1970s constituted an exceptional challenge to the continuing relations between agencies and their environments that had evolved since the end of World War II. The theoretical relevance of the analysis lies in understanding how with the help of the courts these relations were subsequently reconstituted and restored albeit with full and permanent environmentalist representation. Thus, the fundamental and enduring fact of environmental politics in the USA is that, under normal circumstances, the law and the courts only rarely come into play and that most behaviour cannot be explained simply by reference to official legal rules.

If, however, conflict is a central fact of environmental politics and periods of breakdown in the relationships between those involved are normal, then the environmental movement of the late 1960s and 1970s, and the public interest NEPA litigation to which it gave rise, can be seen as just one of a series of encounters in which formal legal forums have been favoured for managing a continuing relationship. Whether the courts are the most relevant or important forum for redefining relationships in environmental politics (which is the conclusion we are drawn to by Taylor's analysis) is, therefore, of much less theoretical and practical interest than the question of who has access to and control of various forums and what conditions favour the use of a particular one. From this perspective, the periodic use of courts and other legal forums is quite normal, even essential for building and maintaining continuing relations.

Taylor focuses so firmly on the attention paid to the courts by actors in the NEPA process that very little can be learned about how other forums have been appraised. Taylor's own evidence, then, is of limited use in examining how a change in the theoretical assumptions about continuing relations might lead to fresh conclusions. At the risk of being speculative and inconclusive, however, we propose to reappraise the 1978 NEPA regulations (Taylor 1984: ch. 13), not in order to describe the regulations or their use but more modestly to assess to what extent they are explained by Taylor's co-evolution thesis.

Throughout the Nixon and Ford administrations the implementation of NEPA evolved with helpful and sympathetic guidance from the Council on Environmental Quality (CEQ). Although President Nixon endorsed NEPA in 1970, and both he and President Ford maintained high-quality appointments at CEQ, the Council showed very little inclination to force the pace of change on the bureaux and agencies of a Republican-led executive branch. The pace of change was forced by environmental plaintiffs, appearing initially before a federal judiciary bearing the imprint of Democratic appointments and working under the protection of a Democratic Congress. CEQ was careful to comment on the implications of NEPA litigation for federal agencies and to amend its advisory guidelines accordingly. Under this accommodating and reactive umbrella, however, there sheltered a wide variety of agency practices governed much more by separate and uncoordinated agency guidelines than by direction from CEQ. As Nixon and Ford judicial appointments slowed the pace of change during the early and mid 1970s it began to look as though the power of environmentalists to discipline every agency's decision making using a federal common law of impact assessment was steadily eroding.

As Taylor properly observes, the election of President Carter in 1976 created an opportunity to realign the politics of EIA that was leading to this erosion. Taylor argues that, from the perspective of the CEQ, the significant choice was how to adapt EIA policy to increasing evidence of judicial restraint and to the general public's more ambivalent attitudes toward environmental issues. 'Bluntly put,' he writes, 'the choice was between attempting to strengthen NEPA by making new courtroom victories possible for the environmentalists, and emphasizing inter-agency cooperation' (Taylor 1984: 277).

In the end, CEQ chose a strategy that, according to Taylor, tried simultaneously to increase its own authority as a forum for interpreting the rules of the NEPA process, and to reduce the courts' self-restraint in NEPA oversight. The viability of this strategy, Taylor concludes, was called into question after the election of President Reagan in 1980 and the subsequent plan (not carried through) to dismantle CEQ although it may yet prove viable if circumstances change. Taylor puts his best hope for the future, however, in the fact that early Reagan budget cuts failed to reduce the number of environmental specialists employed by federal agencies and, hence, failed to dislodge the institutionalization of agency participation in the competitive analysis of environmental impacts created by NEPA.

The most interesting feature of his entire discussion is how much CEQ appears to have been obsessed with appraising the viability of judicial oversight of the EIA process, to the point that repairing the damage done by judicial self-restraint appears to be the only plausible rationale for the regulations. The positive attributes of other forums seem to have been ignored by CEQ, although Taylor's analysis is ambiguous on this score. On the one hand, he describes the regulations as 'clever tinkering' implying that CEQ felt it had to act but could do little. Elsewhere, he describes the events surrounding promulgation of the regulations as an attempt by CEQ to 'seize' the right to review and veto individual agency NEPA procedures.

It is very misleading to portray the 1978 regulations as no more than a defensive reaction to judicial self-restraint. As his first chairman of CEQ President Carter selected Charles Warren, the most successfully entrepreneurial pro-environment member of the California State Assembly over the past twenty years. He was a prime mover of the legislation that created the California Coastal Commission as a permanent agency of state government and, later, the California Energy Commission. Moreover, Warren took with him to Washington, as general counsel of CEQ a young attorney named Nicholas Yost. Previously, Yost had played a major role in rescuing the California Environmental Quality Act of 1970, the state's EIA mandate, from the obscurity to which the administration of the then-Governor, Ronald Reagan, would have assigned it.

Warren and Yost were acutely aware of the opportunities presented by President Carter's strong personal commitment to environmental values and of the chance to introduce some of the measures for improving EIA which Yost had helped to develop in California (Selmi 1984). The inclusion of a provision for referral of irreconcilable agency differences over impact statements to the White House for resolution in the 1978 regulations underlines CEQ's confidence of access to the President in the light of this personal presidential commitment. Warren and Yost were equally aware of the anti-NEPA sentiment that had built up during the election campaign which threatened the indifference to NEPA outcomes Taylor ascribes to Congress. This situation required that the continuance of the indifference be renegotiated for as long as it would take for CEQ to prepare new guidelines for the EIA process.

There was little that could be done immediately and directly to relax judicial self-restraint in NEPA cases. Reversing that trend would take several years of judicial appointments and certainly longer than one four-year presidential term. Consequently, the main chance lay in procedural changes that could revitalize the EIA processs as an instrument of reform so long as the Democrats retained control of Congress and, more especially, the executive. It is in this light, rather than as a reaction to what the courts were doing, that the introduction of certain measures included in the 1978 regulations should be interpreted. Significantly, these regulations brought: enhanced recognition for the professional qualifications of environmental analysts; scoping; and the conferral of quasi-judicial authority for the interpretation of NEPA norms and the review of individual agency procedures on CEQ.

Taylor is correct to emphasize that, although President Carter was quickly persuaded to issue the Executive Order authorizing the replacement of CEQ's advisory NEPA guidelines with legally binding regulations (42 *Fed. Reg.*, 26967), it was 1978 before the regulations were adopted and July 1979 before they took effect. This timing meant that the strategy represented by the regulations barely had time to take effect before another presidential election in November 1980 and that, therefore, it remains untested. It remains potentially viable so long as the basic elements of the insiders–outsiders–forum structure of the EIA process are retained. However, we think the viability of the guidelines strategy came into question before the election of President Reagan. It lost much of its impetus when Warren stepped down as CEQ chairman for personal and family reasons and returned to California in 1979. Warren joined the University of California at Davis, for a time, and this analysis reflects in part discussions with him. Clearly, CEQ's appraisal of the energy and resources available to the environmental movement to exploit the enhanced opportunities of participating in bureaucratic politics provided by the regulations also proved over-optimistic and could not be compensated by appointments to the departments and bureaux of people sympathetic to the environmental objectives.

Thus, we suggest that the origins of the 1978 regulations reflected an attempt to exploit the opportunities provided by the election of President Carter and its success depended on much more than a skilful appraisal of the courts and litigation trends. Strategic assessments were also required of: the probable post-election behaviour of business and industrial critics of EIA; of the capabilities of environmentalists; congressional attitudes and agendas; and the susceptibility of a vast federal bureaucracy to political direction and control. Similarly, we think that those who wanted to use NEPA and the EIA process as instruments of change after the Reagan election victory in 1980 were concerned about much more than possible reductions in analyst–advocates for environmental values within agencies. In fact this did not occur to the degree expected because there was an enormous increase in environmentalist support in the early 1980s. This was not mere chance, but the result of a deliberate political strategy which has probably done more to confound the Reagan administration's plans for environmental policies, including NEPA, than any other factor (see, for example, Belsky 1984, Vig & Kraft 1984, Rose 1986).

Clearly, the idea that continuing relations are the source of the stability, informality and trust essential to the co-evolution of politics and policy under an uncertain mandate is an important one. However, it is very rare in modern environmental politics for continuing relations to be able to exert their influence on such co-evolution for long periods without periodic disturbance that requires their reassessment. The influence of elections, an important, regular and frequent source of turbulence, in the United States, on the evolution of NEPA has received little attention.

With turbulence comes the need to reassess strategically the possibilities for access to and control of the various forums in which conflicts can be aired and norms for their resolution negotiated. Out of such a reassessment, for example, may come a determination to use a legal forum, such as the courts, because it confers important representational advantages on environmentalist plaintiffs in NEPA cases and has more power to shape and sanction collective definitions of relationships than unofficial forums. Irrespective of whichever forum is chosen, entrepreneurs are needed to perform the political work necessary for a successful strategy. This broader view of the influence individual and environmental factors can exert can be readily appreciated in the case of California, which is discussed in the next section.

Entrepreneurship and the impact assessment strategy in California

In late 1982 and early 1983, following the California gubernatorial election in which Republican George Deukmejian defeated Democrat Tom Bradley, speculation about future state policy eventually turned to the fate of the California Environmental Quality Act (CEQA) and its EIA process. However, this was not the first time that the opportunity to appraise thoroughly the costs and benefits of CEQA had arisen.

The first memorable stock-taking followed the California Supreme Court decision in *Friends of Mammoth* (Andrews 1973), a landmark in the evolution of state EIA policy, comparable in significance to *Calvert Cliffs* at the federal level (Anderson 1973). This decision extended the applicability of CEQA to the private development authorization process of hundreds of California cities, counties and special districts.

Early in his first term, Democratic Governor Edmund G. Brown, Jr., let it be known that he might favour the repeal of CEQA. This was shortly after his office had been surrounded by logging trucks, driven by loggers angry at the application of CEQA to timber harvesting. The remark, although probably not altogether serious, prompted another stock-taking (Hill 1975).

There were others. Some were precipitated by court opinions, some by economic trends in the state such as the affordable housing crisis of 1979, and at least one, the massive revision of the state implementation guidelines prior to the end of Governor Brown's second term in 1982, was initiated in advance of broader political change (Wandesforde-Smith 1977, 1981, Hill 1983).

Thus, although one comes away from Taylor's (1984) interpretation of NEPA with the impression that that statute evolved in a relatively stable environment, it would be difficult to form a similar impression in the case of CEQA. With the possible exception of the State of Washington, California has had much the most instability associated with the development of its 'little NEPA' (see Yost 1974 for an early assessment, and more recently Pearlman 1977, the special issue of the *Albany Law Review*, 1982, Renz 1984).

It was suggested previously that Taylor's view of NEPA's history understates the effect environmental turbulence may have had on its evolution. It does this most especially by obscuring the frequent strategic calculations that have had to be made by participants in EIA with respect not just to one but, possibly, to several different forums. The main omission in Taylor's discussion, therefore, is the idea that the evolution of an impact assessment process can be marked by a succession of forums which arises as people who want to direct the evolution of the process try to adjust changes in their access to and control of oversight forums.

In the case of CEQA, evolution of the statute reflects repeated attempts to reassess the significance and worth of the state's EIA process using several forums besides the courts (Wandesforde-Smith 1981). That account emphasizes the increasing importance of the Governor's Office of Planning and Research (OPR) and the office of the Assistant Secretary of the Resources Agency (Norman Hill) as authoritative sources on EIA law and practices in the late 1970s and early 1980s. However, there is no discussion of the extent to which alterations in the environment, such as the replacement of Governor Brown by a Republican, affected the oversight function of the executive branch, particularly OPR and Assistant Secretary Hill.

The earlier account, therefore, does not reveal as much about the dynamics of the continuing struggle of supporters and opponents of CEQA to impose meaning on the Act as it now seems important to know. This can now be done more effectively. Indeed the realization that the election of Governor Deukmejian and the advent of a Republican administration could have major consequences for the politics of EIA and the evolution of EIA policy in California prompted a number of thoughtful commentaries (Eastman 1981, Perlstein 1981, Bass 1983, Nevins 1984, Selmi 1984, Vandervelden 1984, Fulton 1985, Roberts 1985). Between May and July 1985 we interviewed a large number of officials within government and government agencies as well as representatives from outside organizations involved with the EIA process in California in order to provide more insight into entrepreneurial effects. Combining the commentaries listed above, the results of the informal interviews, previously unpublished materials collated by Seyman (1986), and reports (State Bar of California 1983, CEQA/Housing Task Force 1984), it is possible to get a better idea of how elections and regime changes prompt reappraisals of forums for pursuing CEQA disputes and, thus, affect the entrepreneurial dynamics of the ongoing struggle to control the implementation and evaluation of the Act. In this discussion the main focus is on elite entrepreneurship. Accounts of California practice in which the entrepreneurial activities of, for example, project managers, agency leaders, analyst–advocates and interest group leaders are discussed can be found in the cases reported in detail by Eastman (1981), Perlstein (1981), Nevins (1984), Vettel (1985), Sproul (1986) and Seyman (1986).

THE YOST ERA

It is important, first, to recall the conditions under which the early struggle over CEQA was fought. The period 1970–80 is dealt with at greater length in Wandesforde-Smith (1977, 1981). The Act was the brainchild of a committee set up by a Democratic legislature and was signed by a Republican governor, Ronald Reagan, in September 1970. There is no evidence to suggest that either the Democrats in the Legislature or the Republicans in the Reagan administration had thought through the implications of CEQA. The administration's first response, incorporated into the earliest draft guidelines for implementing CEQA, was that the law's requirement for preparing and considering environmental impact reports (EIRs, equivalent to federal EISs) would apply to only a handful of projects proposed by agencies of state government.

The entrepreneurial initiative sweeping away this crabbed interpretation came from the office of another Republican, Evelle Younger, the elected Attorney General of California. Under the vote for constitutional officers in the California system the Attorney General has a measure of political and policy independence from the Governor. The specific architect of the more liberal reading of CEQA, however, was a Deputy Attorney General, Nicholas Yost. First, Yost was able to persuade Younger to support a formal, public challenge to draft CEQA guidelines. Subsequently, Yost won the Legislature over to an expansive reading of the law and eventually, through an *amicus curiae* brief, helped the state Supreme Court to frame its *Friends of Mammoth* decision.

One need not argue that Yost was a hero or a genius to appreciate what he did. He was simply a bright and ambitious young attorney, thoroughly briefed on developments affecting the interpretation of NEPA at the federal level, and able to see how they might be exploited in California. Nor should one suppose that Yost effected the first significant transformation of the meaning of CEQA on his own. Much of the legal groundwork was laid by national public interest law groups initiating NEPA litigation, and in California there was help from sympathetic legal minds in government and in private practice.

Even taking these factors into account, however, Yost deserves credit for taking the initiative, doing the work, and accepting the risks of political entrepreneurship. He took a set of conditions that were originally very unpromising for the vigorous enforcement of CEQA and transformed them (cf. Lewis 1980).

Although Yost left California in 1976 to join the staff of CEQ after the election of President Carter, the conditions he created persisted until the statewide general election of 1978. This must now be seen as an extremely important turning point in the evolution of CEQA, comparable to the *Friends of Mammoth* decision in 1972 and to the election of Governor Deukmejian in 1982.

Under California law, the Attorney General not only represents state agencies in court at their request, but also can intervene in litigation on behalf of the public when fundamental rights are at stake. Either way, litigation initiatives have to be chosen carefully and, if an incumbent governor lets it be known that he does not want his departments and agencies referring cases to the Justice Department, may be quite rare.

It was Yost's accomplishment that, under Governor Reagan and during the first term of governor Brown, the environmental law unit of the Attorney General's Office, working closely with OPR, the Assistant Secretary of the Resources Agency, and with friendly attorneys and others out in the field, became a focal point for monitoring cases raising significant CEQA issues. Cases that might make good CEQA law were kept before the courts, though usually without Justice Department attorneys prosecuting the cases. This momentum was maintained until the elections of 1978 caused another basic change in conditions.

THE HILL ERA

The 1978 election brought George Deukmejian to the Attorney General's Office and resulted in an erosion of much that Yost and his colleagues had accomplished (Bass 1983). Under Younger the environmental law unit often intervened informally in CEQA-related litigation, and sometimes prepared *amicus* briefs. In contrast under Deukmejian, who was inclined to more old-fashioned views of natural resources law, the environmental unit was essentially dismantled, and the state's role in environmental litigation played down.

After 1978, suspected CEQA violations referred to the Attorney General for legal action by OPR and other state agencies were generally ignored. Without the same entrepreneurial skill and drive evident earlier in this key state office, the judicial development of CEQA languished. Previously, the courts had generally responded favourably to suggestions that they adopt an expansive judicial interpretation of CEQA, based on carefully chosen cases brought forward by environmentalist attorneys, often with the tacit or formal imprimatur of the Attorney General. Subsequently, there were fewer such cases and the burden of developing and pursuing them fell almost entirely on environmentalist and local plaintiffs.

Another notable outcome of the 1978 elections, enactment of Proposition 13, a statewide ballot measure limiting increases in local property taxation, also forced a fresh appraisal of the politics of EIA in California. One implication of Proposition 13 was that local government found it increasingly difficult to finance the public services and amenities, such as police and fire protection, sewers, paved streets, schools, parks and open space, normally demanded in association with new residential and commercial development. Local governments, therefore, had a heightened interest in bargaining with developers to provide some or all of these services and amenities as a precondition for granting development approval.

Thus, before 1978 the benefits of strong and vigorous EIA litigation and state EIA policy accrued primarily to groups seeking to represent a broad public interest in environmental quality. After 1978 local governments (or more accurately the local staffs charged with negotiating with developers) became major beneficiaries of the changed situation.

Before 1978, although developers complained about disruptive environmental litigation based on CEQA and about the costs of decision-making delays caused by CEQA compliance, they had an uncertain and locally variable alliance with interests at the local level. Proposition 13 gave them more reliable local allies and, consequently, an even stronger voice than they had been able to buy before in the Legislature.

It was a less than perfect union, however, because local negotiators looking, for example, for concessions on improvements and infrastructure provisions were not interested in seeing the opportunity to exercise discretion and the threat of delay eliminated entirely. While environmentalist and local actors envisage delays in the EIA process as a vital political weapon, developers consider delay generally nothing but an additional and unwanted cost (Sproul 1986). Local government actors, however, were bound to have mixed motives and alliances in the EIA process after Proposition 13.

Whatever alliances they might have formed in pursuit of a revised EIA policy, however, local governments were likely to be even more vigorous and interested participants in the politics of EIA at the state level after 1978. Trading off this effect against the constraining impact of Deukmejian's stance on the judicial oversight of CEQA and the CEQA process, leads to two conclusions.

First, some alternative to the courts as an effective oversight forum would be needed to safeguard the further evolution of EIA in California. Secondly, some way of granting local governments greater autonomy and control over the EIA process would have to be found, otherwise the pressure they were under to bargain rather freely with developers might lead them to support the abolition of the process altogether.

It was against the background of these changed political circumstances that the most comprehensive and ambitious revision of the state guidelines was initiated in the last two years of Governor Brown's second term by Norman Hill, the Assistant Secretary of the Resources Agency for CEQA matters. At that time, it was widely believed that Governor Brown would seek a seat in the US Senate in 1982, with Attorney General Deukmejian likely to try to succeed him as Governor. In fact, this is what happened.

In this situation any attempt to boost the standing of the CEQA guidelines as an authoritative source of EIA norms and to have them reflect the liberal views of the Brown regime, would have to be made before the 1982 elections. We credit Hill with that entrepreneurial insight and, further, with undertaking the enormous political work involved in bringing the revision to completion. The revised guidelines can be found in *California Administrative Code*, tit. 14, sections 15000–387, and appendices.

Hill's contribution goes further than this, however. In the face of severe environmentalist criticism and after protracted consideration, Hill supported and won support in the revised agency guidelines for something called a mitigated negative declaration. The arrangement comes into play (see Fig. 10.1) after a lead agency, typically a local government, conducts an initial study of a project and identifies potential significant impacts. At this point an EIR would normally be required. However, under the revisions, a project proponent can change or accept modifications to the project that will avoid or mitigate the impacts, and thus become eligible for a negative declaration. If this mitigated negative declaration is released for public review and subsequently approved by the decision-making body (typically a city council or county board of supervisors), a decision can be made on the project and a notice of determination filed. In this way, a developer can avoid the expense and delay of preparing an EIR.

One measure of the significance of this innovation is that 95 per cent of all projects that are subjected to the CEQA process (Fig. 10.1) are dealt with through the issuance of a negative declaration (Fulton 1985). Although this figure is probably accurate, it is difficult to make unequivocal statements about EIA practice in California, where so much occurs at the local level. Our experience shows that under the present administration the Governor's Office is unable or unwilling to develop a reliable overview of how EIA is being enforced. This seriously undermines Selmi's (1984) explanation of the evolution of CEQA.

In the context of this discussion, however, a more important aspect of the mitigated negative declaration is that it is a licence for local authorities to bargain with developers within the framework of the CEQA process in order to reach the best deal they can very early in the review process. Essentially, it grants local governments the autonomy needed after Proposition 13 to help make development pay its way, while at the same time preserving the appearance that the state, through the CEQA guidelines, is helping to subject all local development to the same searching statewide standard of environmental review. In effect, the mitigated negative declaration realigned the state–local relationship in EIA policy and practice. It kept the process alive by adapting it to changed political circumstances. Furthermore, its acceptance clearly signifies the extent to which advocates of a vigorous state role in implementing and enforcing CEQA had been forced on the defensive by 1982.

Broader political and economic trends also had notable impacts in the Legislature and the courts. The Legislature felt obligated to underline the need for the CEQA process to produce balanced outcomes when, for example, it approved a bill in 1979 making the provision of a decent home for Californians a goal to rank beside protecting environmental quality. Over this time, there were many such changes (Selmi 1984). The courts were, if anything, even quicker to sense which way the political winds of change were blowing. The 1978 decision in *Laurel Hills* gave perhaps the clearest early indication of deference to local decision making and of an unwillingness to be led much further down the path of finding substantive law to apply in CEQA (Perlstein 1981). The period between 1978 and 1982 also evidenced, in line with national trends, considerable disarray in the California environmental movement, which led eventually to leadership changes and a shift in political direction.

In this broader context, Hill imagined that the mitigated negative declaration and other lesser concessions to local governments and developers could be

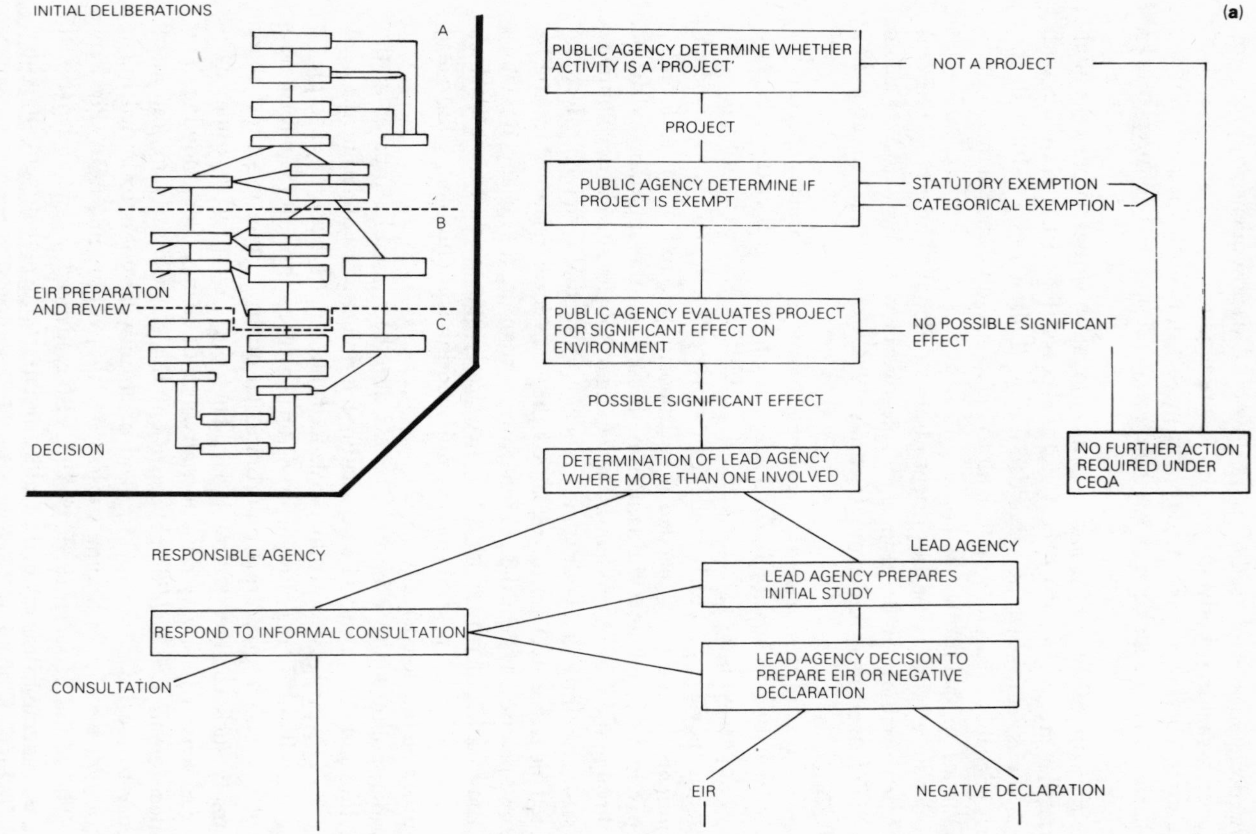

Figure 10.1(a) Environmental impact report procedures in California showing initial considerations.

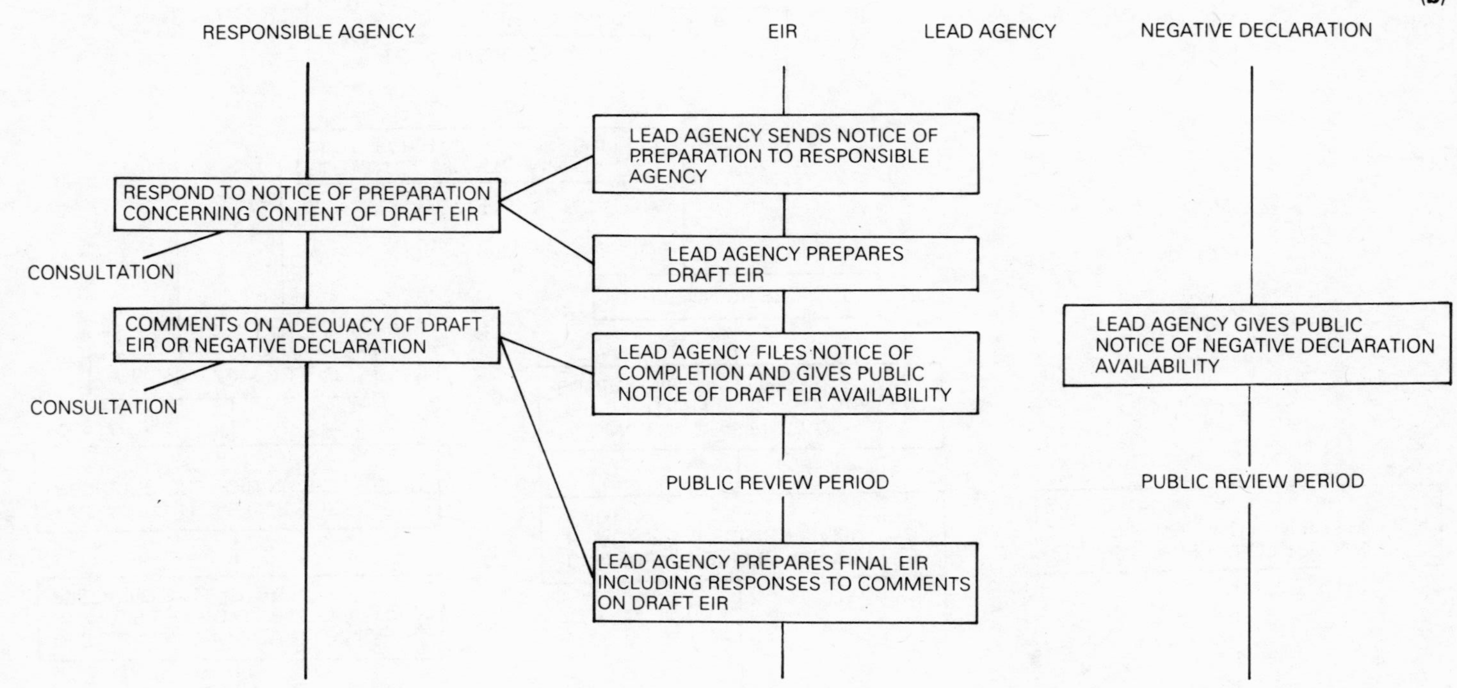

Figure 10.1(b) Environmental impact report procedures in California showing EIR preparation and review.

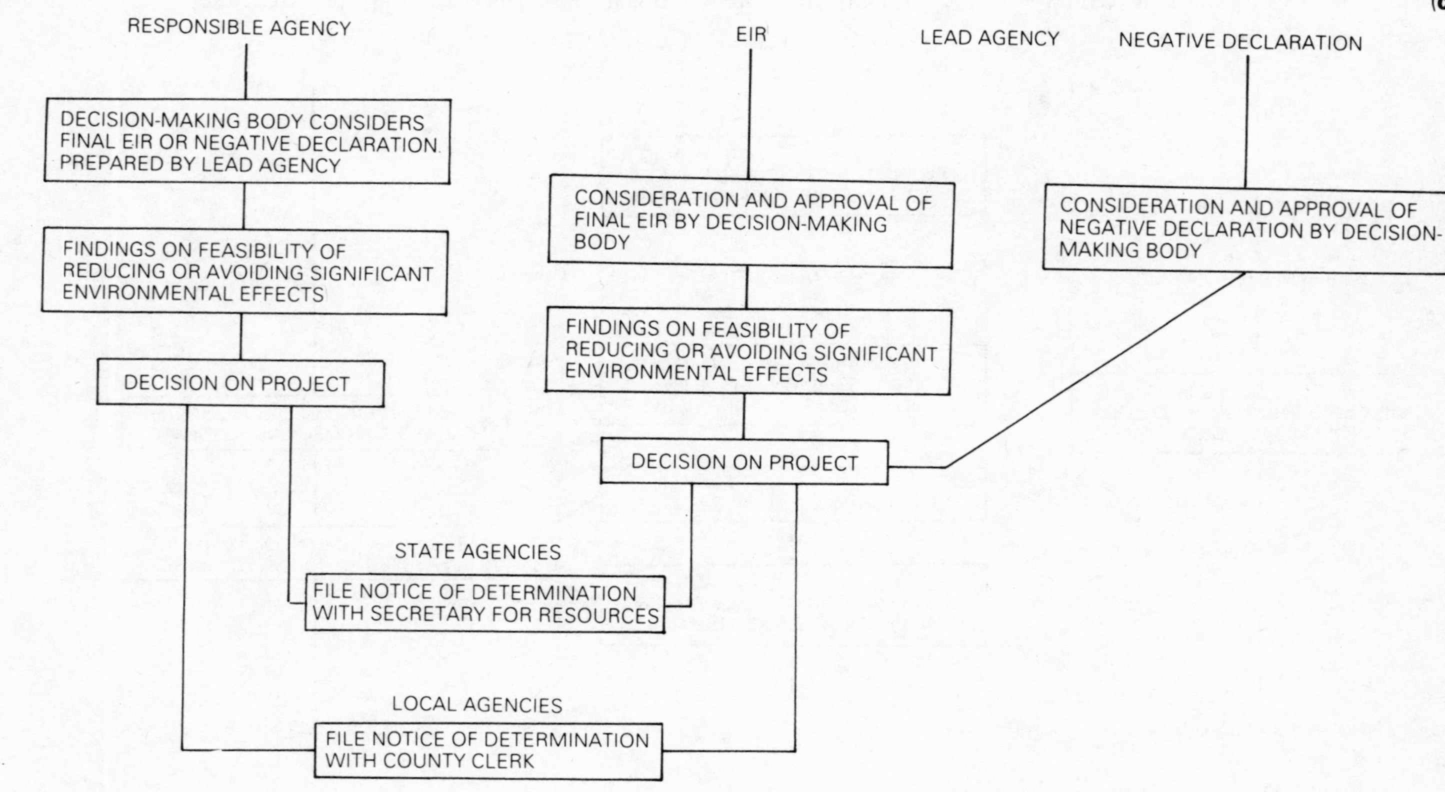

Figure 10.1(c) Environmental impact report procedures in California showing decision.

traded off in the guideline revisions against changes that, with shifts in the balance of access to and control over various forums, might conceivably strengthen state oversight of the CEQA process. Although Hill hoped to secure final approval of the revisions before the end of 1982, this did not happen because of delays in the regulatory review procedure operated by the Office of Administrative Law.

In the wake of the November 1982 elections, therefore, it was possible that instead of adopting Hill's proposals the incoming administration of Governor Deukmejian might substitute a fresh set of guideline revisions. It was this factor, together with the generally uncertain implications of the election results, that made the struggle over CEQA in late 1982 and early 1983 so intense.

THE PRESENT ERA

From the outset, it was clear that Norman Hill would have no part in EIA policy in the Deukmejian regime and he was quickly dismissed. This was far from unexpected, in as much as the post which he occupied was a political appointment within the patronage of the Resources Secretary and, indirectly, of the Governor. What made it remarkable, however, was Hill's generally even-handed and non-partisan approach to his CEQA oversight role and, above all, his widely acknowledged and essentially irreplaceable expertise in the law and practice of impact assessment.

There were other early signs that the Deukmejian administration might be planning a fundamental break with previous policy. For example, the Office of Planning and Research, which had a major role in monitoring local compliance with CEQA, and the State Clearinghouse, a branch of OPR important in ensuring the proper circulation and review of CEQA documents, were reorganized.

Bass (1983) sounded alarms about what these changes might imply for CEQA and for the substantial consulting industry that specializes in EIA advice to state and local agencies. However, it quickly became clear that they were more indicative of a general 'house-cleaning' by an incoming administration than a specific policy thrust. The Republican administration was also constrained by the fact that the Democrats still controlled both houses of the Legislature and had captured the Attorney General's Office.

The first clue to the administration's position on CEQA came in March 1983, when the Governor created a State Development Review Panel and charged it with streamlining the state's processes for issuing permits. This confirmed the generally conservative and pro-business orientation of the administration. It also indicated, however, that there would be no rush to judgement on CEQA. There would be time to study Hill's proposals, determine what else might be accomplished by executive action alone, and consider the possibilities for legislative reform.

This essentially internal review was complete by the end of June 1983, at which point it became quite clear that the administration did not intend to exercise any entrepreneurial initiative with respect to CEQA. The guideline revisions drafted by Hill were approved with no substantial alterations on 30 June 1983, to take effect in August of that year. Six days earlier, on 24 June 1983, the administration announced the formation of the CEQA/Housing Task Force to survey residential builders and developers in the state and to prepare, among other things, a legislative package addresssing changes to the statute. This was an obvious but important acknowledgement that developers had concerns not accommodated by the new guidelines that they would be given help to pursue.

In addition, the Legislature signalled its willingness to respond to the kinds of concerns expressed by Bass (1983) by at least taking a look at CEQA. More specifically, Assemblyman Terry Goggin, chairman of the Natural Resources Committee in the lower house, announced that no bills to amend CEQA would be released from his committee until the second half of the two-year legislative session in January 1984. This would allow time for the Natural Resources Committee to hold joint hearings with the Committee on the Environment of the State Bar of California and to receive a report that that State Bar Committee would prepare.

Goggin had chosen, in other words, to continue the key role of the Assembly Natural Resources Committee as a block to legislation proposing major changes to CEQA. Furthermore, he was calling on the prestige of a study of the State Bar to help him bargain with other legislators, particularly those conservative colleagues proposing bills promoting substantial alterations in CEQA implementation and enforcement.

The result of Goggin's initiative, according to Vandervelden (1984), was to give legislators the courage to vote against bills favoured by developers, who it was felt had failed to establish unequivocally a substantive case for major changes to CEQA. In this respect, the State Bar report was significant because it stood as the only authoritative source of analysis and represented a consensus among a respected cross section of attorneys. It was especially important in persuading certain legislators to drop bills providing for: awards of attorneys' fees to any winning party in a CEQA lawsuit; authorizing bonds to be posted to indemnify defendants for costs and damages caused by CEQA litigation; and establishing new standards of judicial review in lawsuits seeking to enjoin development on CEQA grounds.

On the positive side, Goggin's decision to use his position to shape a new legislative review of CEQA resulted in the enactment of Assembly Bill 2583 at the end of September 1984. However, this was an omnibus bill of individual 'bits and pieces' which reflected no entrepreneurial initiative on Goggin's part comparable to earlier work by Yost and Hill.

The two most significant amendments to CEQA in AB 2583 will probably be that requiring parties to a CEQA lawsuit to meet and confer during the first 20 days of the litigation and that sanctioning tiered EIRs. The former provision is intended to help reveal any hidden agenda on the part of plaintiffs in a CEQA suit. The latter is supposed to encourage environmental assessment of plans for future land use and development, especially local general plans, so that subsequent EIRs for specific projects can be confined to developments raising significant environmental issues not already considered in the EIR for the plan.

For some observers the 1983 CEQA guidelines and the Goggin amendments have brought an uneasy truce to the politics of EIA in California. This truce is likely to be characterized by a continuing rearguard defence of CEQA by environmentalists in the face of unrelenting pressure from developers, but no major policy shifts (Fulton 1985). From this perspective, CEQA will continue to evolve generally in a direction not only responding to developers' financial interest in progressively narrowing and limiting its scope, but also reflecting the belief that an EIR should be construed merely as an information gathering tool (Vandervelden 1984).

For others, however, new guidelines and the 1984 statutory amendments mark the completion of a major shift in policy. The reforms signify the beginning of a new era in the co-evolution of EIA politics and policy in California in which the *laissez-faire* approach to discretionary local implementation has given way to the dictation of local EIA practices by state regulation. From this alternative viewpoint, it is also concluded that policy will continue to evolve, but in a direction that leaves untouched the judicially developed public interest purposes of CEQA. Occasional outbursts of protest from developers and local governments about the burdens they have to bear are unlikely to affect the primacy of the belief that EIRs should be broadly construed as instruments of public involvement in environmental decision making and of citizen enforcement of CEQA's requirements (Selmi 1984).

In our view both of these evaluations read too much into the initiatives taken by the Deukmejian administration and by the Legislature in 1983 and 1984. The fundamental point about both sets of initiatives is that they left essentially undisturbed the arrangement engineered by Hill, whereby locally autonomous EIA practice and non-judicial state oversight exist side by side in California. Neither initiative evidenced the imagination, effort and enthusiasm which reformed policy in earlier eras of CEQA's evolution.

The question to ask about what has been happening in the present era and may happen in the future, therefore, is not whether one or other of the parties contending for control of CEQA has finally got the upper hand. To ask that is to forget how quickly and how often the linkage between politics and policy can change, for electoral and for other reasons, and how unwise it is, therefore, to project present relationships into the future. Rather, the question should be what is required to achieve a more profound transformation of EIA politics and policy than either the Deukmejian administration or Assemblyman Goggin and his staff proved able to effect. To ask that is, of course, to ask how entrepreneurship in politics and the conditions for its exercise combine to produce policy change and evolution.

So much depends on individual circumstances. It seems reasonable to say in the case of California that entrepreneurs for a more vigorous and assertive use of CEQA are to be found. People with the requisite imagination, effort and enthusiasm, however, are not attracted to the Deukmejian regime, whose entrepreneurial tendencies lie in the reverse direction. Furthermore, they are unlikely to find a home in the Legislature, which has dealt with CEQA frequently over the years but almost always in reaction to events elsewhere, for example in the courts, and almost never with any creative spark. The Office of the Attorney General is a more likely base of operations. Indeed, the Democratic incumbent has revived the environmental law unit but is unlikely to have many CEQA cases referred to him by the Deukmejian administration. Similarly, it is always possible that an attorney for an environmental group or in private practice will succeed in persuading a court to open up new pretexts for CEQA litigation. Attorneys in practice, and many still in law school, dream of such a perfect case and it may yet be found.

At the moment, clearly, conditions do not favour vigorous CEQA entrepreneurship. For any of these potential CEQA entrepreneurs to be successful, therefore, conditions have to be right. This is not to say that they should wait for the normal processes of political change to yield an ideal combination of circumstances, namely the co-occurrence of committed environmental liberals as Governor, Attorney General, chairman of Assembly Natural Resources, and Chief Justice. Rather, the right conditions have to be created and sustained by political work. For the foreseeable future in California, this means searching for entrepreneurial opportunities at the state level because, under Proposition 13, the structural constraints on local initiatives are probably too great.

We conclude, therefore, not that recent events have condemned CEQA to a slow demise nor that the state has emerged from the present era as dictator of CEQA implementation and enforcement; rather it appears that there is still an opportunity to exploit the changing politics of California, in order to realign continuing relations in the EIA process and to help reshape the policies of CEQA through an available and appropriate forum. This is, in part, an affirmation that individuals as political entrepreneurs have made and will continue to make an essential contribution to the co-evolution of EIA politics and policy in the state. It is also, in part, a caution against supposing that, because conditions sometimes narrow the opportunities for entrepreneurial initiative, it is the prevailing conditions that force the pace and direction of change, rather than what people make of them.

Conclusions

We began by pointing out that the publication of *Making Bureaucracies Think* (Taylor 1984) is bound to generate new interest in EIA scholarship. It opens new perspectives on implementation and enforcement by pointing to the many conditions that contribute to the effectiveness of EIA in addition to good technical information and sound assessment methods. It lays the basis for a new generation of comparative studies, both across states and nations and between various impact statement strategies. Above all, however, the book deals with and invites further attention to basic questions of policy change, and specifically to the way sustained analytical competition can prod the evolution of policy in a simultaneously effective and acceptable direction.

In the case of NEPA, Taylor's argument is that policy change has occurred as a result of learning, indeed that there has been such a progressive evolution of EIA practice and policy that it is unlikely to be reversed. It is clear that in the beginning, for example, NEPA litigation succeeded in disrupting a pattern of continuing relations that had grown up around federal agencies like the Corps of Engineers and the Forest Service since the end of World War II. It was the familiar pattern of closed subgovernments revealed by all the great, postwar pluralist analyses of federal resource management (see, generally, Wengert 1955). Agencies and their closest political allies negotiated the balance of judgements that went into decision making, including the extent to which decisions would be seen to rest on analysis, among themselves.

It is also clear that a measure of stability had returned to continuing relations by the end of the 1970s and that the new pattern was one in which environmentalists were well represented. Agencies no longer resisted EIA nor merely tolerated it for the sake of laying a paper trail *en route* to decision making and keeping up environmental appearances. Rather, Taylor aruges that the agencies welcomed EIA, even came to love it, because it restored much of the informality in their deliberations and negotiations that the era of litigation swept away. EIA, in short, proved itself useful to agencies as a device for negotiating answers to hard choices even in situations where all conflicts could not be resolved. Judgements were balanced in such a way that the answers were perceived to be both effective and acceptable.

On the one hand, Taylor perceives EIA as a mechanism that works because of its structure. In this respect, the key to the success of EIA in making policy subsystems better at producing knowledge, and in making that knowledge the basis for decisions, is its ability to make a market in analysis. It takes what would otherwise be a mere redundancy of agency critics and puts them in analytical competition with each other and with insider analyst–advocates. This competition, properly regulated by an authoritative oversight forum, is the agency of learning. One of the optimistic messages conveyed by this conception, therefore, seems to be that learning and, hence, the evolutionary progress of policy and decision making on the basis of better knowledge, will continue as long as analytical competition can be kept healthy and vigorous.

By this account, the value of EIA rests on what Taylor (1984: 37) calls its internal architecture. It requires some strategic choices by Congress and the White House to maintain the foundations, and an authoritative overseer, probably the courts. By and large, however, the value of EIA ought to be realizable despite the comings and goings of particular individuals, the political ups and downs of election returns and, to range into even more distant contingencies, general social and economic conditions.

On the other hand, there is the possibility that EIA promoted adaptive co-evolution of politics and policy not so much because of its special internal structure but because it was embedded in a pluralist political system that, given enough time, tends to converge on equilibrium solutions to social problems. In other words, there may be forces in the system that contribute to the development (or decline) of consensus on policy goals and which originate independently of improvements in knowledge (Heclo 1978). From this perspective, the argument that the benefits of EIA were produced without regard to individual- and regime-level variables would be even stronger than in the structural explanation.

Unfortunately, as we noted in our careful review of his interpretation of the history of NEPA, Taylor missed an opportunity to test the applicability of his structural explanation of EIA learning by saying too little about the impact of the Reagan administration on the structure created by the 1978 NEPA regulations. The broader thesis based on the natural pluralism of the US system is difficult to evaluate even over a period of at least a decade (Yngvesson 1985). The same basic questions about the relationship of EIA to policy change arise in the case of CEQA and, indeed, in every other EIA mandate in the United States and abroad. From the experience of CEQA, it seems that correcting the disabilities of Taylor's views may be far less compelling than exploring factors he has chosen not to stress.

Indeed, we have argued that in California, far from being independent of individual- and regime-level variables, the ability of EIA to produce better knowledge and to contribute to the progressive improvement of policy and practice is difficult to comprehend except in terms of individual entrepreneurial and strategic responses to a changing political world. It is hard to imagine EIA in California surviving so long and being accepted as useful by so many without the creative imagination and entrepreneurial hard work of people like Nicholas Yost and Norman Hill. The suspicion that a similar conclusion might be reached for NEPA if its history were re-examined is, of course, buttressed by the fact that Yost went to Washington with much of the entrepreneurial spirit that California had taught him could be productive.

We want to close by again pointing to the strange world that, on the basis of Taylor's information on the Corps of Engineers and the Forest Service, actors in the EIA process seem to inhabit. It is not a world from which political judgements are totally absent. However, the range and frequency of such judgements are extremely limited, allowing an astonishingly intense and single-minded concentration on the business of EIA. It is an environment where regular and frequent elections never seem to intrude, and where the relevance and contribution of EIA to a larger world outside the process itself never seems to be checked or questioned.

In the face of a system that seems to do such a good job of keeping bureaucratic 'noses to the grindstone' one is lulled into thinking that it runs itself. If that is so, the impact statement strategy of administrative reform is indeed a remarkable policy instrument, perhaps one that can make policy subsystems better at producing knowledge for decisions no matter who is in charge of preparing and evaluating EIAs, or who designs the guidelines.

Experience of EIA illustrated by the evolution of CEQA suggests that the impact assessment process is not self-sustaining and self-regulating. In the real world that is inhabited by people involved in impact assessment, learning and improvement cannot be taken for granted by virtue of the process used in analysing and making choices. Learning and improvement must be worked for or, more accurately, created by people whose motivation extends beyond making knowledge production the principal value in a policy-making subsystem.

It has long been a central tenet of the EIA literature that impact analysts, and even assessment project managers and authors of manuals and guidelines, have no entrepreneurial function. Now that Taylor has revealed the widespread existence, indeed the necessity, of entrepreneurship in analytical competition, that tenet is no longer viable. With it goes the neat distinction between trained assessors who merely assess and lay politicians who decide. It is time to acknowledge the prevalence of political entrepreneurship and to recognize that, without entrepreneurs who can first see opportunities for learning lessons and then come forward to do the necessary political and analytical work the co-evolution of politics and policy is difficult to imagine and impossible to attain. When it does happen, it bears the imprint of the individuals responsible.

11 *The EIA directive of the European Community*

P. WATHERN

Introduction

Between 1977 and 1980, 'Brussels watchers' amongst the environmental impact assessment (EIA) fraternity could gauge their standing in the hierarchy by whether they were privy to the most recent version of the proposed EIA directive as these documents diffused out only slowly from an inner circle of luminaries. Indeed, there were so many drafts of the directive over this period that even the pundits seemed to lose count. Estimates of how many were produced ranged from 'over twenty drafts' (Haigh 1983) to 'no fewer than 50' (Milne 1986). Not only was there a long gestation period before the draft directive was formally published in 1980, but there were also protracted deliberations before a final text was agreed by the constituent member states of the European Community (EC) in July 1985. In all, a decade elapsed between the initial discussions on EIA as an element of EC environmental policy and its realization.

It would be correct, but far too simplistic, to say that the recalcitrance of certain member states, particularly the UK, was responsible for these inordinate delays. Indeed, the EIA directive merely provides one of the more extreme examples of the difficulties involved in formulating and adopting EC policy. To see how these difficulties arise, it is important to understand how EC policy evolves and to consider the role of various Community institutions within this process.

Community policy, however, is not created in a vacuum, as each member state has a range of domestic provisions which may be enhanced, nullified or even countermanded by proposed EC legislation. Thus, the EIA directive must be set within the context of national planning law. National perceptions of priorities concerning the natural environment influence the evolution of EC policy and even the political relationships between member states determine the agreed Community stance which is finally adopted. National perceptions also dictate the way in which Community policy is implemented within each member state.

In this paper the influence of Community institutions and national planning law within individual member states on the evolution of the directive are reviewed. The main provisions of the directive, representing a minimum package of measures acceptable to all member states, are described. Finally, implementation in the UK is reviewed to assess the extent to which these measures facilitate realization of the objectives of an EC policy or merely seek to reinforce the national position.

EC policy and Community institutions

EC environmental policy is enunciated in very generalized terms in three Community environmental action programmes which have been adopted since 1973. Most aspects of this policy are reactive provisions which aim at curing specific environmental ills, particularly those caused by pollution. In contrast, EIA is preventive and seeks to anticipate and resolve in advance potential environmental problems. Although preventive policy was mentioned in the 1973 action programme, the first explicit reference to EIA as an objective of EC policy was included only in the second action programme (Council of the European Communities 1977).

There are a number of legislative means available for translating the general statement of intent in action programmes into specific provisions, namely regulations, directives, decisions, recommendations and opinions. Regulations are the most forceful, being laws which are directly applicable within member states. Directives specify binding policy objectives, but leave the means for achieving them to each member state. Decisions are binding only upon those specified, whereas recommendations and opinions carry no mandatory obligations. In the field of environmental policy, directives have been the dominant legislative device (Haigh 1984).

The formulation of Community policy is theoretically a simple process, with a number of institutions fulfilling clearly defined roles. In practice, however, the procedure is complex and much of it is conducted behind the scenes.

Draft directives are formulated by civil servants within the Commission of the European Communities (the EC bureaucracy) in the light of agreed Community statements, the results of contract research and pending national legislation. The period involved in formulating a draft directive may be a protracted one during which various experts within the member states are usually consulted. In the case of the EIA directive, five years elapsed between the commissioning of a research project on EIA in 1975 and publication of the draft directive in 1980 with at least 21 versions during this period.

A directive has no power until it has been adopted. Generally, before this takes place more formal negotiations between the member states occur and the provisions are refined by the Commission's Scientific Advisory Committee, comprising expert technical representatives from each country. The work of this committee has been described as a mystery, not least by its members (Anon. 1985a). In addition, the draft must be submitted to the European Parliament. Although the views of the Parliament are not binding, they are normally taken into consideration when the final draft is formulated. This draft is presented to the Council of Ministers for adoption. Formally, the Council comprises the foreign ministers of each member state, although in practice the position is delegated to an appropriate minister depending on the matter under discussion. The Council of Ministers must be unanimous in a decision to adopt legislation, which for the EIA directive finally occurred on 3 July 1985.

A number of factors influence the stance of the member states over proposed directives. Philanthropy is not often evident as member states actively promote their own priorities, try to contain measures which might have high domestic political or economic costs, and seek alliances which carry scope for future national advantage. The deliberations prior to the formal adoption of a directive provide one opportunity for member states to modify the scope of a proposal. As a directive does not specify the mode of implementation, but leaves this to each member state, another opportunity to modify or even nullify its effect is provided. Thus, a government antagonistic to the objectives of a policy has the potential for pre-adoption emasculation of a directive and post-adoption deflection of its intent (Wathern *et al.* 1983).

Environmental assessment provisions in member states

The commitment that any member state is likely to demonstrate towards individual EC legislative proposals is a direct reflection of the priority afforded to it within domestic policy. This perception is important as it influences the brief given to national representatives in formulating the proposals, the stance of ministers in negotiating the agreed provisions and finally governments in implementing policy. National attitudes may range from active support through indifference to outright opposition. Many of the responses of member states to the EIA directive become clear when national provisions for the environmental assessment of development proposals are examined. Although each is in some respect unique, certain countries have been selected for more detailed consideration in the following discussion to show the range of approaches that have been adopted. In the case of the UK, however, the description is intended to provide sufficient background information to explain its protracted opposition to mandatory EIA.

BELGIUM

In Belgium, EIA seems to have been a victim of the increased devolution of powers to the regions after 1980. In 1977, the Belgian Minister for Public Health and Environment identified EIA as a major priority of his administration and envisaged legislation by 1980–1 (Anon. 1978). The minister's aspirations, however, have not been realized and there have been no unifying proposals emanating from the regions which exercise devolved powers over environmental protection. The attitude of the major linguistic groups, the Flemish and the Walloons, towards environmental concerns differ markedly with the result that, paradoxically, the EC directive is likely to be a major impetus for standardized EIA provisions.

Until such times as the EIA directive is implemented, environmental appraisal of development proposals rests with existing statutes relating to safety, pollution control and land-use planning. Of particular note are the laws on the control of dangerous, dirty and noxious establishments, the 1946 amendment to the General Regulations for Protection of Labour, as well as land-use planning and building authorization legislation, embodied in the 1962 law on Land Development and Town Planning (Lee & Wood 1985).

DENMARK

At present, there are no specific legislative provisions related to EIA in Denmark. Some elements of an EIA system for both projects and plans, however, exist under present legislation. The main provisions relate to the pollution certification system operated under the Environmental Protection Act 1973 and to the preparation of regional plans required by the National and Regional Planning Act 1969. Of particular note, with implications for the EC directive, is the authorization procedure for many major development projects. These are authorized through the legislature.

FEDERAL REPUBLIC OF GERMANY

Within the federal republic of West Germany there is a formal division of responsibility for development and its regulation between the national government and the individual states. EIA procedures exist at both levels of government, but only for large projects in the public sector. Although some national pollution control and other environmental legislation provide elements of an EIA system, as yet no formal EIA legislation exists at the federal level. In place of a draft bill formulated by the Ministry of the Interior in 1974, but never placed before the Bundestag, a cabinet resolution was adopted on 12 September 1975. This resolution includes recommendations concerning not only the procedures to be adopted in carrying out an EIA for federal actions but also details of its content. Bunge (1984) indicates that these are considered minimum requirements which, in practice, are generally exceeded.

The provisions of the resolution need not apply if other regulations achieve the same objectives. Given the range of other legislation which exists, this is an important caveat which has been used as the basis for non-implementation by certain federal agencies (Kennedy 1981). Haigh (1983) considers that the lack of effective public participation within the system has resulted in the procedures having little discernible effect. Kennedy (1981), on the other hand, considers that the lack of accessibility merely obscures a commitment to environmentally sensitive planning and decision making, at least within certain agencies.

Individual states have given serious consideration to EIA provisions applicable to their own activities. In 1976, Saarland was the first to adopt provisions, a direct copy of those formulated by the federal government. Subsequently, West Berlin and Bavaria have adopted procedures, and Hamburg and Hesse are in the process of doing so. The deliberations of others, for example, Schleswig-

Holstein and North Rhine Westphalia, are contingent upon the EC directive (Lee & Wood 1985).

FRANCE

To date, France has adopted the most formalized system of EIA within the Community, embodied in the law on the *Protection de la Nature* which became operational in 1978. *Etudes d'impact* are required for a range of developments, although certain minor infrastructure provisions are specifically exempted. Under the terms of the legislation, impact studies should contain an environmental analysis of the area; should review the environmental impacts of the proposal; should include an analysis of the alternatives considered and a justification of the selected option; and should indicate any mitigating measures which have been investigated (Monbailliu 1981). Lee & Wood (1985) estimate that approximately 8000 *études d'impact* are prepared each year.

Haigh (1983) has criticized the public participation procedures that operated in France in parallel with the system in the past. The *enquête publique* is little more than a consultation exercise. The documents are merely made available for the public to make written comments on the proposal in a formal register. The proponent subsequently has the opportunity to respond, without redress, to these comments (Macrory & Lafontaine 1982). Since October 1985, however, new procedures, more akin to the British public inquiry, have been in operation for major developments. Prieur (1984) considers that the major benefit of the EIA process in France has been to effect an increase in environmental awareness.

GREECE

Lee & Wood (1985) report that there are certain provisions which provide for an elementary type of environmental study in Greece. These provisions exist in regulations related to urban development, forest protection, mining and quarrying, the protection of the marine environment, and the licensing of new industrial operations. It is anticipated that these fragmentary provisions will be replaced by a formal system as part of a new environmental legal framework. When the draft law was made public in February 1986, the responsible minister indicated that it would be at least two months before the legislation could be presented to parliament, following a period of the widest possible discussion (Anon. 1986).

ITALY

In January 1984, a bill which would introduce formal EIA procedures was presented to the Italian parliament. By the middle of 1986 it had not been enacted. Until this comprehensive law is operative, only fragmentary provisions exist at the national level. Within the various regions, however, some elements of an EIA system have been introduced (Lee & Wood 1985).

LUXEMBOURG

EIA procedures have existed in Luxembourg since July 1978. The 1978 law on the protection of the natural environment and the 1979 law relating to dangerous, dirty and noxious installations require the assessment of development projects. In addition, planning law, in particular the 1937 law relating to amenity plans and the 1974 law on the management of land use, integrate some elements of EIA into land-use planning (Lee & Wood 1985). It is estimated that between 5 and 20 projects per year are submitted to some form of environmental evaluation under these provisions.

THE NETHERLANDS

The Dutch approach to EIA has been perhaps more deliberate and reasoned than that of any other state within the EC since official interest in the Netherlands began in about 1974. The early history of Dutch experience is reviewed in Jones (1980) and the detailed proposals for a national system of EIA are described in Jones (1983) and Brouwer (1986). The deliberations on EIA have been characterized by a combination of commissioned research and case study EIAs. In 1977, nine trial EIAs were initiated encompassing both industrial development proposals and a forward planning appraisal. Experience from these studies did much to help frame the subsequent provisions. In addition, a series of research studies covering such topics as the ways of assessing impacts on the physical environment, scoping techniques, the content of EIAs, and impact assessment guidelines was commissioned. The programme of research laid great emphasis on evaluating the practical experience gained in other countries.

In the light of this practical experience and these research findings, as well as a wide range of consultations, a draft bill was presented to parliament in May 1981. Since that time, an interim EIA policy has operated which has allowed further research and impact studies to be undertaken. The bill has been described by Jones as 'extensible' in that it sets the basic procedural and substantive aspects of EIA, but retains a facility to be supplemented by specific pieces of legislation, formulated in response to future experience. Brouwer (1986) indicated that the law should be fully operational by the end of 1986 and anticipated that 10–15 EIAs would be produced each year.

PORTUGAL

Portugal, with Spain the most recent state to join the EC, is in the process of harmonizing its own domestic legislation to conform with existing EC provisions. At present, there are no EIA requirements, although an environmental law (Project of Law No. 79/IV), which includes provisions for the evaluation of environmental impacts, is under discussion in the Portuguese parliament. Until this legislation is implemented, the impact of a proposed project is regulated under various sectoral laws.

The most significant law concerns the 'regulation of location and functioning of industrial settlements' dating from 1966. According to this legislation, industry is classified into three categories depending upon its potential for inflicting environmental damage. The law also specifies the technical rules and norms with which industry must comply. The most damaging activities, the explosives and phosphorus industries, mining and the extraction of radioactive ores, are subject to more specific constraints.

Impacts on specific environmental media are regulated under separate laws. In 1980, legislation recommending the establishment of commissions for the preservation of air quality in critical geographical regions was adopted. A number of these commissions have now been established. Water quality has long been an issue of concern and basic legislation for the protection of water was enacted in 1933. During 1986, a project on the River Ave hydrological basin, designed to provide the basis of a comprehensive policy for the management of water resources, was initiated. In addition, specific legislation and regulations for chemical wastes and noise have recently been adopted.

REPUBLIC OF IRELAND

There has been a system for the appraisal of new industrial projects in the Republic of Ireland since 1970. The main response to the need for EIA was to superimpose impact assessment procedures contained in the Local Government (Planning and Development) Act 1976 upon this system. The EIA process requires the production of a report detailing the environmental effects of a proposed development project to be submitted with a formal application for planning permission. The impact statement must also be made available to the public. Dalas (1984) considers that the inclusion of a Ir£5M threshold on the capital value of projects requiring assessment and the exclusion of specific projects, such as the works of national and local government, are major deficiencies of the system. Lee & Wood (1985) claim that few EIAs have been completed in the Irish Republic.

SPAIN

The EC EIA directive was adopted before Spain signed the treaty of accession. On joining the Community, therefore, Spain accepted a number of elements of environmental policy which it had not been party to formulating. Thus, as there is no national mandatory procedure for EIA, provisions will have to be enacted in order to comply with the directive (Acre 1986).

This is not to say that environmental assessment does not exist in Spain at present. As in many other countries, a piecemeal approach has been adopted with provisions scattered amongst various sectoral regulations. For example, a 1961 by-law on troublesome, unhealthy, harmful and dangerous activities requires applications for the development of certain categories of industrial plant to contain documents detailing amongst other things, the environmental and health consequences of the development and any remedial measures that are proposed. The regulations operate at a local government level and cover a wide range of private and public sector developments. An order of the Department of Industry on the prevention and correction of industrial atmospheric pollution issued in 1976 contains comparable provisions related to atmospheric emissions. The 1985 Water Law requires EIA for certain types of development affecting water resources. Mining legislation also indicates a growing commitment to

EIA; thus, for instance, legislation dating from 1973 requires studies to protect the environment during mining operations, while the Royal Decree of 1982 on restoration and that of 1984 on opencast coal mines make explicit reference to environmental evaluations (Fuentes 1985).

UNITED KINGDOM

Recent experience with the EC directive suggests antipathy towards EIA within the UK. In fact, there has been a marked polarity in attitudes towards EIA with significant increases in the number of proponents in recent years. Even the attitude of central government departments appears somewhat ambivalent. The early 1970s were characterized by a spate of novel developments within the UK as North Sea oil exploration and exploitation presented planners with the need to appraise a whole new industry. In a situation where all of the information, and consequently the initiative, seemed to lie with the developer, the possibility that EIA in some guise might redress this imbalance was actively pursued particularly by the Scottish Development Department (SDD). The influence of these early development proposals is reviewed in Clark *et al.* (1981a).

In 1973, SDD and the Department of the Environment (DoE) funded a research project undertaken by a research group (Project Appraisal for Development Control – PADC, latterly the Centre for Environmental Management and Planning – CEMP) at Aberdeen University to produce a manual for the assessment of major development projects (Clark *et al.* 1976, 1981b). Subsequently, a study of the practicality of introducing EIA into the planning system was commissioned (Catlow & Thirlwall 1976). It is interesting to note that the findings of both studies were published as DoE research reports. This device effectively distanced government from formal endorsement of EIA, while at the same time allowing it to commend the PADC manual to developers and planning authorities.

Developers in some ways have been more enthusiastic advocates of EIA than government. British Gas, Shell, British Coal, British Petroleum and the North West Water Authority are but a few of the major developers within the UK who have adopted EIA in the last decade. In addition, despite central government's stated opposition to the EIA directive, the Ministry of Defence prepared an EIA for the Faslane Trident base. This EIA was written in such a way that it complied with the draft EC EIA directive which was then being contested by the UK government (Foster 1984).

While individual developers have appreciated the need for EIA, there has been collective opposition from industry as a whole, at least as expressed through the Confederation of British Industry (CBI). The CBI has retained a fixed position of opposition to mandatory EIA procedures, for example, in evidence to the House of Lords Select Committee (House of Lords 1981). The CBI has argued consistently that EIA causes delays to development, a view which appears to have exerted considerable influence over an increasingly receptive government.

The UK government has maintained that the elements of EIA are already present in a flexible guise in existing provisions under town and country planning legislation. Under planning statutes, all development requires prior approval unless specifically exempted. Amongst the major exemptions are agriculture and forestry (these are not considered development under the planning acts) as well as development by statutory undertakers (generally projects such as power stations undertaken by public utilities) and by the Crown, for example, military installations. The responsible agencies, however, are encouraged to follow comparable procedures. As planning authorities have the right not only to request appropriate information from the developer concerning the proposal, but also to initiate any studies necessary to formulate a decision, the system is considered to be sufficiently flexible to appraise simple development proposals as well as, for example, complex industrial or civil engineering schemes.

In the UK, many major developments are subject to an inquiry in public, although this is not obligatory. Public inquiries have a quasi-judicial structure with the right to legal representation. The adversarial nature of the proceedings means that evidence presented by a developer or an objector can be contested under cross-examination.

Evolution of the EC directive

Comparison of the draft directive published in 1980 (Commission of the European Communities 1980) with the text finally adopted by the Council of Ministers in 1985 (Council of the European Communities 1985) reveals a plethora of minor and major modifications. These changes combine to produce 'a less powerful – if in places more flexible instrument than that sought by the Commission' (Anon. 1985b). The changes reflect the compromise achieved between member states, although it is impossible to determine which countries were responsible for individual changes as the discussions occurred mainly behind the scenes. From 'inside information' and a scan of the environmental press over the period, however, it is possible to discern the influence of certain member states in the final form of certain of the provisions.

For example, Danish representatives prevented the directive from being adopted in November 1983 on the grounds that the directive would undermine the sovereign power of the Danish parliament to approve development projects. This objection has been accommodated in the final text by the expedient that the provisions should not apply to 'projects the details of which are adopted by a specific act of national legislation'.

French opposition centred upon the extensive and rigid provisions concerning consultations related to trans-boundary pollution included in the draft directive. These provisions required *inter alia* the assessment of impacts upon the environment within another member state and the need to send information on the project to the appropriate authority within any country affected for comment. Both long-standing, trans-boundary pollution problems associated with discharges from mineral workings to the Rhine, and controversy over nuclear power stations in border areas, for which EIA would be mandatory, made France particularly sensitive to this issue. In the final text, consultations between member states concerning trans-boundary impacts have been placed within the framework of normal bilateral relations, while the requirement to consider impacts in neighbouring countries has been dropped.

The reservations of the United Kingdom to the directive were often voiced after the draft directive was published. In 1980, an under-secretary at the DoE advised the House of Lords Select Committee that EIA should not even be a matter for EC legislation (House of Lords 1981). The UK stance has been that mandatory provisions for EIA are not acceptable and that the directive should be more 'pragmatic and flexible' than the draft proposed, see for example, Anon. (1980) and House of Lords (1981).

It appears that the UK has achieved its main objective in containing the possible effects of the directive. Thus, the original draft proposed that member states had to obtain the prior agreement of the Commission to exempt projects from the provisions of the directive. Although projects can be exempted 'in exceptional cases' the responsibility for this decision has passed back to the member states. They are now merely required to advise the Commission of their reasons for doing so.

The major omission from the final text, however, relates to the use of EIA in forward planning and policy making. Lee & Wood (1978), who carried out the first research project on EIA for the Commission, argued for project EIA as the first stage of EC preventive environmental policy, only because the experience of EIA in plan and policy making was rudimentary. It is unlikely that they envisaged this first meagre step taking a decade. With publication of the final directive it is clear that the attempt to commit the member states to further directives containing provisions for EIA in plan making and policy formulation have failed. It is clear that a Community preventive environmental policy will not be achieved using this device.

There is certainly more flexibility in the directive than originally proposed, but the UK has also had to make some concessions. For example, the DoE was clearly opposed to the Commission drawing up lists of projects for mandatory EIAs (House of Lords 1981), yet such lists have been included.

While some member states may feel content at having contained the potential effects of the EIA directive, others are less sanguine. Brouwer (1986), for example, considers that from a Dutch perspective 'this EC-directive, like so many others, is a very weak compromise. It is more the result of the cumulative resistance from the development promoters and bureaucracies in the member countries than a synthesis of the best ideas for the protection of the environment'.

Content of the directive

The provisions of the EC EIA directive can be grouped into four categories. These relate to the specification of projects requiring EIA, the scope of an assessment, consultation and the role of the Commission. In the following discussion each of these items is given separate consideration.

PROJECTS REQUIRING EIA

Member states are required to assess the effects of both public and private projects which are likely to have significant impacts on the environment as a consequence of their nature, size, or location. Of these criteria, nature and size are given more detailed consideration in the annexes to the directive, while the implications of project location as a determinant of environmental impact for a particular project is not addressed further.

Projects for which an EIA is mandatory are specified in Annex I of the directive (see Table 11.1). Annex II containes additional project types which may be subject to an assessment 'when member states consider that their characteristics so require'. Although member states may specify criteria or critical thresholds defining the circumstances in which an assessment would be required, they are not compelled to do so. Defining such thresholds would create, in effect, further classes of mandatory assessments. The categories of projects included in this annex are listed in Table 11.2 and the project types

Table 11.1 List of Annex I projects requiring environmental assessment.

Crude-oil refineries (excluding undertakings manufacturing only lubricants from crude oil) and installations for the gasification and liquefaction of 500 tonnes or more of coal or bituminous shale per day.
Thermal power stations and other combustion installations with a heat output of 300 megawatts or more and nuclear power stations and other nuclear reactors (except research installations for the production and conversion of fissionable and fertile materials, whose maximum power does not exceed 1 kilowatt continuous thermal load).
Installations solely designed for the permanent storage or final disposal of radioactive waste.
Integrated works for the initial melting of cast iron and steel.
Installations for the extraction of asbestos and for the processing and transformation of asbestos and products containing asbestos; for asbestos-cement products, with an annual production of more than 20 000 tonnes of finished products, for friction material, with an annual production of more than 50 tonnes of finished products, and for other uses of asbestos, utilization of more than 200 tonnes per year.
Integrated chemical installations.
Construction of motorways, express roads and lines for long-distance railway traffic and of airports with a basic runway length of 2100 metres or more.
Trading ports and also inland waterways and ports for inland waterway traffic which permit the passage of vessels of over 1350 tonnes.
Waste disposal installations for the incineration, chemical treatment or landfill of toxic and dangerous wastes.

Table 11.2 Categories of projects included in Annex II.

Agriculture
Extractive industry
Energy industry
Processing of metals
Manufacture of glass
Chemical industry
Food industry
Textile, leather, wood and paper industries
Rubber industry
Infrastructure projects
Miscellaneous
Modifications to Annex I developments

within the agricultural category are included in Table 11.3 in order to illustrate the level of specificity in the directive.

From Table 11.1 it can be seen that threshold criteria also help to define some of the projects requiring mandatory EIA. Member states are empowered to exempt, 'in exceptional cases', specific projects from the provisions of the directive.

SCOPE OF AN ASSESSMENT

The directive requires developers to supply information on the proposal, which must be considered by the competent authority in arriving at its decision. This information, generally in the form of a report or environmental assessment (the directive does not use the term environmental impact statement) must include a description of the site, the design and size of the project, remedial measures, and the data necessary to assess its main environmental effects.

Annex II of the directive indicates those aspects of the environment which are likely to be affected by a development and which, therefore, should be addressed in an assessment. Two provisions are of special note for this

Table 11.3 Agricultural projects included in Annex II.

Agriculture:
Projects for the restructuring of rural land holdings
Projects for the use of uncultivated land or semi-natural areas for intensive agricultural purposes
Water management projects for agriculture
Initial afforestation where this may lead to adverse ecological changes and land reclamation for the purposes of conversion to another type of land use
Poultry rearing installations
Pig rearing installations
Salmon breeding
Reclamation of land from the sea

discussion. First, developers are required to describe the main alternatives of the project which have been assessed, but no indication is given as to whether this relates simply to alternative sites or should encompass, for example, alternative technological means for realizing the same objectives. Secondly, developers must specify the forecasting methods used, providing an opportunity for impact projections to be scrutinized independently.

CONSULTATION

In the directive, consultation is specified in general terms. First, member states are required to ensure that authorities with special responsibility for the environment are given an opportunity to comment on a proposal based upon information supplied by a developer. This information is also the basis for public consultation. Although the obligation for public participation is explicit, member states are responsible for determining how this shall be achieved. In particular, the definition of the public to be consulted, the means for notifying the public and reviewing assessments as well as the form of the consultation have been left open to the governments of member states.

THE ROLE OF THE COMMISSION

The Commission has a clearly defined role with respect to the directive, namely co-ordinating information exchange. Member states are required to inform the Commission of any criteria and thresholds used to determine Annex II projects which should be subject to mandatory assessment. In addition, information rationalizing a decision to exempt a project from the terms of the directive must be supplied. Although this information will be of value to the Commission in the preparation of a report on the operation of the directive, required by 1993, its major utility appears to be in formulating further proposals aimed at harmonizing EIA practice within the Community.

Implementation in the United Kingdom

Member states are required to comply with the directive by 3 July 1988. At present, the only guides to future implementation are the public pronouncements by governments and the stances adopted during negotiations on the directive. Between 1980 and 1985, some member states, such as Germany, indicated a high priority to the adoption of the EIA directive. It can be assumed, therefore, that the directive will be implemented quickly and smoothly within such countries.

In countries with a small constituency for environmental issues, EC legislation already provides a major impetus for domestic reform and the same is likely to be true of EIA. In the case of Belgium, for example, the EC requirement for EIA is likely to force uniform action by the regions. Lee & Wood (1985) consider that the Republic of Ireland may need to broaden its EIA provisions in order to comply with the directive. The system for EIA developed in the Netherlands appears to be in conformity with the directive (Brouwer 1986). Some member states, however, particularly Greece and Italy, seem to have practical difficulties and sometimes even apparent reluctance, in incorporating EC provisions into domestic legislation, which may lead to delays in implementation of the directive.

The case of the UK needs special consideration because of the protracted opposition of government to the directive. In addition, some elements of a UK response are beginning to emerge which appear to conform to a pattern previously seen in the implementation of other EC directives.

Certain facets of the directive may yet prove 'hostages to fortune' with respect to its implementation within the UK. Thus, the provisions designed to exempt projects 'adopted by a specific act of legislation', may provide an opportunity to circumvent the requirements of the directive in at least one important case. There appears to be a commitment on the part of the UK government to push the Channel Tunnel project through to fruition as quickly as possible. The UK Transport Minister is already committed to there being no planning inquiry for this development. Under the terms of the EC EIA directive, participation would be obligatory. The project, however, is likely to be approved using the parliamentary device of the hybrid bill (previously used to approve the Windscale nuclear waste reprocessing plant), thereby placing the most important UK civil engineering project of this decade outside the scope of the EIA directive.

An essential feature of all EC directives is that detailed implementation is left to the member states. Consequently, shortly before the EIA directive was adopted, the DoE established a working group to oversee implementation, but only with respect to developments covered by planning law. The initial remit of the group was to draw up guidelines concerning the type of information to be covered in an environmental assessment and to determine pre- and post-assessment procedures for handling an appraisal. The results of the group's deliberations will be published as a consultative document.

When the group was established in April 1985, it was anticipated that the consultative document would be available in 'late Autumn' 1985. The group met for the last time in November 1985, but difficulties were encountered subsequently. Thus, on 30 January 1986, a minister at the DoE indicated that 'there are one or two points that we are still considering' (Waldegrave 1986). When this chapter was written (mid-April 1986) the document had not been published.

The working group comprised some twenty members drawn from government departments, local authority associations, the planning profession and industry as well as one member from an amenity society. The Ministry of Agriculture, Fisheries and Food (MAFF) refused to participate in the group, apparently on the grounds that most agricultural practice is outside the scope of planning legislation and, therefore, not affected by the directive.

It should be noted that the farming lobby, powerful and effective within the UK, has successfully withstood all attempts to bring agriculture under town and country planning law since its initial enactment in 1947. MAFF's stance is an attempt to maintain the *status quo*. The directive is a threat to this position, as poultry and pig-rearing installations and the intensive cultivation of semi-natural areas are included in Annex II of the directive. A recent proposal to bring intensive livestock units under planning control following a recommendation of the Royal Commission on Environmental Pollution has been restricted to those that are close to residential areas, which constitute, therefore, a public nuisance rather than an environmental hazard. This again indicates a policy of containment on the part of agricultural interests. A blanket exemption of agricultural projects in Annex II or of those not subject to planning law would do much to reduce the potential of the directive. This appears to be MAFF's objective.

The second issue which is of note is the controversy which has arisen over public participation. DoE's original proposal was to introduce a formal requirement that public consultation should commence prior to the submission of a planning application, arguing that this would help define the scope of an appraisal. Industrial representatives within the working group opposed this suggestion. In the draft consultative document prepared by this group it was proposed that a new notification system should be adopted. This would require developers to notify the planning authority of the intention to submit a planning application for a development for which an environmental assessment would be necessary. This should occur at least 2 months, but not more than 5 years, before submission of the planning application. Subsequently, the environmental assessment would have to be submitted with the planning application. This modification to present practice would require an amendment of UK statutory procedures, but it remains to be seen whether the minister will accept this change.

The greatest controversy in the UK, however, is likely to attach to the treatment of Annex II projects. The civil servant chairing the working party has stated that it was understood in the UK that 'it would be open to the Government not to make Assessment compulsory for Annex II projects' (Fuller 1986). The unpublished draft consultative document prepared by the working group goes further and states categorically that 'the Government has decided that no such extension' of mandatory provisions to Annex II projects should be made. A senior official of the EC Commission indicated in September 1985 that Annex II is not considered by the Commission to be optional and that any decision by a member state to exclude all types of project listed in Annex II would contravene both the letter and the spirit of the directive. Therefore, it seems that this issue will only be resolved when the UK government officially notifies the EC Commission of the measures taken to implement the directive.

The final recommendation included in the final draft consultative document relates to the consideration of alternatives; a very restrictive viewpoint has been adopted. A developer only has to include reference to alternative sites in an environmental appraisal, and need not mention, for example, other ways of achieving the same objectives.

Thus, within the UK it appears that an attempt is being made merely to absorb the EIA directive into current practice, with few substantive changes in approach. The directive could have provided a vehicle for reform of the UK planning system with, for example, mandatory requirements for the assessment of agricultural and forestry developments, an objective of the environmental lobby for many years. What is likely to be achieved will be far more modest. Undoubtedly, some classes of Annex I project, such as waste disposal facilities, are likely to be subject to more detailed scrutiny than previously. Yet, having substantially reduced the potential of the EIA directive prior to its adoption, the UK government appears to be continuing to nullify its impact in formulating the implementation procedures. This approach to containment, however, is nothing new. It has also been adopted, for example, with the shellfish directive (Young & Wathern 1984) and the less favoured areas directive (Wathern *et al.* 1986).

In most member states, the EIA directive will have beneficial effects. For some, it will provide a means of overhauling land-use planning systems which are fragmentary. To a large extent, however, its impact, as in the UK, will depend upon domestic political interests. Without the necessary commitment to anticipatory preventive environmental policy, individual member states may do no more than just formally comply with the directive. Considering the fate of these relatively modest provisions for project assessment which already exists in some guise or other in most member states, the prognosis for the aspiration that there should be further directives concerned with plan and policy appraisal is bleak.

Conclusions

A series of directives requiring environmental impact assessment not only for projects, but also for plans and policies, would do much to advance the cause of preventive environmental policy within Europe. After a decade of deliberations, however, the EC has been able to adopt only the most meagre of provisions for project assessment which do no more than formalize those that already exist in most member states. In Belgium, Greece, Italy, Portugal and Spain the need to comply with the directive may be a spur to formulating national legislation, but for the remainder little material change is likely. This is not totally unexpected, however, as some member states, particularly the UK, certainly approached the discussions over the EIA directive intent on containing its potential effects. The need for the Council of Ministers to be unanimous in a decision to adopt legislation ensures that Community policies are the minimal provisions acceptable to each member state.

A directive is not the most effective means for introducing EC environmental policy, primarily because formulation of the detailed implementation provisions is left to individual member states. The result is that policy may be applied unevenly across the Community, further underlining the wide discrepancies which are evident in the priority afforded to environmental protection in different member states.

Sufficient discretion over the directive remains with member states to ensure that compliance can be achieved with widely varying implementation procedures. From detailed consideration of the situation in the UK, it is clear that at least one member state has ensured that the directive will not impinge upon domestic environmental policy. Yet, this has been achieved without the UK being in breach of the directive except, perhaps, over the Annex II projects.

UK government treatment of major development proposals, however, gives an even stronger indicator of intent concerning EIA. The Channel Tunnel proposal is the first Annex I project in the UK since the EIA directive was adopted. Therefore, it should require mandatory environmental impact assessment under the terms of the directive. By taking the proposal out of the established development control procedures, however, the UK government has placed it beyond the scope of the directive in order to avoid the protracted public consultations which almost certainly would accrue.

From UK experience, it is clear that the national perception of priorities concerning development and the environment, is likely to be the major factor in applying EIA, rather than the existence of an EC directive. Furthermore, those who look to EIA as a means of balancing the legitimate, but competing, demands of development and the environment, particularly when government has a pre-stated interest in the outcome, have been afforded little encouragement by the provisions of the EC directive.

Postscript

With publication of the draft guidelines for consultation in late April 1986 the uncertainties associated with the deliberations of the working group were resolved (DoE 1986). The published guidelines differ somewhat from the recommendations agreed as the final draft by the working group. That the 'requirements of the directive can be met within the context of the existing planning system' is a reflection of a government decision to reject a number of the group's recommendations. The most important issue, however, that 'the Government does not foresee that it will be necessary to make the carrying out of formal assessments mandatory' for Annex II projects is broadly in line with the interpretation of the working group. Rather, the secretary of state will be empowered to require the preparation of an assessment for 'particular Annex II projects which are so substantial in their environmental impact that a formal assessment ought to be carried out'. Furthermore, local planning authorities will be advised that 'it is not appropriate' for them to require developers to undertake formal assessments other than those listed in Annex I. Effectively, a central restraint on the proliferation of EIA is being retained by government. This will ensure, for example, consolidation of MAFF's position.

Adopting the recommendations of the working group with respect to prior notification would require modifications to planning law. Industrial representatives on the working group, however, were opposed to the insertion of another formal stage into the development control system, arguing that commercial interests might be prejudiced and highlighting the possible lengthening of the assessment process with concomitant delays in development authorization. In a political climate which favours development and deregulation it is not surprising that their views have prevailed. The only change suggested in the consultative document is that planning authorities will be given 16 weeks in which to consider such developments, rather than the existing 8-week statutory period.

In the consultative document, the consideration of alternatives has been broadened somewhat from the narrow recommendation of the working group. Developers are advised that in the assessment they should detail the 'main alternatives considered and reasons for final choice'. There is no requirement to consider only alternative sites. Presumably, therefore, the adequacy of an assessment could be questioned by objectors at an inquiry for the failure to consider, for example, policy issues. It might even prove possible for a planning authority to request additional information on this topic under existing powers which permit it to request the data necessary to determine a planning application.

In 1986 member states adopted new procedures over the adoption of many of its provisions. Henceforth, majority voting will operate in many situations which may act against individual member states who have used unanimous voting as a veto to resist reform in the past.

12 *The legislative framework for EIA in centrally planned economies*

A. STARZEWSKA

Introduction

One of the fundamental functions of a socialist state is to secure proper living conditions for its citizens. Consequently, in countries with centrally planned economies, there is a legal framework which ensures a particular place for environmental protection in the economic activity of the state. To achieve this objective requires a range of provisions which include, besides the strictly social ones, a great number of conditions generally described as 'ecological' or 'environmental requirements'.

In order to determine how environmental impact assessment (EIA) could fit within the overall planning process in socialist countries, it is important to consider three aspects. These are the constitutional and legal framework within which EIA would have to operate; current practice in development planning; and the scope for using EIA in centrally planned economies. In this chapter, each of these facets is given separate consideration, along with a detailed review of recent experience of EIA in Poland.

Constitutional foundations of environmental protection

The constitutions of socialist countries determine that environmental protection and regulation of the use of natural resources for present and future generations are basic functions of the state. The complexity of environmental–social interrelationships requires a particular form of constitutional device. Thus, certain constitutions contain articles dealing directly with environmental protection, for example, by identifying responsibilities with respect to the environment, while others specify the socioeconomic objectives of environmental policy. The most important principles of environmental protection incorporated within the constitutions of individual socialist countries, and some examples of relevant articles, are listed in Table 12.1

To understand how constitutional principles relating to the environment operate in some socialist countries it is advisable to examine some examples. For instance, the Czechoslovak constitution regulates environmental protection above all by Article 15 which states that:

Table 12.1 Constitutional devices governing the utilization of natural resources.

Constitution form	*Example*
Codification of the independent function of the socialist state in nature conservation, rational use of natural resources and environmental protection.	Article 15, GDR Article 31, Bulgaria Article 18, USSR
The right of exclusive state (socialist) ownership of the components of nature and natural resources.	Article 11, USSR Article 12, Poland Article 8, Hungary
Principle of planning in the development of the national economy and the use of natural resources, including national economic planning, raw material requirements analysis and rational allocation of resources.	Article 7, Hungary Article 193, Yugoslavia Article 16, USSR
Formulation of rights and duties with respect to the environment. A right to live in a favourable environment is paralleled by a duty to protect the environment.	Article 87, 1926, Yugoslavia Article 34, GDR Article 42, 67, USSR
Definition of the role of public organizations and movements advocating nature protection.	Article 29, GDR Article 27, Romania
Codification of the unity of home and foreign policy and the environmental responsibilities of the state and society.	Article 20, Hungary Article 28, 29, USSR

> the socialist state shows concern for the improvement and all-round protection of the nature of the homeland and preservation of its beauties in order to create more and better conditions for the well-being of the people, the health of working people and for their rest and leisure.

The constitution also defines the aims of environmental protection, thereby establishing conformity between the economic and environmental interests of society. This should ensure an increase in the well-being of the people along with the provision of healthy living conditions (Lisitsin 1985).

The Hungarian constitution specifies the socialist attitude towards nature. In this sense, attitudes towards the environment are framed by needs implicit in the state (socialist) ownership of the main means of production. Thus, the components of nature and natural resources provide the material basis for development which unites the interests of environmental protection and economic development. Any environmental measures, however, may be constrained by needs implicit in the state ownership of the means of production.

The planned character of state activities in environmental protection and the comprehensive nature of measures necessary for the protection and use of the environment are also noted in the Hungarian constitution. Collectively, these provisions establish the basic means for co-ordinating development of the economy and environmental protection with due regard for any immediate and long-term consequences that may appear in the relationship between people and nature. Article 57 is of particular interest. This article formulates the right of citizens to a clean environment as a fundamental human right, a feature which is becoming increasingly common in present-day legislation.

'With the aim of improving and protecting man's environment' Yugoslavia's constitution seeks to 'ensure conditions for the preservation and improvement of natural and other values of man's environment, which are necessary for healthy, safe and effective living and working conditions for present and future generations'. Another important constitutional provision specified is the need to prevent and eliminate 'harmful consequences which, on account of the pollution of the air, soils, inland water bodies and the sea, may jeopardize these natural values or create a threat to the life and health of the people'. The provision of conditions for exercising the right to a healthy environment is vested in the 'public community'. Each person is obliged to use the 'natural riches . . . in such a way as to ensure conditions for man's work, life and leisure in a healthy environment'.

Legal aspects of environmental protection

All legislation is an extension of the primary provisions implicit in constitutions. Thus, within any country, the same constitutional principles provide the basis of both environmental policy as a whole and current legislation specifically dealing with environmental protection.

A short review of environmental protection legislation in socialist countries provides a starting point for considering not only the possibilities of introducing EIA procedures in such a system, but also the form that they should take. It is interesting to note that environmental protection legislation has undergone a parallel evolution in different socialist countries.

Historically, the first objective of environmental law in socialist countries was nature conservation. Under such laws, individual sites and natural features of special aesthetic, cultural and historical value which should be preserved intact, are removed completely or partially from economic activity. In many instances, however, legislation on environmental protection has developed concurrently with laws related to the use of natural resources.

Thus, the laws and regulations on nature conservation frequently include provisions governing the use of natural resources. This feature, for example, characterizes the first law on nature conservation passed in a socialist country, namely that adopted in Poland in April 1949. In it, nature conservation was interpreted not only as the protection of different types of flora, fauna and natural areas, but also as the preservation and proper use of natural resources. Similar approaches have been adopted, for example, in Romania, the German Democratic Republic (GDR) and in the two constituent parts of Czechoslovakia.

The Czechoslovakian legislation, adopted in 1950 and 1954, stands apart somewhat from other laws characteristic of this period, in that a new approach concerning the importance of 'environmental–social relationships' was already emerging. This concept has reached greater prominence subsequently and is discussed in more detail below.

The approach to nature protection in the narrow sense of nature conservation is also characteristic of the early period. The Hungarian law of 1961–2 also adopts a narrow approach to nature conservation covering, for example, biotopes of scientific and cultural value, endangered species and natural and historical relics. This legislation, however, operates in conjunction with regulations governing the protection of individual biotopes, such as the laws on forest and land protection and on water protection adopted in 1961 and 1964 respectively.

Threats to the natural environment resulting from economic growth and industrialization have brought a fundamental change in environmental policy towards preventing environmental degradation. This change is reflected in a number of legal acts governing not only the protection of such environmental elements as air, water and soil, but also the rational use of natural resources.

Thus, in Czechoslovakia, for example, there are laws related to the protection of the purity of the air, agricultural land, state forests, wildlife and water quality, as well as rational use of the subsoil. In addition, there are two pieces of Czechoslovak legislation concerning the connection between resource use and environmental protection. The 1976 law on territorial planning and regulation of construction provides an important judicial means for regulating urban areas and for protecting the entire human environment. A law dating from 1966, pertaining to the health of the population provides for the regulation of the air, water, land and other elements of the environment from considerations of sanitation and hygiene. In the other socialist countries of Eastern Europe, there is much legislation regulating the use of various elements of the environment, comparable to that which exists in Czechoslovakia.

In recent years, however, laws concerned with regulating various facets of the natural environment have been incorporated into more comprehensive legislation establishing standards for environmental protection. The emergence of these more comprehensive laws reflects the recognition not only of the integral unity of natural systems and their importance for economic activity (the 'environmental–social relationships' noted above), but also the need to preserve ecological equilibria. This understanding is the basis of comprehensive legislation which deals with the protection of the complex formed by the interaction of nature and the social environment. Such laws are evidence of the tendency towards maximum coverage of all aspects of the 'nature–society' system by environmental legislation in socialist countries. This progressive tendency towards integration, however, is offset by the relative isolation of the legal systems related to different components of the natural environment. This fragmentation of the legal system makes consolidation of environmental protection within a single piece of legislation difficult.

One of the first national laws reflecting this comprehensive approach to nature and environmental protection was the law on the planned application of socialist culture in the use of resources, adopted in the GDR in 1970. This law emphasizes the:

> planned development of the socialist culture in the use of resources as a system of conscious [that is, guided by society] formation of the environment and environmental protection with the aim of keeping in unity . . . the natural and productive foundation of society – the land, the water, the air, flora and fauna.

It should be pointed out that similar legislation exists in other countries. For example, a comprehensive law on environmental protection was adopted in Romania in 1973. This law provides, for the first time, a definition of the environment which includes the totality of natural and man-made factors which together 'affect the ecological equilibrium and determine the conditions of human life . . . [and] . . . society's development'. Besides identifying the components of the environment which are protected this law also details the functions of central and local government bodies, including the health, agriculture, food and industry ministries, with respect to environmental protection in Romania.

A higher degree of consolidation of environmental regulations is found in the law 'On the Protection of the Environment' adopted by the Hungarian People's Republic in 1976. This is the first law in Hungary which provides an all-embracing, general regulation concerning basic aspects of environmental protection. It should be emphasized that the Hungarian law has a wider significance than merely establishing environmental regulations, as it also identifies the means that have to be adopted for environmental protection.

This legislation sets the context in which the law envisages both a prohibitive and a permitted order in the relationships between society and nature. This 'permitted–prohibited' approach gives enterprises and organizations using natural resources much latitude, but with one exception. They must observe the maximum permitted concentrations of pollutants. Accordingly, the main aim of the law is not protection of the environment *per se*, but rather a preventive, long-term regulation and planned restructuring of the human environment. The Hungarian law imparts a dual meaning to the concept of 'environmental protection'. First, it aims at protecting the environment from both existing pollution and disturbances of the natural equilibrium. Secondly, it envisages the creation of an environment which ensures proper living conditions for people. Thus, in this context, there is no great difference between environmental protection and the planned restructuring of the environment, as these concepts are inexorably interlinked.

The characteristic features of this type of legislation can be seen from a description of one of the latest examples, the Polish law on the protection and modification of the environment adopted in early 1980. This is probably the most comprehensive of all such acts within socialist countries. The Polish law contains a detailed characterization of the general principles of environmental protection together with a description of the concepts underlying environmental protection and the control of pollution. It sets out the basic elements of environmental management for different natural resources such as soils, minerals, surface and coastal waters, the air, flora and fauna. The law then indicates practical steps to be taken in accomplishing the aims of environmental protection, formulates the functions and competence of various economic and managerial bodies protecting and using the environment, states restrictions on the use of natural resources, and lists specifications for the equipment to be used in economic activities. Furthermore, it deals extensively with questions about the responsibility for damage caused to the environment. The principle of compensation for the use of natural resources is emphasized with any charges and fines constituting a special fund for environmental protection. The law deals in detail with organizational and legal questions of environmental protection, including the creation of protection bodies, state inspectors, local and public organizations and institutions.

From this wide review of existing legal provisions for environmental protection in the countries of Eastern Europe, it is clear that a comprehensive approach has been adopted, particularly in recent years. This is a result of the realization that more components of the environment should be afforded protection and that many economic activities need to be limited and managed more carefully in the future.

Environmental protection in the planning system

In considering the opportunities which exist for introducing EIA within socialist countries, it is important to establish the relationships between planning and environmental protection. Figure 12.1 illustrates the connection between socioeconomic planning and the natural environment according to Bochniarz & Kassenberg (1985).

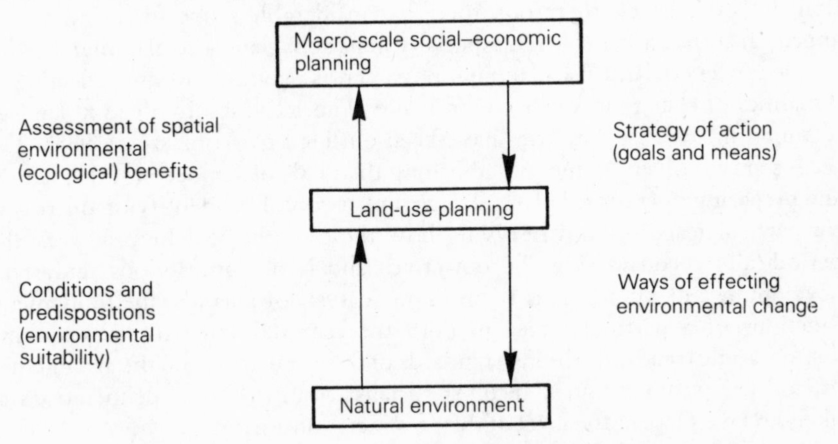

Figure 12.1 Relationship between socioeconomic planning and the natural environment.

On the basis of this figure, it is clear that there is a role for EIA within the planning system. In the context of project planning in socialist countries, this means primarily the application of some existing methods which together constitute an approach comparable to EIA. For example, it is possible to consider a development within an area by simulating its likely subsequent effects. Such an approach contains the essential elements of EIA within the planning process, namely the identification of potential effects, their assessment and the eventual comparison of alternatives.

This application of EIA is obvious. It is essential, however, to consider EIA within planning at the level of the whole national economy. In socialist countries this means within the context of the socioeconomic plan. Moulding the environment to the present and future needs of society must take place within a predefined space whose essential component is the natural environment. Thus, it is the natural environment which determines the development potential of any state. Consequently, the quality, arrangement and accessibility of reserves and natural resources, as well as the structure of ecological systems define the limits of planning freedom within which the development of a country can proceed. Within any particular state, spatial differences in the available resources create regional variations in the opportunities for economic development. The function of the socioeconomic plan is to optimize these opportunities.

There are various ways of viewing the problem of environmental protection in the socioeconomic plan. While the environment may act as a constraint on development, it can also be considered a component of development or as an element of the investment programme. Establishing the extent to which the environment acts as a constraint involves determining the objectives of socioeconomic development with respect not only to the national economy, but also to land use. Thus, appraisal must involve consideration of the available resources, the value of the natural environment and the extent of its contamination. In this respect, therefore, there is considerable scope for using an EIA approach at the strategic level, especially in national land-use planning.

The process of land-use planning, however, is a continuous one reflecting the dynamics of change in socioeconomic life. The activities involved in land-use planning in socialist countries have been outlined by Tomaszek (1985). Two phases are involved, namely establishing the goals of socioeconomic planning and preparing operational plans. The first phase involves identifying the reasons for particular actions and resolving how these might be achieved, as well as periodically reconsidering the objectives and basic assumptions underlying development planning. This requires an analysis of *inter alia* the fundamental functions of a particular area in both the regional and national economy, demographic trends, the living standards of the population and the arrangement of land use within an area. In most socialist countries these deliberations are likely to take place at the level of the provincial authority.

During the second phase, operational plans are prepared. These plans identify the changes necessary to achieve the predetermined goals and, concurrently, fulfil a number of other tasks. Such plans aid the formulation of recommendations concerning the location of projects and modification of the present pattern of land use. Simulation studies show the impact of alternative locational variations of an investment on the natural environment. Similarly, detailed land-use proposals incorporating appropriate modifications for particular parts of an area would be established. Possible suggestions for updating other land-use, socioeconomic and departmental plans together with programmes for environmental protection and other actions affecting land use throughout the country would be provided. Finally, an indication of current knowledge on the state of the environment and foreseen modifications to it would be given.

Clearly, within such a system the essential requirement is for accessible information on the present and projected state of the environment. One advantage of this system is that it allows an immediate reaction to change, either by undertaking actions to prevent undesirable processes, or by modifying land use within the area affected by them.

In socioeconomic planning within socialist countries, therefore, the place of EIA becomes distinct; it is involved at the stage when operational plans are devised. In practical terms, therefore, EIA will be most useful when output is expressed in conventional land-use planning terms, such as maps of threatened elements of the environment. Assessment of the effects of a project which, for convenience, can be termed 'classical EIA', takes place within the overall decision-making process shown in Figure 12.2 (Tomaszek 1985). A schema of

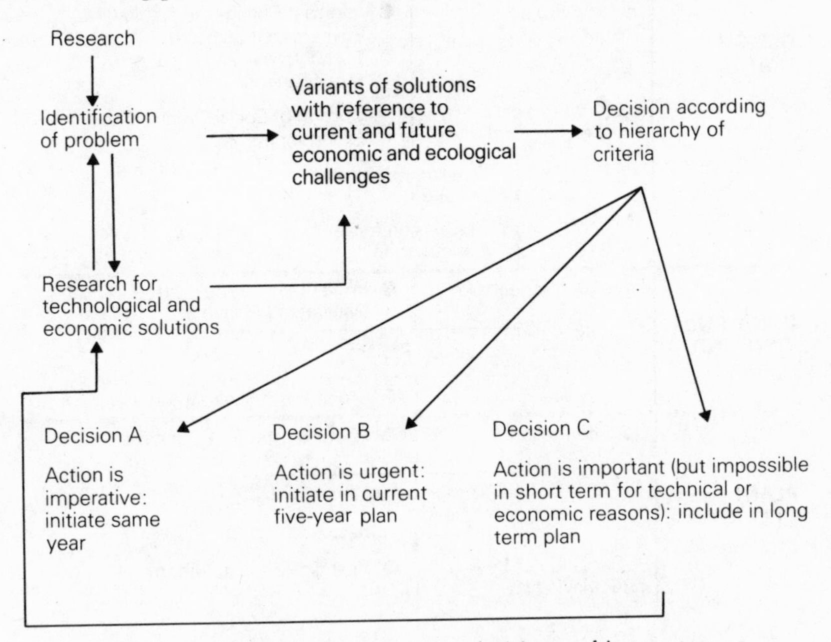

Figure 12.2 EIA in the investment decision-making process (modified after Tomaszek 1985).

the investment process which includes specific provisions for ecological investigations is featured in Figure 12.3 (Kozlowski 1985).

Within the process of land-use planning, forecasts of changes in the state of the environment consequent upon realization of a plan are made. EIA, here, is

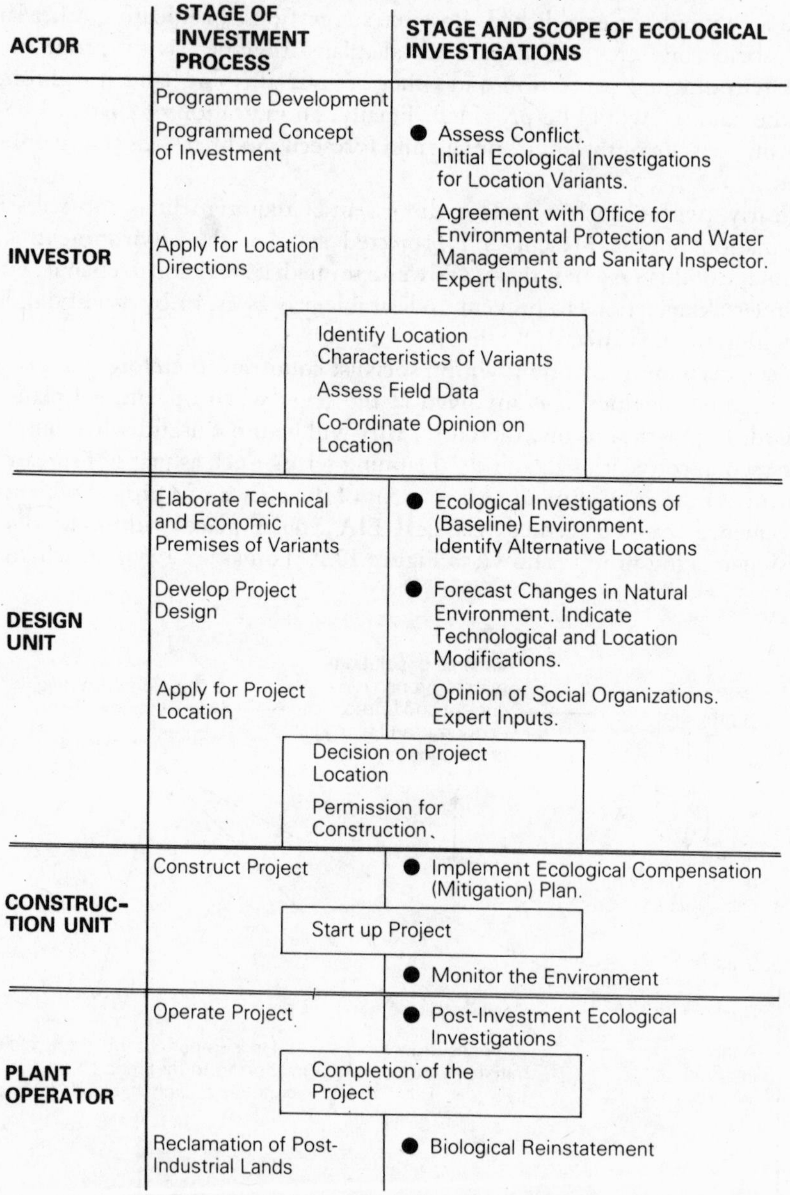

ACTOR	STAGE OF INVESTMENT PROCESS	STAGE AND SCOPE OF ECOLOGICAL INVESTIGATIONS
INVESTOR	Programme Development	
	Programmed Concept of Investment	● Assess Conflict. Initial Ecological Investigations for Location Variants.
	Apply for Location Directions	Agreement with Office for Environmental Protection and Water Management and Sanitary Inspector. Expert Inputs.
	Identify Location Characteristics of Variants Assess Field Data Co-ordinate Opinion on Location	
DESIGN UNIT	Elaborate Technical and Economic Premises of Variants	● Ecological Investigations of (Baseline) Environment. Identify Alternative Locations
	Develop Project Design	● Forecast Changes in Natural Environment. Indicate Technological and Location Modifications.
	Apply for Project Location	Opinion of Social Organizations. Expert Inputs.
	Decision on Project Location Permission for Construction.	
CONSTRUCTION UNIT	Construct Project	● Implement Ecological Compensation (Mitigation) Plan.
	Start up Project	
		● Monitor the Environment
PLANT OPERATOR	Operate Project	● Post-Investment Ecological Investigations
	Completion of the Project	
	Reclamation of Post-Industrial Lands	● Biological Reinstatement

Figure 12.3 Ecological investigations (●) in the investment process (modified after Kozlowski 1985).

suitable for formalizing forecast operations and for assessing alternatives. This assessment of alternatives may involve ranking particular kinds of influence and weighting various spatial elements.

In the long-term planning horizons involved in strategic planning, there is also a role for EIA. The objective of such planning exercises, should be to determine desired land structures based upon clearly identified demands for energy and water, as well as upon limits on environmental contamination. Consideration of the imposition of rigid limit values is outside the scope of this discussion as it belongs to the legal instruments of planning. However, similar objectives can be achieved by a variety of other measures. Thus, general bans on location within predetermined areas and on the production of substances hazardous to the environment, as well as fines for transgressing contamination standards could be used in the preparation of a plan. As a number of variants will probably be considered in devising a plan, operational plan formulation is also an appropriate component of the planning system where EIA should be adopted.

The scope for adopting EIA in centrally planned economies

The legal situation in most countries with a centrally planned economy shows an attitude to the environment which is consistent with that implicit in EIA. While the laws described above provide the legal framework for the adoption of EIA procedures, only certain socialist countries already have the conditions necessary for the introduction of such a system. In this respect, the GDR and Hungary are the most advanced. The case of Poland is somewhat different, in that recent decisions have provided an opportunity to gain practical experience of EIA. Work on the application of EIA in Poland occurred simultaneously with considerable legal and organizational changes with respect to environmental protection; this enabled positive steps towards the introduction of EIA to be taken which are discussed in more detail in the next section.

EIA is at a beginning in Hungary. Orders governing investments are contained within a regulation of the Council of Ministries, which has been modified several times since its introduction in 1974. According to these orders, every investment decision must be based upon an inquiry supported by appropriate economic and technical documentation. Each investment decision has to be based upon the demands of environmental protection as well as land-use planning. A 1983 regulation of the National Council for Environmental Protection and Nature Conservation expressed the Council's intention to make it obligatory for the planning of energy production investment to include a consideration of environmental effects. A government order to implement this provision is expected (Enyedi & Zentai 1985).

In the GDR there is, as yet, no counterpart to 'classical' EIA procedures (Kotyczka 1984). However, environmental considerations are widely taken into account in the planning and decision-making process. Utilization of the environment is planned under the responsibility of national government authorities, regional and district councils, industrial groups, companies and other institutions, each in charge of its own delimited area.

Decisions on possible courses of action are taken on their merits, according to a list of criteria defining an order of priority laid down under planning guidelines. Other indicators, such as assessments of effectiveness, cost and revenue performance, are also taken into account. All efforts are geared towards finding the most favourable variant which satisfies both economic and ecological requirements. The decision-making process can be described in general terms by the diagram in Figure 12.4.

Additional importance is attached to complex economic and ecological assessments, with reference being made to environmental effects likely to

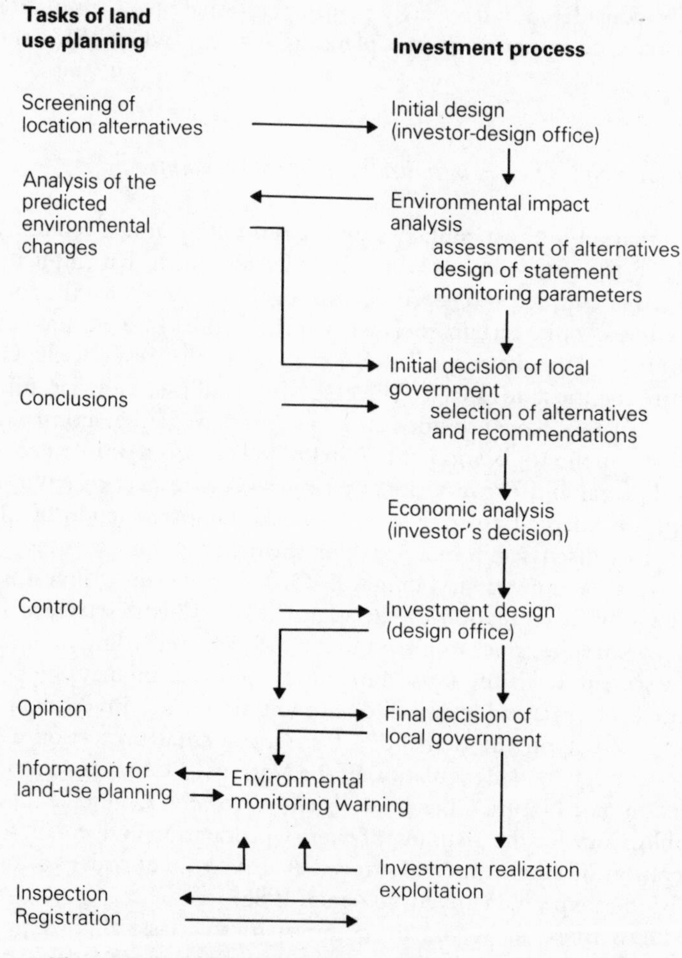

Figure 12.4 General view of the decision-making process (modified after Kotyczka 1984).

emanate from economic actions. Within the GDR, however, it is admitted that there are constraints upon how effectively this can be done. These constraints reflect a need for more detailed knowledge on, for example, long-term effects, beneficial and adverse environmental impacts of human activities, the existing flora and fauna, and the nature of the built environment. A number of studies have been started to remedy some of these deficiencies.

In summary, therefore, countries with centrally planned economies have all of the necessary formal legal conditions for EIA procedures to function. In certain countries, notably the GDR and Hungary, considerable advances have been made to gain acceptance of the principle that the approval of a particular development project shall be based, amongst other considerations, upon an assessment of its effects on the environment. Elsewhere, the comprehensive approach to the environment, reflected in recent legislation, indicates that the ecological consciousness in these countries is high enough to recognize the need for environmental impact assessment.

An interesting legal and administrative approach to environmental problems exists in Czechoslovakia (Madar 1985). First, the procedures endeavour to optimize the relationship between economic production and the environment, including protection against the consequences of human activities, particularly pollution. Secondly, the procedures address the care of both natural and man-made components of the environment. At the same time, there are individual decrees concerning protection of specific components of the environment and various aspects of its pollution.

Collectively, this legislation indicates that there is relatively comprehensive legal coverage of environmental aspects. There is, however, still no comprehensive, systematic approach to project assessment and plan making, an important attribute of EIA. Indeed, Czechoslovakian specialists themselves point to the need for the system to satisfy certain fundamental criteria. First, legal rules are required to codify the environment completely. Secondly, the system must also accommodate the complex interactions between individual components of the environment. Thirdly, the system should be able to react quickly to any factor likely to influence the state of the environment. Finally, the system for safeguarding the environment has to define accurately the competence of individual authorities, organizations and responsible persons (Madar 1985).

There can only be one conclusion from such statements. The introduction of EIA procedures in socialist countries, together with all of the necessary legal and organizational changes, would be appropriate given the present high environmental awareness, the current wide consideration of environmental aspects by the legal system and the appreciation that a different, more comprehensive approach to planning is required.

EIA procedures in Poland

Significant changes in the integration of environmental considerations into economic activity within Poland occurred in 1983 with the establishment of a special ministry, the Office for Environmental Protection and Water Management (OEPWM). The responsibility for environmental protection has been transferred to OEPWM from the Ministry of Administrative Territorial Management and Environmental Protection (MATMEP) which, in turn, has become the Ministry of Administration and Spatial Management. Consequently, responsibilities formerly dispersed amongst a few ministries, in particular MATMEP and the Ministry of Agriculture, have been concentrated in a specialized ministry which will have no other responsibilities apart from environmental protection and water management.

Besides the emergence of OEPWM, two legal acts are of particular importance for the introduction of EIA procedures. These are the law on protection and development of the environment and the law on land-use planning. Article 20 of the law on environmental protection states that 'the state administration responsible for environmental protection can demand from the investor presentation of the opinion drawn up by an approved expert on the impacts of the project . . . on the environment'.

The July 1984 law on land-use planning imposed a requirement that both OEPWM and the Main Sanitary Inspector must agree on the location of investments which have adverse effects upon the environment and human health before they can be approved. This requirement extends not only to the consideration of alternative variants of the proposal, but also to the subsequent decisions establishing the location of the preferred alternative. For projects determined by the law as particularly noxious, assessment of environmental impacts must be undertaken by an expert from a panel indicated by the ministry. A decree of the OEPWM Minister issued soon after the law was enacted defines the criteria by which a project is defined as particularly noxious and, therefore, subject to an assessment of its impact on the environment. These criteria are listed in Table 12.2.

These legal and organizational changes in the environmental management system coincided with work on the application of EIA in Poland. First, the EIA approach was applied in the assessment of development plans within two pilot areas, in Upper Silesia and in the Legnica–Głogów copper mining region. The project, carried out by the Institute for Environmental Development (IED) in Katowice, was sponsored by the World Health Organization. The results of the study (IED 1983) clearly indicate the valuable experience that was gained by the Polish research workers concerning the theory and practice of EIA. An account of these studies is given in Janikowski & Starzewska (1986). One outcome of this project has been the development of recommendations for the introduction of EIA into the Polish planning and decision-making system.

The work at IED on this project, led to the formation of a group of well-trained and experienced people who have since started the introduction of the EIA approach to development appraisal at various scales. The approach has been used subsequently for a number of developments including a major ore mining and processing development, a regional development complex, several sewage treatment plants, steelworks and power plants. For each case study, the

Table 12.2 Criteria for triggering an assessment of the environmental impact of development proposals in Poland.

Emission of atmospheric pollutants:
- (a) > 5000 tons/year dust and gases to area immediately surrounding installation, or on to specially protected or ecological hazard area.
- (b) > 20 000 tons/year dust and gases to other areas.

Discharge of liquid wastes requiring treatment:
- (a) > 2000 cu. m/day to flowing water or > 1000 cu. m/day if ecological hazard or specially protected area.
- (b) > 100 000 cu. m/day to Baltic Sea, reservoirs or ground water.

Waste materials produced and dumped:
- (a) > 5000 tons/year of specified noxious waste or > 20 000 tons/year of other noxious waste on areas of ecological hazard or specially protected area.
- (b) on other areas, > 10 000 tons/year of specified noxious waste or > 100 000 of other noxious wastes.

Deterioration in water relations of lands with particular socioeconomic value.
Noxious impact on soils, agriculture or forest plantation over an area > 50 ha.
Abstraction of water:
- (a) > 40 000 cu. m/day from surface water.
- (b) > 4800 cu. m/day from underground sources.

Power supplied by transmission lines > 400 kV.
Noise generated above limits specified by Cabinet decree.
Electro-magnetic fields in the frequency 0.1–300 000 MHz at intensities above limits specified by Cabinet decree.

role of the group appointed to advise on appraisal was to provide a scientific consultation service, rather than to complete the routine work concerned with project assessment. At the same time, the organizational framework for routine preparation of EIA is just being developed; an EIA review bureau is likely to be established in the near future.

Intensive activities aimed at familiarizing Polish planners with EIA procedures are being undertaken in the form of lectures, publications and training courses. Participation at seminars abroad is an important way of learning from the experience which has been gained of EIA elsewhere. Simultaneously, the leading scientific group of IED workers is developing the most appropriate methods and techniques for use in the appraisal of future projects, which may be commissioned by investors. This scientific work is also making a significant contribution to increasing the store of practical and research experience. The results are being published and utilized, amongst other things, for lecturing on the training courses mentioned above.

Conclusions

The adoption of EIA in the socialist countries of Eastern Europe seems the logical extension of the evolution of measures aimed at nature conservation and environmental protection. As the consequences of environmental degradation resulting from rapid industrialization become apparent, the adoption of EIA seems to be inevitable. Almost certainly, it will soon be applied, at least, for the most noxious projects. Increasing consciousness of the need for an integrated approach to development and the environment and, in particular, for EIA is reflected by the recent dynamic evolution of environmental legislation.

The establishment of specific EIA procedures, however, would require a large staff of devoted and efficient workers who must be thoroughly trained. The next few years will see the realization of a comprehensive training programme based mainly upon the Polish experience of EIA. The training programme will also be able to draw upon the experience of systems analysis methodologies which have been developed at various scientific centres within the socialist countries, particularly in Czechoslovakia.

13 *The EIA process in Asia and the Pacific region*

NAY HTUN

Introduction

There is a significant and increasing awareness of environmental problems in Asia and the Pacific region. This is underscored by the fact that, during the past decade, most countries in the region have established institutional mechanisms to protect the environment, by setting up specific ministries, offices, or departments with environmental responsibilities. The media have also played an important role in heightening public awareness. Newspapers regularly and frequently carry local and international articles on environmental issues as do radio and television. Major environmental events such as the Ixtoc 'blow-out', the sinking of the *Amoco Cadiz* and the Bhopal and Chernobyl accidents were reported speedily and communicated to the homes of the vast majority of the population. The 'only one earth' perception is gaining ground with the advent of the age of telecommunication, so much so, that such incidents raise concerns as if they were happening nearby and not events occurring in far-off places in other parts of the world.

With this growing awareness, there is an increasing realization that the potential impacts of proposed development activities need to be assessed, so that appropriate mitigating measures can be adopted. Furthermore, the concept that the environment and development can be mutually enhancing and do not inherently conflict is beginning to gain ground. In this context, the environmental impact assessment (EIA) process is seen as a means not only of identifying potential impacts, but also of enabling the integration of the environment and development.

While EIA is being increasingly applied in the region, this is still a relatively recent phenomenon. Only within the last decade have countries been concerned with EIA. The major focus throughout has been at the project level, with attention paid, primarily, to techniques and tools used in assessment. There has been relatively little investigation of the whole EIA process. Many of the EIAs that have been carried out have attempted to identify and predict the potential effects that might result from the proposed activities. Very few have attempted to consider what would be the net positive and negative impacts of these effects on human health and welfare and on the environment.

Another observation on these EIAs is that in many studies, particularly the earlier ones, the major emphasis was on assessing effects on the physical and natural environment with relatively little consideration given to social, cultural and economic aspects. When included, these components have been studied by separate teams of experts and their results, generally, have been segregated from other aspects in the report. There have been few, if any, attempts to integrate physical and natural studies with socioeconomic considerations.

This chapter contains a review of experience of EIA within Asia and the Pacific region. This discussion places the main emphasis on: the range of legislative and non-legislative means that have been adopted for instituting a system of EIA; practical experience of different facets of the EIA process; and prognoses for the future development of EIA in the region.

Environmental machinery

In most countries in the region, the major components of the machinery for managing and protecting the environment are beginning to be put into place. This has greatly facilitated the introduction and application of EIA. Institutionally and organizationally, most countries have established specific divisions, sections, or groups responsible for EIA within an appropriate organ of the central government administration.

The main functions of these organizations are to: provide terms of reference or delineate the scope of an assessment; review and comment on an assessment report; and finally to co-ordinate and liaise with the project proponent and, sometimes, with the study team. Only very seldom are these organizations empowered to carry out an EIA study. As an increasing number of bilateral and multilateral funding sources require EIA studies for proposed development loans and aid projects, significant impetus has been given to the institutional machinery within countries receiving such assistance.

EIA LEGISLATION

An increasing number of countries have enacted legislation that specifically requires an EIA study to be undertaken and approved before a project can start. The criteria that determine whether an EIA is required are generally based upon: the nature of the potential impacts; the type of project or activity; size; and the location of the proposal. The desire to establish legislative requirements, yet at the same time to avoid judicial involvement in the procedures to the extent that has occurred in the United States, is noticeable in most of the EIA legislation enacted in the region.

A number of countries have enacted specific requirements for EIA and have established criteria to be used in deciding whether EIA should be undertaken. These are detailed below.

Australia: The purpose of the 1974 Environmental Protection (Impact of Proposals) Act is to ensure that, to the greatest practicable extent, matters significantly affecting the environment are fully examined and taken into account. To rationalize the application of the Act, a 'Memorandum of Understanding', executed by the department responsible for environmental matters in conjunction with the relevant minister and public authorities, is being formulated to arrive at an 'understanding' on the type of proposals which, normally, would be classed as 'significant'. The memorandum, however, lacks any legal status.

Indonesia: Article 16 of the 1982 Act Number 4 concerning 'Basic Provisions for the Living Environment' urged the adoption of an EIA study for any proposed activity that will have significant impacts upon the environment. On 5 June 1986, Regulation PP No. 29/1986 on Environmental Impact Assessment was issued. The regulation requires the project proponent, in the first instance, to conduct an environmental assessment (PIL) based upon guidelines prepared by the Ministry of Population and Environment. After evaluation of the PIL a decision is made as to whether a full EIA (called an *Andal*) is needed. For projects which are likely to have significant impacts, proponents may elect to proceed directly with the preparation of an Andal instead of a PIL. Proponents are also requested to prepare an environmental management plan (RKL) and an environmental monitoring plan (RPL).

Republic of Korea: The 1979 Environmental Conservation Law, revised and enacted in April 1983, incorporates provisions for EIA. Since March 1981, government agencies and government-funded institutions implementing developments related to: energy; water; apartments and tourism complexes; the construction of industrial areas or complexes; ports; roads; railways; airports; reclamation and dredging have to prepare an environmental impact statement (EIS) and submit it to the environmental administration. These provisions do not apply to developments in the private sector.

Malaysia: The Environmental Quality Act of 1974, as amended in January 1986, requires EIA for all projects that will have major environmental impacts. The amendments apply to both private and public sector projects. Specific categories of projects that will need an EIA are being drafted.

Pakistan: The 1983 Ordinance No. XXXVII requires every proponent of a project which is likely to adversely affect the environment, to file a detailed EIS with the Pakistan Environmental Protection Agency at the time that the project is planned.

Papua New Guinea: The fourth of the five goals of the constitution recognizes environmental responsibility: thus, 'We declare our fourth goal to be for Papua New Guinea's natural resources and environment to be conserved and used for the collective benefit of us all, and to be replenished for the benefit of future generations.' The 1978 Environmental Planning Act enables the minister, if in his opinion a proposal may have significant environmental implications, to serve a requisition on the proponent to submit an environmental plan.

Philippines: The 1977 and 1978 Presidential Decrees 1121, 1151 and 1586 formally established not only the requirement for EIA, but also a system for the preparation of EISs. Proclamation No. 2146 of 1981 identified a number of areas and types of projects as being environmentally critical and, thus, falling within the scope of the EIS system. First, environmentally critical projects, comprising specific named industrial sectors and project types within the categories heavy industry, resource extractive industry and infrastructure provision, have been determined. Secondly, environmentally critical areas consisting of eleven different types of locality including watersheds, recharge areas for aquifers and mangroves have been designated.

Sri Lanka: The 1980 National Environmental Act enables the authorities to require an EIA to be carried out for all public and private development projects. In January 1984, this was made mandatory. In 1986, the Act was strengthened with the result that EIA should now be conducted for all major impacts on the environment.

Thailand: The 1975 Improvement and Conservation of National Environmental Quality Act, as amended in 1978, established the legislative framework for EIA. It provides the powers necessary to issue notifications prescribing categories and magnitudes of projects or activities for which an EIA report is required for consideration and approval prior to the commencement of a project. The first notification, issued in 1981, contained the ten project categories listed in Table 13.1

Table 13.1 Categories of projects subject to EIA in Thailand.

Dam or reservoir (storage volume greater than 100 000 000 m^3 or storage surface area greater than 15 km^2).
Irrigation (irrigated area greater than 12 800 ha).
Commercial airport.
Hotel or resort facility in environmentally sensitive areas such as adjacent to rivers, coastal areas, lakes and beaches or in the vicinity of national parks (greater than 80 rooms).
Transit systems planned and operated by Expressways and Rapid Transit Authority of Thailand (ETA).
Mining.
Industrial estates.
Commercial port and harbour (capacity for vessels greater than 500 gross tonnes).
Thermal power plant (capacity greater than 10 MW).
Industrial plants:
Petrochemical (greater than 100 tonnes/day raw materials required in production processes of oil refinery and/or natural gas separation);
Oil refinery;
Natural gas separation or processing;
Chlor-alkali industry requiring NaCl as raw material for producing Na_2CO_3, NaOH, HCl, NaOCl, and bleaching powder (production capacity for each, or combined, greater than 100 tonnes/day);
Iron and/or steel industry (requiring iron ore and/or scrap iron raw materials greater than 100 tonnes/day or using furnaces with combined capacity of greater than 5 tonnes/batch);
Cement;
Smelting other than iron and steel (greater than 50 tonnes/day production capacity).
Wood pulp (greater than 50 tonnes/day production capacity).

OTHER EIA PROVISIONS

Other countries in the region have adopted a range of EIA provisions which, as yet, have not been codified into statutory requirements. The measures which have been adopted are reviewed below.

China: The 1979 Environmental Protection Law Articles 6 and 7 provide the basis for EIA requirements. Within the People's Republic of China a number of development types, namely: industrial projects; mining; irrigation works; port construction; development of large and medium-sized cities and regional development activities, can be subject to the preparation of an EIA.

India: The 1977 Constitution (Forty-Second Amendment) Act, Article 48A specifically places an obligation on the state to protect and improve the environment and to safeguard the forests and wildlife of the country. Consequently, several programmes to incorporate environmental protection into development projects were included in the sixth five-year plan covering the period 1980–5. The Environment (Protection) Act, No. 29, 1986, provides the Central Government with the powers to take all such measures as it deems necessary or expedient for the purpose of protecting and improving the quality of the environment and preventing, controlling and abating environmental pollution. Any major project requires the approval of the Planning Commission and the preparation of a review report by the Department of the Environment. Similarly, major industrial developments in the private sector which require federal government approval, will only be granted an installation licence after review by the Department of the Environment. Guidelines and checklists for undertaking EIA have been issued for hydroelectric, irrigation, thermal power generation, industrial development, harbour, mining, rail and road construction projects. Draft guidelines are also being prepared for the environmental assessment of new towns and for the planning of military facilities.

Japan: The context for EIA in Japan was set in August 1984, when the Cabinet issued a decision on the implementation of EIA. This was followed in November 1984 by the promulgation of the so-called Principles for Implementing Environmental Impact Assessment by the Environmental Agency. Under the terms of these provisions, EIA is required for the projects listed in Table 13.2

Nepal: In 1982, the Environmental Impact Study Project of the Department of Soil Conservation and Watershed Management was established to prepare some studies on EIA for renewable and non-renewable energy resources.

New Zealand: Environmental law in New Zealand is not confined to one specific piece of legislation, but is found in a number of Acts relating to pollution control, resource management and the protection of species and habitats. The Environmental Protection and Enhancement Procedures set out the process of EIA that applies to projects and developments which either require statutory approval or have government involvement.

Singapore: A vetting process is used by the Ministry of Environment, in close collaboration with other government authorities, for all project proposals

Table 13.2 Projects subject to EIA in Japan (with appropriate qualifying legislation).

Construction/reconstruction of national expressway, trunk road, or other roads.
Construction or improvement of railways.
Construction of airports and changes in airport facilities.
Reclamation and dumping works.
Projects implemented by corporations to prepare land for residential development, factories, business establishments and other purposes.
Construction of dams and river management facilities, e.g. weir, watergate, revertment (prescribed by the River Law).
Land readjustment projects (Land Readjustment Law).
Residential development (the New Residential Development Project Act).
Industrial estate construction projects (various legislation covering projects in suburban development and redevelopment areas in the National Capital and Kinki Regions).
Urban infrastructure development projects (Law for Development of the Basis of New Cities).
Centres for distribution businesses (Law of Development for Business Centres).
Other similar projects specified by the competent ministry in consultation with the Director-General of the Environment Agency.

such as land development or the allocation of new industries. All project proposals involving any land development must first obtain the necessary planning approval, under the Planning Act, from the Ministry of National Development (MND). Acting as co-ordinator, the MND channels the proposal simultaneously to various relevant authorities, such as the Ministry of Environment, for comment. Planning approval will only be issued when the various authorities give clearance. Similarly, building plans need to be vetted and cleared. As in land development, planning and building approval must be obtained for factories. In addition, there are further controls on proposed industrial activities through zoning, as well as vetting on possible water and air pollution.

EIA practice and application

Most of the legislation, implicitly or explicitly, requires the application of the EIA process at the planning stage so that environment and development can be integrated. In practice, however, the process is normally applied when decisions on a certain set of project options such as siting, raw material requirements and the type of processes to be employed, have been made. It is not surprising, therefore, that the findings of an EIA study, generally, tend to endorse the options chosen and attempt to allay concerns that the potential environmental impacts, as originally suspected, are minimal and that the benefits of the project outweigh the damage costs. Thus, EIA is not part of the overall project planning process. Furthermore, EIA usually takes place at the project level. As yet, there have been very few programme or plan EIAs, although the appraisals for the Songkhla Lake in southern Thailand and the Eastern Seaboard development activities in the eastern part of the country, for example, were attempts to do so.

One means of promoting greater integration would be to require that proposed development plans for a country should be subject to an EIA at the formulation stage. This, however, is not yet part of current practice, although a few countries such as Thailand, Malaysia and Indonesia have included a specific section dealing with the environment in their national development plans.

INITIAL ENVIRONMENTAL ASSESSMENTS

Several countries in the region have instituted initial environmental examination or preliminary environmental assessment procedures. The basic aim of these provisions is to screen proposals to determine whether a full-scale EIA is required. If an EIA is not needed, the initial assessment will form the basis for a decision. Such initial assessments are widely practised and encouraged in Thailand and the Philippines. They have been found to be low-budget evaluations, based on readily available information or on professional judgement. In Thailand, as many as 50 initial assessments per month have been carried out. Some are produced and reviewed quickly, while others necessitate field trips and studies. In the Philippines, a matrix which rates impacts and project activities is also prepared.

Initial assessments are normally carried out by the government authority responsible for the environment. In the event that a full-scale EIA is required, the results of the initial assessment are used to determine and identify key issues that merit further detailed study. Throughout the region, full EIA studies are usually carried out by experts commissioned by the project proponent.

REVIEW OF EIA STUDY

Only a few countries in the region have legislation which clearly specifies the details of the review process. Thus, in Thailand the National Environment Board (NEB) reviews an EIA study within 90 days of submission and forwards its considerations to the authorizing agency and to the project proponent. If the NEB rejects the study the proponent revises the EIA, which is reviewed by the NEB within 30 days, with its final deliberations again being forwarded to the agency. Similarly, in the Philippines, all agencies having jurisdiction over, or special experience of, the subject matter of an appraisal must comment upon a draft EIS within 30 days.

Mandatory provisions do not necessarily guarantee a public input into the review process. In Australia, for example, the minister, in determining whether to make public a draft EIS, is required to 'take into account any views expressed by the action Minister or the responsible authority on whether the draft environmental impact statement, or any part of it should be made available for public comment' (Fowler 1982). No other factors upon which the minister might base his decision are specified. This provision could prompt a negative approach towards publication, since the attitude of the action minister or responsible authority (in effect the proponent) could normally be expected to be against publication. When the EIS is made available for public comment, the procedures require that a notice must be lodged in the official Gazette and 'in such newspapers and on such occasions that the Department approves'. Any written comments from the public must be supplied to an address given in the notice within a specified period of 'not less than 28 days'. Copies of any comments must be supplied to the department and the proponent within 7 days of expiry of the period for public comment.

The review of an EIA study is normally done by the government agency with responsibility for the environment, taking into account comments provided by other competent agencies as well as by outside experts retained to review specific aspects. In Malaysia, however, the review is conducted by an independent, multidisciplinary, review panel which provides recommendations to the authorizing authority.

In Japan, the draft EIS is made available for public review for a period of one month. During this period, the proponent should hold explanatory meetings about the project within the area likely to be affected. This is not a mandatory provision as it may not be possible to arrange such meetings for reasons beyond the control of the proponent. In such circumstances, the proponent must make efforts to publicize the draft EIS by other means. Should residents of the project area wish to comment upon the draft in writing, they are given a further two weeks beyond the end of the review period in which to do so. The proponent must summarize all comments and forward them to the relevant prefectural governor and mayor for further comment, which must be given within a three month period. When these comments have been received, the proponent must prepare a final EIA, whereupon, the cycle of review and comment is repeated. Subsequently, the EIS is accepted by the national administration and sent to the competent ministries and agencies including the Environmental Agency. These procedural steps are shown in Figure 13.1.

SCOPE AND COSTS OF EIA

One of the most important aspects of EIA is determining the scope of an assessment which involves the consideration of temporal, spatial and technical criteria. These criteria and their respective subsets of issues are interrelated and both influence, as well as being affected by, cost considerations.

With respect to temporal factors the need to carry out the study over at least a one year period is generally considered necessary so that seasonal influences can be discerned. The expected persistence of an effect is another time-related factor which should be taken into account.

The geographical size of a study area depends upon a number of factors, including, for example, the dispersion pattern of potential pollutants, the relative importance of the primary, secondary and tertiary impacts, and the location of any unique ecological, archaeological and cultural sites in the vicinity of the proposed project. Administrative boundaries are also known to be determining factors in delineating the spatial scope of an appraisal.

Technical criteria have probably been the major consideration in defining the scope of an EIA undertaken within the region. The nature and types of, for

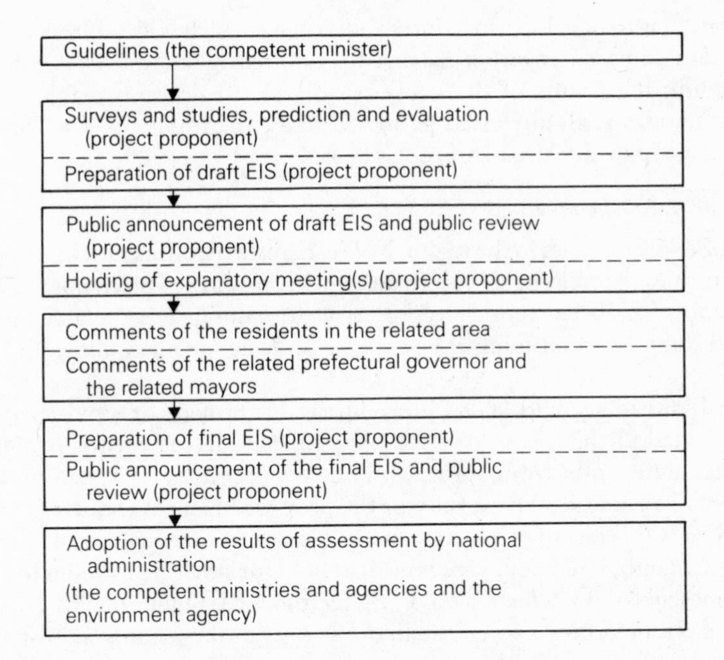

Figure 13.1 EIA procedures in Japan.

example, air pollutants, flora, fauna and social indicators that should be assessed are generally the first sets of parameters that have been taken into account in determining the scope of a study.

The level, number and type of expertise that is needed for an assessment depends upon its scope. Hence, it is clear that the cost of an EIA is also related to the scope of an assessment. Most, if not all, developing countries lack reliable baseline data which is one of the major impediments to the adoption of a more empirical approach for determining the scope of an EIA. Generally, expediency and the availability of financial resources for carrying out an assessment are the main determinants. Some estimated costs of EIA in Thailand, taken from UNEP (1982), are shown in Table 13.3.

While the industrialized countries have developed procedures and guidelines for scoping, these are still lacking in the developing countries in the region. The

Table 13.3 Estimated cost of EIA studies in Thailand.

Project	*Project cost (US$m)*	*EIA cost (US$m)*	*EIA cost % of total*	*Type of study*
Pa Mong Hydro Power & Irrigation	300	0.89	0.30	EIA
Songkhla Industrial Estate	18	0.087	0.48	EIA
Natural gas pipeline	434	0.049	0.14	EIA
Pulp mill	75.5	0.026	0.034	IEE

major need is to develop procedures for scoping which will: address potentially important impacts; ensure a balanced, comprehensive, detailed and optimal study; link the results of the study to the related decision making; provide opportunities for all interested parties to make their opinions and viewpoints known; and expedite the EIA process.

INCLUSIVE/EXCLUSIVE LISTS FOR EIA

A number of countries in the region have established notification lists of projects and activities that will be subject to an EIA. According to Snidvongs (1985) this procedure has been found to be easy to administer, flexible and least cumbersome for modifying the scope of projects included under EIA provisions.

The disadvantage with such a procedure is that projects or activities which are not included in the list, but which, nevertheless, have potentially significant impacts, are exempt from appraisal. A case in point was a tantalum smelting plant in Thailand. As the capacity of the proposed plant was only 6 tonnes per day (even at this rate of production it was capable of meeting over 40 per cent of world demand), it did not exceed the threshold for smelting facilities (more than 50 tonnes per day) which would trigger the EIA requirement. There were massive public protests when the first test runs of the plant were just about to commence. Eventually, these protests became politicized and led to the mob burning down the new smelter. It has been surmised that, if an EIA had been prepared at the project planning stage, the incident would not have occurred. An EIA would have addressed the opinions and concerns of the local people and would have included a consideration of alternative site locations.

ROLE OF THE PUBLIC

The role of the public is not generally defined in most countries in the region, exceptions being Japan, Australia and New Zealand. In this context, it is useful to differentiate between 'involvement' and 'participation', although these two terms are frequently, but mistakenly used synonymously. Public involvement should be seen as being concerned with technical issues, since the intention is to improve the effectiveness of an EIA study by providing the means and the opportunity for gathering information from, and exchange views about the perceptions and concerns of, the public likely to be affected by the proposals. There are also other, equally important, reasons for widening the role of the public in EIA, because of the need to anticipate potential conflicts of interest. The opportunity for views to be represented and for the assessment study process to be scrutinized by the public may serve to improve understanding and acceptability not only of the assessment results, but also the final decision on the proposal. Since the state of the environment has become a matter of growing concern, the public increasingly wants to know about the issues that have to be examined and the process used for that examination (Nay Htun 1984).

Public participation takes the process one or more steps further. It implies that the public participates in the decision-making process both with regard to the review of the results of an EIA study and with respect to the decision on whether the proposal should proceed based upon those results.

The role of the public clearly depends upon the institutional and political structures existing within a country. There is an increasing consensus that the technical aspects of public involvement are useful. The main issue to be resolved, however, is to determine how the public can be involved in a constructive and meaningful manner. The mechanisms used are seen to provide genuine opportunities for the free and frank exchange of views that will be considered and reflected in an assessment.

METHODOLOGIES

A variety of methodologies such as checklists, impact matrices, networks, overlays and simulation models have been used in the region. Checklists have been found to provide for a rapid assessment of potential impacts, particularly for initial environmental evaluations. Full EIAs have then been carried out on those impacts expected to be significant. In the Philippines basic checklists for describing the proposed project and for compiling the description of environmental conditions are available. Various checklists have also been prepared for rapid assessment of development projects in Thailand.

Impact matrix approaches have been found to provide a useful visual summary of the impacts which can help in communicating the results of a study. An additional advantage is that a high degree of training and expertise is not required in order to carry out an assessment using this approach. It was employed, for example, for the Kujang fertilizer project in Indonesia. This experience indicates that the matrix, like all other methods, has certain defects. The most serious deficiency is that it does not bring out the intermediate relationships that exist in complex systems.

The network method was used for the Saguling hydroelectric power plant in Indonesia. The investigators found the method to be flexible and considered that it was able to show secondary, tertiary and subsequent impacts for each development phase. In Japan, the Principles for Implementing Environmental Impact Assessment recommend that pollutant impacts are predicted from calculations based on simulation models, simulated experiments and references to previous cases.

All of the methods mentioned above have certain advantages and disadvantages. Lohani & Halim (1982) reviewed the utility of these methods and their resource implications when used for EIA studies within the region. This experience has been brought up to date in a recent publication by the United Nations Economic and Social Commission for Asia and the Pacific (ESCAP 1985).

There are two major shortcomings in the methodologies that have been used. The first is that there are, as yet, no procedures or techniques for the systematic integration of physical and other natural components of the study with the social and economic aspects. The second concerns the very limited experience in the application of the EIA process at the programme and plan levels. Nearly all of the studies have been undertaken for specific projects.

It is clear that one of the major underlying factors contributing towards methodological problems is the lack of reliable baseline data. This has made objective evaluations of possible changes in environmental quality, at best, difficult and, often, impossible. This limited baseline data also has major cost implications. Acquiring baseline data is generally one of the most expensive and time-consuming activities in the EIA process.

PRESENTATION OF RESULTS

With growing experience in the application of EIA in the region, improvements in the presentation of assessment results, so as to aid decision making, are beginning to be made. The most important issues relate to the format, structure and content of EIAs; suggestions for mitigating measures; and proposals for alternatives.

With regard to the first issue, the utility of a number of excellent comprehensive assessments has been significantly reduced because of shortcomings in communicating the results of an EIA study to decision makers and the media, as well as to the general public, in a clear, concise, meaningful and useful format. From the studies that have been undertaken, various deficiencies and their likely causes can be identified.

Study teams normally consist of experts with specialist knowledge in science, technology and engineering. Very seldomly do the study teams include a person with expertise in writing and audio-visual communication. Most of the studies conducted in the developing countries within the region are designed to meet the requirements of foreign investors, as well as bilateral and multilateral funding and aid agencies. Hence, the study results are usually written in English. While many of the experts undertaking the assessment read and understand English well, they do not have sufficient command of the language to enable them to write clearly and succinctly.

Often, insufficient thought is given at the outset of a study as to who will be the target user of the results, so that the expected contents and presentation can be planned and structured accordingly. Finally, study reports are generally not widely available or are circulated for review and comment to only a restricted group of experts with detailed knowledge of the various aspects of the proposal. Without the opportunity, or the requirement, for open scrutiny by peers and by the informed public, there is a tendency to present the results in a highly technical and specialized form.

With respect to the second major deficiency, it is clear that most EIA study reports are voluminous documents containing long descriptions of the project and detailed accounts of the location and the condition of the local environment with inadequate consideration of mitigation measures. Any recommendations concerning mitigation are generally confined to 'end-of-pipe' pollution control measures and not preventive approaches based on either the application of resource conservation and recycling concepts or the reuse of residues. As experience has demonstrated that preventive measures are more cost-effective than corrective solutions, the inclusion of such advice will enhance the efficacy of the EIA process.

The third shortcoming is that very seldom do the studies include advice on alternatives with regard to, for example, siting, raw materials, processes and finished products that could have less adverse impacts on the environment and utilize resources more efficiently. Decision making is improved when there are various options, with details of their respective environmental and natural resource implications, from which to choose.

MONITORING AND AUDITING EIA STUDIES

There are very few countries in the region that have specific mechanisms for monitoring and auditing environmental impacts. Thus, there are no provisions which require the quality of those aspects of the environment and natural resources included in an EIA to be systematically monitored after the construction phase or while the activity is in operation. Consequently, there is no way of assessing whether the predicted effects and impacts actually occur. Similarly, there are very few institutional mechanisms designed to determine whether the recommendations made in an assessment study to reduce, mitigate and prevent potential impacts were implemmented by the project proponent.

Only through monitoring and auditing can experience be consolidated. Incorporating such mechanisms into the EIA process will not only strengthen it, but will also enable the various procedures, techniques and methodologies used for identification, prediction and evaluation to be improved.

Evolving trends in the EIA process

There has been a significant increase in the use of EIA within Asia and the Pacific region during the last decade, with a major acceleration occurring in the last five years. As a rough estimate, probably a few thousand EIAs have been carried out. Valuable lessons have been learnt from this experience and there is now an emerging core of national personnel who are beginning to have the necessary expertise in using various methodologies for identification, prediction and evaluation. The frustrations and difficulties of working in a multidisciplinary team are also beginning to be accepted and overcome.

The constraints that reduce the benefits that can accrue from using the EIA process are also increasingly being realized as a result of this experience. This was evident from the conclusions of a workshop concerned with assessing the application of EIA within the countries of the Association of South East Asian Nations – ASEAN produced in collaboration with the Carl Duisberg Gesellschaft – CDG (ASEAN-UNEP-CDG 1985).

A number of considerations are being addressed which should improve the efficiency of the EIA process. These include a critical review and assessment of the various components of the EIA process not only to show where the constraints in the system are, but also to indicate the mitigating measures which could be adopted to overcome them. Furthermore, it is recognized that increasing application of the EIA process during the early phase of project identification and planning, as well as the extension of the EIA process to the programme and plan levels of development, are still needed within the region.

There are several methodological shortcomings which should be improved. Developing methodologies and procedures for integrating natural and physical aspects with social and economic components in assessment studies is one of the priority needs. Baseline data availability remains a recurrent problem in the region. Systematic and structured collection and collation of information would not only improve the availability of baseline data, but also aid prediction and evaluation as well as making more cost-effective use of resources. The need for better data management is recognized, particularly in view of the increasing use of remote sensing and computerized information storage and retrieval systems. In order to improve the utility of EIA both as an environmental management tool and as a means for identifying the interaction between development and the environment, incorporating auditing and monitoring mechanisms within the EIA process is seen to be an important requirement.

There is increasing recognition of the need for establishing efficient and effective procedures for reviewing the results of EIA studies. A major consideration in this respect is a greater recognition of the importance of, and the need for, communicating EIA results in clear and concise language in a useful and understandable format. This should facilitate greater involvement of the public, particularly the informed public, in defining the scope of an assessment as well as in evaluating and reviewing the results of a study in order to facilitate decision making for sustainable development.

Note

The views expressed in this chapter are those of the author and do not necessarily represent the policies of UNEP.

14 *EIA in Latin America*

I. VEROCAI MOREIRA

Introduction

Like Asia or Europe, Latin America is not a homogeneous region in many respects. The name Latin America itself conjures up a jigsaw puzzle whose pieces would be extremely difficult to fit together on account of the historical vicissitudes of these countries and the present situation which exists within them. Any attempt or intent to treat the recent advances in EIA in all Latin America together would prove to be impossible and could lead to the most base generalizations or to a repetition of the preconceived judgements found sometimes in the literature on EIA in developing countries. Therefore, this chapter is restricted to the provision of some information showing how these countries have been dealing with their environmental problems and how EIA has been adapted to suit the national situation in a few of them.

The twenty independent countries that form what is known as Latin America cover an area of 20 019 000 km^2 (7 729 344 square miles) with a population of about 300 million inhabitants. Eighteen countries are former Spanish colonies, whereas one, Haiti, was colonized by France and one, Brazil, by Portugal. The largest and most populated is Brazil comprising 8 511 965 km^2 and 130 million inhabitants. El Salvador is the smallest (20 935 km^2) and Panama is the least populated with about 1.5 million people. These countries have gradually evolved to form a diverse range on the basis of their different geographical situations, the various origins of their settlers and immigrants and their individual political and cultural circumstances.

Although their social and political organizations are at different stages of development, they share in common not only socioeconomic characteristics which are the consequence of colonization, but also the model of development that has prevailed for one and a half centuries of politically independent history. Latifundia, monoculture for export, and pre-capitalist forms of labour division are still a large component of agricultural exploitation in Latin American countries. On the other hand, it is a well-known fact that industrialization in these countries came about as a response to the economic needs of Western industrialized countries. Therefore, industrialization has been concentrated precisely in those sectors of the economy of international market interest, namely agriculture and mining, rather than in local consumption or local resource utilization. This concentration in the primary sectors of the economy eventually turns these countries into easy prey to international economic crises (Furtado 1970).

In general, the dominant classes in the region are composed of agro-exporter oligarchs, major traders and bankers allied to international capitalism. Middle classes are weak and unable to start up autonomous development programmes. Lower classes, the largest section of the population and the one that grows at the fastest rate, have very limited access to the benefits of development.

Another common characteristic of the region is the political instability resulting from the susceptibility of political leaders and military forces to manipulation by conservative classes and international capitalism (Aquino *et al.* 1984). A slow and inefficient bureaucracy that disregards the needs of the civilian population and local concerns is only to be expected in most of these countries.

The decline in environmental quality that has been observed throughout the world in recent decades has taken a particular form in Latin America. Within the same country, large undeveloped lands exist side by side with modern industrialized areas, and the negative effects of an accelerated economic growth have been added to the well-known social and environmental problems associated with poverty. This has made the environmental situation within Latin America more grievous than that which exists within the developed regions of the world. Air pollution from urban traffic, chemical and organic water pollution from unplanned urban and industrial areas, destruction of important ecosystems, contamination by pesticides and toxic substances are exacerbated by emigration from rural to urban areas lacking basic infrastructure and social services.

In this context, the measures for mitigating the harmful effects of existing activities and the recovery of depleted environmental resources are much more pressing than the implementation of preventive measures. Almost all of the technical and financial resources of environmental agencies in Latin American countries have been allocated to pollution control equipment and to regain water and air quality in order to reduce the harmful effects on, and to facilitate protection of, human health. Cubatão in São Paulo, Brazil, is one of the most famous cases of pollution which has had dramatic consequences for public health. It is also a clear example of how local environmental agencies have to devote time and resource consuming efforts to defend and assist a large and deprived population.

As well as dominating the attention of scientists and the resources of the public administration, this critical environmental situation also served to arouse public and government concern not only to the urgent need for a preventive environmental policy, but also to the relevance of procedures for its implementation. Thus, since the mid-1970s, preventive regulations have been included in the legal and insitutional reorganization promoted for environmental purposes in many Latin American countries (Ballesteros 1981, 1982).

As has happened in other developing regions, the initial demand for EIA in Latin America came formally from development aid agencies. The past and present role of these agencies is the central theme of another section of this book. However, the subject deserves a few comments on the response of some Latin American governments to external financial help and technical assistance for EIA provided in close co-operation with international organizations such as the United Nations Environment Programme (UNEP) and the Pan-American Health Organization (PAHO).

The external support for EIA was crucial and represented almost the triggering initiative for the emergence of government and public environmental awareness in Latin America. It has been a powerful influence on the establishment of environmental policies and laws, the enhancement of scientific capabilities and the mobilization of community environmental interests. It is important to consider, therefore, the reasons why these efforts have not sufficed to make EIA a fully adopted policy instrument in these countries and implemented, not only for development projects dependent upon external aid, but also for those requiring internal decisions.

Much has been said in the literature about the application of EIA in developing countries, the different approaches adopted, the causal factors responsible for these efforts and their uneven results – see, for example, the special issue of EIA review on developing countries (Wandesforde-Smith *et al.* 1985). Certainly there is still a need for external co-operation and technical assistance in terms of either advice or financial help for baseline data collection; environmental monitoring programmes; the improvement of methods to meet local needs; institutional development; and different forms of training for government officers, politicians and representatives of social groups. However, other factors which cannot be influenced easily by external actors are also important. These include, for èxample, the inability of 'political will' to internalize EIA into the planning and decision-making system, as well as administrative impediments to, and the difficulties involved in promoting, a suitable interaction among sectoral government authorities responsible for project and programme approval and implementation.

At this point, the democratic character of EIA must be stressed. As an environmental policy tool, a legitimate EIA process demands, in addition to the factors already cited, a broad participation of social groups in decision making which implies the free availability of information, as well as full discussion of a proposal and its likely effects.

One of the strongest obstacles to the institution of a comprehensive EIA process in certain Latin American countries is the authoritarian character of their governments which are neither concerned with the democratic management of environmental resources nor willing to make known or discuss the development actions they have decided to undertake. For projects dependent upon financial aid, assessments of limited scope, preferably performed by foreign experts ignorant of the social relevance of environmental components likely to be affected, are a hypocritical means of meeting formal requirements without any commitment to the adoption of internal EIA procedures.

In the few Latin American countries where EIA has already been legally instituted, the effectiveness of enforcement and the delays in implementation are also related to the degree of centralization of power and the extent of societal democratization. Advances in the application of EIA,. as measured by increased population representativeness and environmental awareness have been accomplished.

The performance of scientific groups concerned with the environment and the economy in Latin America, though a factor of a different sort, is also relevant to the appraisal of EIA improvements. At an early stage, these groups started to discuss policy strategies for promoting sustainable economic development consonant with adequate use of environmental resources. Theoretical discussions have passed from the broader questioning of development models and their environmental consequences, to a more specific debate on how environmental concerns could be incorporated into the development of Latin American nations. The main consensus is that environmental considerations should be one more issue to be introduced into economic and planning systems. This means that global evaluations of development tendencies, followed by reorientation of planning at national, regional and local levels towards environmental protection objectives, would be more advisable than the adoption of EIA, which is envisaged as a conventional tool for the appraisal of specific projects and programmes more suited to developed countries (Giglo 1982).

On the other hand, the internal efforts to institute and implement EIA have been supported by criticisms of planning experience in Latin America. A number of well-devised plans have not been put into practice because of, amongst other reasons, a poor understanding of economic and social structures and processes, the lack of clear objectives and priorities and the discontinuity of public administration.

Of course, the best strategy to guide decision making in order to accomplish environmental goals depends on both institutional opportunities and the political context that is operating. Thus, EIA, as well as any other available environmental policy instrument should be considered equally to suit national and local situations. In Latin America, the fact is that prejudgement against EIA may limit scientific contributions and restrain the enhancement of procedural and conceptual skill. Most Latin American countries have not yet formulated national environmental policies or promoted integrated environmental legislation. Legal provisions for specific environmental aspects such as air quality, sanitation, forest management, wildlife protection and water resources can be found in a series of enactments under the responsibility of appropriate government authorities in many countries. Some countries made early attempts at environmental management, but without much subsequent progress. This is the case, for instance, with the Dominican Republic, where the government created a commission to analyse environmental pollution in 1972, but has made little subsequent progress. Similarly in Panama, there has been no further initiative towards the creation of a coherent legal and administrative system for the environment even though the political constitution of 1972 established the principle of environmental management by declaring that the state is responsible for the conservation of ecological conditions in the country, for preventing environmental pollution and for the maintenance of ecosystem stability, in full harmony with national economic development (Azuela 1982).

Other countries, although lacking specific regulations, have constituted new administrative units or organizations to deal with environmental management. In this way, Nicaragua seems to be developing integrated environmental management efforts. As part of the revolutionary process, the institutional reorganization which is under way has considered the environment a priority of national development, emphasizing the direct management of natural resources rather than the creation of a legal framework. From the 'Statute on the Rights and Guarantees of Nicaraguans' (Estatuto sobre Derecho y Garantías de los Nicaraguenses) of 1979, the state is obliged to adopt measures for the enhancement, in all respects, of labour and environmental conditions (Azuela 1982). Since then, however, nothing has been added to the existing laws. Important advances have been made with the creation in 1979 of the Nicaraguan Institute for Natural Resources and Environment (Instituto Nicaraguense de Recursos Naturales y Medio Ambiente – IRENA), an autonomous organization to promote integrated management and rational exploitation of environmental resources. IRENA is responsible for: co-ordinating all actions, plans and projects related to the use of natural resources and the generation of energy, promoted by a number of agencies; the formulation and implementation of a global plan for management and conservation of natural resources and the environment; promotion, development and co-ordination of related research; formulation and recommendation of norms and procedures for the regulation and approval of all actions that affect natural resources and environmental quality (Ballesteros 1982). In recent years, IRENA staff have been working hard to accomplish these demanding responsibilities. Though advances have been slow, the general belief amongst local professionals is that as the revolutionary process progresses, they will be able to implement efficient mechanisms to prevent environmental degradation and to promote sound development.

In the remainder of this chapter, a number of other countries are considered, systematically, in more detail. Special attention is focused on mechanisms for environmental protection and experience of EIA-related activities. Uruguay and Peru have no system of mandatory EIA, so these countries are discussed briefly first. The remaining countries are dealt with in a chronological order related to the introduction of mandatory EIA. Argentina is dealt with amongst this group as current legislation is likely to make EIA mandatory in the immediate future.

Uruguay

In Uruguay, the institution responsible for environmental control is the National Institute for Environmental Preservation (Instituto Nacional parar la Preservación del Medio Ambiente – INPMA) created in 1971 under the

Ministry of Education. Its major objective is to attend to every aspect related to the conservation and protection of the human environment. The institute has normative, executive and co-ordination functions, including research and studies concerning the environmental consequences of development. Although there is no legal provision for EIA in Uruguay, public and private organizations may, at their discretion, ask for project evaluations which are co-ordinated and performed by INPMA staff. The first experience with EIA was in 1975, when the development aid agencies funding the dam and hydroelectric complex of Salto Grande on the Parana River, a bilateral project shared with Argentina, demanded an appraisal (Ferrari 1983). Several evaluations have been performed since then for agro-industry and tourist developments. There is no agency responsible for reviewing EIA reports.

Peru

The Peruvian National Office for Natural Resources Evaluation (Oficina Nacional de Evaluación de Recursos Naturales – ONERN) was created in 1962 as a unit of the Ministry of Public Works and Development with the remit to survey and evaluate natural resources. In 1973, ONERN became an autonomous organization of the Presidency of the Republic, with the responsibility of assisting the National Institute of Planning (Instituto Nacional de Planificación – INP) formulate development policies and of performing studies on the interaction of man and the human environment in order to propose actions for environmental conservation (ONERN 1978). In later years, the Ministries of Health, Mining and Housing have also included environmental directorates within their organizational structures.

There is no environmental policy law in Peru. Current legislation is ineffective in dealing with the environmental problems of the country, mainly due to pollution from copper mining and processing, which affects both urban and rural areas (Olano 1985). Environmental impact assessment, however, has been required by development aid agencies or has resulted from technical uncertainties about the consquences of specific activities. EIA has been promoted by project proponents and undertaken by foreign experts, often with the participation of ONERN and domestic consultancy firms. The EIA initiatives for major developments number about 20 and include 10 for mining and metallurgy plants and 6 for water resource management projects. As there are neither procedures nor guidelines for EIA implementation, environmental assessments focus on single ecological elements, such as water or air quality, being limited to surveys and semi-detailed analyses of impacts on natural resources.

In the early 1980s, the government of Peru centralized decisions on development projects within the Corporation of Economic Development (Corporación de Desarrollo Económico – CONADE). Project assessment is limited to the consideration of engineering and financial issues, which has constrained the role of INP, ONERN and other development institutions that had previously considered environmental aspects, prior to project approval.

Professionals of ONERN have been engaged in a large number of research projects such as ecosystem mapping, inventories and process analysis of geological, soil, forest and river basin resources, thereby preparing information and data bases for EIA. In their view, the evident harmful consequences of major activities on environmental quality stress the urgent need for an environmental management policy. They are also conscious that implementation of an EIA process for decision making in Peru will require, as a minimum, an organic environmental law able to complement legislation and to support EIA; co-ordination of government and public actions along with the strengthening of ONERN and health authority responsibilities; development of the operational and technical capability of government organizations to ensure that environmental constraints and EIA results are considered in project planning and implementation (Olano 1985).

Colombia

The first country to institute a formal EIA process in Latin America was Colombia. In 1973, the National Congress conferred to the President of the Republic the authority to enact an environmental policy law and to determine the principles and procedures for pollution control, environmental protection, and the management of renewable natural resources. Thus, in 1974, the National Code of Renewable Natural Resources and Protection of Environment was issued. It consists of 340 articles comprising general policy directives, as well as extensive arrangements detailing: the lease of each kind of resource; the application of permits and licences for a large range of activities; the protection of health; and the organization of community associations, both for exploitation of natural resources and for the defence of environmental quality. Measures for policy implementation were invested in the national government through central and regional public organizations (UNEP 1984).

The general policy directives are comprehensive. They include a broad list of factors causing environmental degradation which should be considered; principles governing the use of natural resources and the environment; provisions for trans-boundary pollution control and the management of resources shared with other nations; financial measures; and tools for environmental policy development and implementation. There are dispositions for the presentation of 'environmental impact statements' by all public and private agents who intend to promote any work or activity likely to produce environmental damage. This statement is to be based on a prior 'environmental and ecological study', which must consider not only physical factors, but also social and economic impacts on the region caused by the activity.

In the meantime, the government of Colombia promoted an administrative reorganization in order to implement this policy. The National Institute of

Renewable Natural Resources and Environment (Instituto Nacional de Recursos Naturales Renovables y el Ambiente – INDERENA), originally created in 1968, acquired new functions and responsibilities. These were to assist the government in matters related to the protection of the environment and natural resources; to co-ordinate the actions of other public organizations concerning environmental policy; to execute forest recovery actions and river basin management; to propose and maintain environmental protection areas; and to regulate and control exploitation and use of renewable natural resources, including issuing of permits and licences for any kind of interference with the environment. INDERENA, under the Ministry of Agriculture, maintains a central office for environmental evaluations and for issuing permits, as well as 22 regional offices for inspections, surveys and monitoring of environmental quality.

The Regional Autonomous Corporations (Corporaciones Autónomas Regionales), created in 1976 under the National Department of Planning, are in charge of promoting social and economic development in 16 country areas. They are also responsible for the use of natural resources related to housing, sanitation and energy generation. The Environment Sanitation Directorate (Dirección de Sanaemiento Ambiental), under the Ministry of Health, is in charge of water, air and soil pollution control, with competence to establish norms and standards. There are also units for environmental protection in several other government organizations.

Efforts to establish the necessary procedures to apply EIA, however, have not yet been fully accomplished. As part of the measures already established to complement the 1974 code, the regulations for water resource management issued in 1978 reinforced environmental impact statement requirements and specified the procedures for issuing permits and licences for activities involving the use of water for domestic supply, energy generation, effluent disposal and any kind of marine exploitation. Only in 1984, however, did sanitary and public health regulations identify the cases in which an environmental impact statement would be required. These are activities involving the discharge of effluents dangerous to human health; dams and energy generation plants; exploitation of non-renewable natural resources; ports and airports; industrial complexes; housing developments; and any activity that affects water quality, marine resources and topography.

In each of these cases, INDERENA staff have developed comprehensive terms of reference for the preparation of EISs and for project review. Since 1976, these guidelines have been followed for the review of a large number of private projects (Pérez 1981). Also, several major developments concerning coal mining and power generation financed by foreign aid agencies have been subjected to EIA. Impact statements have been prepared by domestic consultancy firms with the assistance of INDERENA and foreign experts.

In spite of this rather long period of practising EIA, current procedures for selecting projects to be submitted to EIA have warranted the attention of INDERENA. At present, there are no specific guidelines and selection is based on the common sense and experience of professional staff. Furthermore, there are no mechanisms for assessing project alternatives or for promoting public involvement; approvals by different agencies, in particular health and environmental authorities, are not integrated; most public projects and works are not subject to EIA; and the monitoring of impacts as often recommended in EIAs is not implemented on account of the shortage of resources. Thus, in 1985 INDERENA submitted to the government a draft proposal, along with the necessary provisions to complement existing EIA regulations.

Venezuela

The creation of a legal system and an institutional framework for the protection, conservation and enhancement of the environment has given a fundamental impulse to government actions in Venezuela (Delgado 1983, Arocha 1985). The legal system, composed of a number of sectoral laws and regulations, is directed by the Organic Law of Environment promulgated in 1976. This law established the guiding principles for environmental planning and management; determined the measures for the control of development activities; and provided directives for scientific research, environmental education and the preservation of representative areas and ecosystems. The institutional framework is headed by the Ministry of Environment and Renewable Natural Resources (Ministerio del Ambiente y de los Recursos Naturales Renovables – MARN) which, since 1976, has been responsible for planning and implementing all related government actions, devising the necessary regulations and norms and issuing permits for public and private projects that affect the environment. In 1983, MARN was given the additional function of planning and administering land-use development.

In this context, the EIA process is but one of a series of government instruments for environmental policy. Permits and authorizations for the use of natural resources, though conceptually similar to EIA, are envisaged as tools for a lower level in the hierarchy of decision making. On the other hand, regional plans for land development are regarded as environmental evaluations at a higher level, even though effective demands on land-use cannot be seen precisely in most cases. These plans are one outcome of a large project entitled 'Venezuelan Environmental Systems' which surveyed the natural environment of the whole country, updated regional socioeconomic studies, devised a number of inventories, carried out analyses of environmental problems, and formulated a national system for data processing. This also resulted in the enunciation of the fundamental concepts of policy and legislation to be developed for the rational use of natural resources, land-use planning, environmental management and control of development activities in the country.

The legal basis for EIA was established by the Organic Law of Environment. Thus, since 1976, MARN has been working to implement EIA and to improve its legal and technical capabilities. In 1977, an inter-directorate commission was organized with the task of co-ordinating EIA procedures within the ministry. The first problem concerned a decision on the criteria for determining which activities would be subject to EIA.

Environmental law in Venezuela defines a large number of permits to be obtained and requirements to be met by project proponents. Thus, predominant opinion has been that EIA must be applied only to those projects likely to produce either qualitatively or quantitatively significant impacts on the environment, ordinary permits being applied to the others. Considerations about the cost and timing of the preparation and review of EIAs made the inter-directorate commission produce a broad list of activities that, depending upon size and location, could be subject to EIA as a precondition of implementation. Terms of reference have also been prepared for EIA content and review. Both the submission and scoping of EIA have been left to MARN.

In early cases, assessments were required when site and project characteristics had already been decided and the results were useful only in mitigating and correcting harmful impacts. In spite of the comprehensive guidelines provided by MARN, only recently has EIA begun to be applied to projects prior to implementation and for particular environmental components. Most reported cases concern water management projects, mining, oil exploitation and coastal development. However, only a few of them seem to have produced effective results.

Therefore, although EIA application in Venezuela has made important advances, there are several aspects that still need considerable improvement. The main preoccupations seem to be: the enforcement of procedures and technical norms, especially objective criteria for project screening; mechanisms to ensure the application of EIA from inception to the end of decision making; and the application of the procedures to public development projects which have been, more or less, outside MARN's control. Also, the enhancement of the administrative capability to implement the EIA process is an issue that must be addressed.

Mexico

The basis for the adoption of EIA in Mexico was set in the middle of the last decade with the creation of several environmental units within the government administration. These include the Directorate of Environmental Impact in the Ministry of Agriculture and Water Resources, where a number of assessments were performed for forest and agricultural development; the Intersectoral Commission for Environmental Sanitation, whose design of a procedural framework resulted in the first legal provisions for EIA with the adoption of the Law of Public Works in 1980; and the Assistant Secretary of Ecology in the Ministry of Human Settlements and Public Works, who promoted a large number of environmentally oriented regional plans (ecoplans) as well as the project 'Environmental Impact Assessment of Development in Mexico Valley'.

These initiatives helped to promote technical capabilities in environmental subjects (La Garza 1984, Lámbarri 1985).

In 1982, promulgation of the Federal Law of Environmental Protection and the institution of a new Ministry of Urban Development and Ecology (Secretaría de Desarollo Urbano y Ecología – SEDUE) concentrated sectoral legislation and government responsibilities concerning the environment. The law encompassed directives for the protection of water, air and the marine environment, and determined that proponents of public and private projects likely to produce environmental damage or to exceed norms and standards have to present an environmental impact assessment (manifestatición de impacto ambiental – MIA). These MIAs must be reviewed by SEDUE prior to any approval or decision on implementation. SEDUE was given the responsibility to establish complementary norms and procedures, implement actions for environmental protection and co-ordinate the EIA process.

Since September 1983, SEDUE, through the General Directorate of Ecological Planning and Environmental Impact (Dirección General de Ordenamiento Ecológico e Impacto Ambiental – DGOEIA), has developed a number of technical and procedural guidelines which have been applied but, so far, only experimentally. These guidelines have a broad environmental scope and comprise provisions for project submission and review, as well as detailing formats for the preparation of preliminary assessments and EISs. These are to be applied in accordance with the type, location and potential impact of the proposal. By the end of 1984, 88 public projects and 11 private ones had been subjected to this experimental scheme. However, only ten developers had presented the required statements and received the corresponding approvals. Fifty-five others were required to carry out further studies.

To identify the development actions which should comply with EIA, SEDUE first examined the project programmes of several government organizations and public companies. Then, agencies responsible for issuing licences were asked to instruct the proponents of projects likely to affect the environment to submit their proposals to SEDUE. A large programme for the diffusion of EIA procedures to other government authorities was also carried out resulting in agreements being signed with agencies responsible for mining, industrial development and public works projects. However, these arrangements do not appear to have resulted in an effective commitment to EIA.

From the beginning of 1985, a new strategy for improving the efficiency of the EIA process has been tried by SEDUE. It includes negotiations with other ministries for the submission of projects; participation in the committees responsible for project authorization and financing; the intervention of local SEDUE offices in the licensing procedures of other agencies at state level; and the dissemination of information on the legal, administrative and conceptual basis of the EIA process. At the same time, DGOEIA has promoted the reformulation of EIA guidelines in terms of specific project characteristics and started to improve the personnel and structural capability of the unit in charge of EIA review. These efforts are expected to contribute to EIA being applied more regularly, though an additional legal enforcement seems to be needed. Consequently, new regulations have been drafted and will be put forward in the near future.

Brazil

In 1973, the government of Brazil instituted the Special Secretariat of Environment (Secretaria Especial do Meio Ambiente – SEMA) as a response to the recommendations of the United Nations Stockholm Conference on the Human Environment. At that time under the Ministry of the Interior, SEMA was made responsible for public actions with respect to the environment and the rational use of natural resources. At that time, the emerging concern about environmental problems was clearly subordinate to economic development which continues to be the main government goal. SEMA has issued environmental control programmes, which detail the application of federal legislation on specific aspects of the environment. These programmes are to be carried out by the state environmental protection agencies created since 1974 (Wandesforde-Smith & Moreira 1985).

In 1981, however, promulgation of the National Environmental Policy Law brought the consideration of environmental issues to other public sectors. Moreover, policy was reoriented towards the protection and enhancement of environmental quality to the benefit of human life, in order to ensure harmonious socioeconomic development. Environment is considered a public heritage to be protected in the use of natural resources. To implement the national environmental policy goals, the law created the National System of the Environment (Sistema Nacional do Meio Ambiente – SISNAMA) to integrate the activities of federal, state and local government organizations and institutions involved with the environment. The leading organization within SISNAMA is the National Council of the Environment (Conselho Nacional do Meio Ambiente – CONAMA) whose function and initial composition was defined in 1983. Initially, CONAMA was composed of representatives of several ministries, some state governments and civil associations, presided over by the Minister of the Interior, assisted by the Secretary of SEMA. In 1985, with the creation of the Ministry of Urban Development and Environment, the composition of CONAMA was enlarged to include representatives of all states and territories, government agencies representing 17 ministries and 16 non-government organizations. In spite of their size and consequent problems associated with operating both organizations, the institution of CONAMA and SISNAMA was a clear opportunity to decentralize actions and responsibilities related to environmental management and to harmonize government actions at different levels.

In this political and institutional context, state environmental agencies were given the primary role, as they were responsible for implementing both state policies and the CONAMA directives. Federal action in implementation is restricted to a supportive role, particularly to supplement the inadequacies of individual state authorities. Thus, state agencies are responsible for the provisions related to the environmental control of development activities, including the application of EIA.

The first environmental assessment performed in Brazil, the EIA for a hydroelectric power plant, dam and reservoir financed by the World Bank, dates from 1972. Since then, a considerable number of projects dependent upon external aid have been subjected to environmental analysis. However, the results have had hardly any effect in preventing harmful impacts nor been of much use in making decisions on implementation. At the beginning, assessments were made by foreign consultants. Gradually, groups of Brazilian experts, research institutions and consultancy firms have been involved, later studies being performed with little or no foreign assistance.

Legal provisions for EIA came first at the state level, when the government of Rio de Janeiro instituted a permit system for pollution control in 1977. Regulations for implementing the permit system detailed a requirement for EIA reports in order to instruct developers making permit applications. These provisions could be varied at the discretion of the environmental agency. These provisions, however, have had little influence on either decision making or project implementation, having been applied to only a small number of private urban development projects and for the construction of pollution control equipment (Wandesforde-Smith & Moreira 1985).

The national environmental law of 1981 considered EIA to be one of the instruments for policy implementation. The regulation decree of 1983, which implemented this aspect of the legislation, directed the application of EIA to the permit authorization process. This decree established the basic contents of EIA studies and reports; indicated that the cost of EIA should be met by proponents; noted that EIA reports should be made available to the public; and determined that CONAMA should devise directives and criteria for EIA implementation. Thus, according to these provisions, EIA could be applied to the licensing of a broad range of actions, but only after CONAMA had formulated the necessary directives.

However, these directives were not issued until the beginning of 1986, after protracted negotiations between environmental organizations and other governmental sectors. They provide a list of the activities for which an EIA must be submitted and detail the procedures to be followed, as well as the responsibilities of proponents and environmental authorities. There are also provisions for community involvement and public hearings. State environmental agencies, in addition to the task of devising more detailed guidelines, are required to harmonize the permit system under their responsibility with the current process for decision making related to each kind of activity. The directives confirm that EIA studies are paid for by the proponent and prepared by independent multidisciplinary teams.

Efforts to implement these regulations began soon after their approval by CONAMA, priority being given to government actions with the most significant environmental consequences, namely mining, hydroelectric generation and petroleum and gas production. However, the procedural and technical capability of state environmental agencies is uneven. In several states the permit system has been regularly applied to private projects, mainly in the industrial sector, whereas in others, the permit process was instituted late and only with the aid of SEMA. At the moment, therefore, problems concerning EIA implementation are related to the technical and administrative difficulties within environmental agencies. An equally, and perhaps more decisive, factor is the need for a suitable understanding between these agencies and the government institutions responsible for major development programmes. Traditionally, these government organizations have never been subject to any interference in their decision making.

For other aspects influencing the EIA process, such as the availability of data, resources, technical expertise, scientific knowledge and skilled professionals to prepare statements, the situation is more favourable than in other developing countries. The general belief is that once demands are clear and supported by public opinion and political interest, needs will be gradually met.

Argentina

At present, current environmental legislation in Argentina relates only to certain aspects and even these provisions are not in force in all regional and local jurisdictions. Nevertheless, the federal environmental organization has perhaps the most technically accomplished staff for the application of EIA within Latin America.

The situation in Argentina has been reviewed by Balderiote (1985). In 1975, requirements imposed by the World Bank for financing the Salto Grande Development, a river basin project involving water resource management for power generation shared with Uruguay, led the bilateral committee managing the project to request the assistance of environmental organizations in carrying out the appraisal. Research on previous EIA cases led to the development of methods, techniques and procedures which could be applied to the Salto Grande Development.

This experience was extended to the development of other environmental research and gave rise to the Programme for EIA of Major Infrastructure Works by the staff of the National Directorate of Environmental Planning (Dirección Nacional de Ordenamiento Ambiental – DNOA). Originally established in 1975, under the Assistant Secretariat of Environment, DNOA is presently under the Secretariat of Housing and Environmental Planning of the Ministry of Health and Social Development. In 1985, six river-basin multiple-use developments and a railroad network project were being assessed and managed by DNOA.

None of these cases, however, can be regarded as a valid EIA, since assessments began after the decisions to implement had already been taken and the construction work contracted out. As there is no mandatory requirement for EIA in the country, DNOA decided to persuade the proponents of major works to enter into a voluntary commitment. Most probably, the favourable response reflects more the pressure of public opinion and international aid agencies than the persuasion of the environmental authorities.

The tactic chosen to implement EIA, namely to learn on a case-by-case basis, was probably the only realistic option available to DNOA. In later years, however, DNOA staff have devised a comprehensive set of guidelines for applying EIA to major infrastructure developments. These guidelines comprise directives on project-screening criteria, scoping formats, methods, procedures and monitoring. Also based upon this experience, the Secretary of Environmental Planning submitted a Bill to the present government proposing directives for environmental management, including the use of EIA for actions supported, financed, or promoted by government organizations. Other constraints on EIA relate to the lack of both an administrative framework at regional and local level and mechanisms for public involvement. These can certainly be solved once the environmental organizations are provided with the appropriate powers.

Conclusion

Great debtor nations like Brazil, Argentina and Mexico certainly need environmentally sustainable development which requires pragmatic policies and management actions. A number of environmental professionals in Latin America regard EIA as an efficient and credible instrument for implementing such policy actions and objectives. While an adequate legal framework and a competent and flexible administration are necessary to carry out EIA, even though responsibilities for environmental matters still need to be clarified and shortages of resources for scientific research and technology is a recurrent problem, the fact remains that EIA has already been put into action in these countries. Moreover, once issues related to environmental quality concerns not only the public administration, but also the whole of society, the practice of EIA can provide an opportunity for meeting the compelling societal demand for information and participation.

Acknowledgement

The author would like to thank Mr Jaime Hurtubia from UNEP's Regional Office for Latin American and Caribbean Countries for providing copies of the unpublished papers prepared for the workshop on EIA and Health Effects of Major Projects in Latin America held in Mexico City in early 1986 under the sponsorship of UNEP and PAHO.

Part V

EIA AND INTERNATIONAL AGENCIES

15 *WHO interest in environmental health impact assessment*

E. GIROULT

Introduction

Official backing for the involvement of the World Health Organization (WHO) in environmental health impact assessment was sanctioned by Resolution WHO/35.17, approved by the World Health Assembly in May 1982. This resolution recommended that environmental health and health impact assessment studies should be carried out and developed prior to the implementation of all major economic development projects, with special reference at a global level, to water resources development projects. There are geographical variations in priorities with respect to environmental health impact assessment which are dependent upon the relative state of development within a particular region. At the European level, for example, priority is given to hazardous industrial developments with special attention focused on the control of toxic chemicals and on the risk of major technological disasters (see, for example, Gilad 1984). The general framework for WHO involvement in environmental health impact assessment at present is set by the Seventh General Programme of Work covering the period 1984–9 (WHO 1983a). This states that WHO

> will pursue the study and analysis of situations in which ecological changes, particularly those resulting from urban and rural development, might give rise to health hazards. For this purpose, it will promote the study and analysis of, and the collation of information on, types of ecological changes that might create such hazards; research on the prevention of communicable diseases that are spread by deficient sanitation and are associated with rural and urban development as well as on factors that might promote and adversely affect the quality of life that are associated with such developments; and the participation of health experts in the planning of rural and urban programmes to make the control of hazards, due to ecological changes, an integral part of such plans.
>
> (Objective 11.2, para 285)

In practical terms, WHO's policy towards environmental health impact analysis has two main aims. The first is to strengthen health and safety considerations in impact assessment. Secondly, WHO seeks to encourage member states to undertake such assessments for each major development project. WHO has four basic functions in relation to its main role as technical adviser to the ministries of health of its constituent member states. These are the collection and dissemination of existing knowledge and national experiences related to health; transfer of this knowledge into national policies and programmes; co-ordination and mobilization of other organizations involved in health activities; and the development of new knowledge through the stimulation of research. Each of these functions has implications for environmental health impact assessment.

The collection and dissemination of information has been achieved through the organization of courses and seminars, as well as by the preparation and distribution of guidelines and other documents. Thus, an EIA course at Aberdeen University is sponsored annually by WHO. In addition, a specialized course on water resources development projects as well as national seminars in a number of countries including Greece, Poland and Turkey have been held. The Regional Office for Europe has produced guidelines for assessing the environmental health impacts of agricultural irrigation projects (Environmental Resources Ltd. 1983), petrochemical developments (D'Appolonia SA 1982), and urban development. (Environmental Resources Ltd. 1985). Similarly, WHO headquarters has published a training manual (WHO 1983b), guidelines on risk assessment (WHO 1983c) and advice on the rapid assessment of pollution (WHO 1982).

WHO is also aiding the transfer of this knowledge into national policies and programmes. Thus, the WHO Regional Office for Europe, in collaboration with the United Nations Development Programme, is assisting in the development of environmental health impact assessment procedures in a number of countries including Greece, Poland, Portugal and Turkey. In addition, closer co-operation is being sought with other organizations involved in this area. Within Europe, liaison with the Commission of the European Communities and the United Nations Economic Commission for Europe, Division for Human Settlements and the Environment are particularly significant. Finally, WHO is promoting research aimed at methodological development within the field of environmental health impact assessment.

Present experience indicates that there are considerable methodological advances to be made in environmental health impact assessment before it reaches a comparable status to that which has been achieved in other aspects of the assessment process. In this paper, deficiencies in current practice in environmental health impact assessment are reviewed. Secondly, the interrelationships between environmental factors and human health are discussed. Thirdly, an approach which could be adopted in assessment is described. Finally, an example is analysed to show the application of this environmental health impact assessment procedure.

Present status of health impact assessment

Environmental impact assessment (EIA) methodologies and procedures were developed, first in North America, then in Europe, mostly by ecologists concerned more about natural resources and the conservation of cultural heritage than about human health. Efforts were made, nevertheless, to develop health impact assessment methodologies and procedures. However, two kinds of difficulties were encountered. First, the lack of epidemiological knowledge regarding dose-response relationships frequently made it difficult to identify the health implications of environmental changes such as the release of pollutants. Secondly, government authorities responsible for authorizing development or reviewing impact statements have not been prepared to release documents referring directly to the health impacts of a development, expressed in terms of morbidity, disease incidence, mortality, or projected death rates. Governments prefer to treat such data as confidential in the rare cases that they are computed.

These difficulties have resulted in the fact that the health component has been the main weakness of many EIA studies. A study of the health component of past EIAs, commissioned by WHO, has been undertaken by the Centre for Environmental Management and Planning (CEMP), Aberdeen University and the Instituto Superiore di Sanitat (ISS), Rome. The results of this study showed that off-site health effects generally have been given a low priority and have not been treated systematically (CEMP/ISS 1986). This study contained an analysis of 13 EIAs produced between 1973 and 1982, covering petrochemical complexes, bulk and intermediate chemical manufacturing facilities, fertilizer plants and a natural gas terminal, located in Canada, Hong Kong, the UK and the USA. Only one EIA contained a section on health impacts. Even though each development handled or manufactured known or suspected carcinogens, only one EIA contained a comprehensive inventory of toxic and other materials – a fundamental requirement for assessing the health effects of a chemical plant. A formal hazard appraisal was included in only two EIAs.

Another important prerequisite for a health impact assessment is appropriate baseline information. Thus, data on the characteristics of the population which may be affected by routine or accidental emissions from a chemical plant, information about the distribution of homes and other buildings, as well as data concerning local lifestyles, which are relevant for identifying exposure to adverse effects and pathways of potentially harmful materials, are required. Eleven of the EIAs reviewed in the study were found to contain few baseline data appropriate for health impact assessment.

WHO has an obvious commitment to act in order to develop additional methodology to achieve stronger consideration of the public health component of EIA. To overcome the difficulties described above, work has been aimed at developing a new approach to the consideration of health aspects in EIA. It is proposed that a new concept be adopted, namely environmental health impact assessment. This involves the assessment of a proposed action with respect to its effects upon environmental parameters which are known to have an important influence upon health. For convenience, these parameters are referred to as environmental health factors. Clearly, an understanding of the relationships between environmental factors and human health is essential for the development of an appropriate approach to environmental health impact assessment.

Interrelationships between environmental factors and human health

Reviews of the current state of knowledge with respect to the interrelationships between environmental factors and human health can be found in Cohen (1983), Parke (1983), Robinson *et al.* (1983) and Donaldson (1984). In most cases, environmental health factors are not, *per se*, agents of disease. However, they may facilitate human contact with agents of disease or may weaken human resistance to infection. Certain environmental alterations may create direct contact between some agents of disease and human beings. For example, urban air pollution results in people breathing potentially toxic gases or suspended particles which directly infect the lungs. Other changes, however, have an indirect impact. Reservoir eutrophication, for example, may affect human health because the water treatment process used to produce drinking water will not remove all organic micropollutants resulting from lake eutrophication which, in combination with chlorine from drinking-water purification, leads to the formation of carcinogenic polychlorinated aromatic compounds.

Additional considerations which influence potential impacts relate to the conditions necessary for exposure. Certain changes only have an impact on health if people are exposed to them. Thus, for example, a poorly drained agricultural scheme will increase the habitat available to *Bulinus* (snails acting as the intermediate host of schistosome parasites). This will not necessarily increase the incidence of the disease schistosomiasis (bilharzia), as the snails locally may not be infected by the parasite. Similarly, even when infected snails are present, there may be no health impact if the agricultural work has been mechanized. If workers do not walk barefoot in the fields, there will be little likelihood of their contracting the disease, because infection generally occurs through the parasite penetrating the skin of the foot.

The main agents of disease can be classified into three groups, namely biological, chemical and physical. Biological agents are found amongst a large number of taxa including helminths, protozoa, bacteria, mycobacteria, *Rickettsia* and viruses as well as specialized parasites from a range of other groups. Chemical agents can be classified most realistically according to their effect. Chemicals may influence environmental health factors through toxic, carcinogenic or mutagenic effects. Finally, physical agents include materials such as dust and other irritants as well as energy in the form of harmful radiation and vibration, especially noise.

Biological, chemical and physical agents exert an effect upon health in a variety of ways. These are listed in Table 15.1. Certain of these effects may occur in a number of direct or indirect ways. Thus, for example, people may be

Table 15.1 Possible effects of environmental health factors.

Contact with agents of disease:
exposure to agents of communicable diseases
intensive exposure to toxic materials or agents causing immediate acute disease
low-level exposure to toxic materials or agents which may cause acute or chronic diseases long after exposure
Exposure to agents which may cause genetic changes
Lowering resistance to infection
Producing subclinical irritation, nuisance and discomfort
Contributing to aggravation of existing disease
Creating conditions incompatible with, or derogatory to, the achievement of physical, mental and social well-being

exposed to agents of communicable diseases by favouring an increase in the population of either the pathogen or its intermediate host, by facilitating direct contact between human beings and the pathogen, and by increasing the incidence of pathogen infection of food, consumable liquids, or the air.

All human beings are not equally sensitive to the impacts of the agents of disease or to environmental health factors. Sensitivity may vary at different scales. The regional variation in disease resistance is well established. Certain age groups within a population may be particularly at risk, although all individuals within an apparently uniform group, for example, infant males, may not be equally sensitive. It is important in environmental health impact assessment to identify risk groups, that is, the group of people most sensitive to a particular health hazard. The likelihood of an environmental health impact occurring is also a direct consequence of exposure, that is, the intensity or duration of contact between people and the disease agent or the environmental health factor.

Appraisal of the health impact of exposure to a disease agent or to an environmental health factor is generally described as 'risk assessment'. Figure 15.1 shows a schema of the overall risk assessment process. There are, however, certain problems associated with this term in the literature. The term 'risk assessment' has been used not only for assessing the consequences of exposure to a hazard in environmental health impact analysis, but also for determining the probability and consequences of accidental hazards. There are considerable parallels in the two processes, although it should be noted that environmental health impact assessment should cover not only the planning, construction, operation and decommissioning stages of a project, but also the accidents or disasters which may arise during its construction and operation. It is important that both aspects are considered; indeed the discussions related to health impacts are equally applicable to the consideration of accidents.

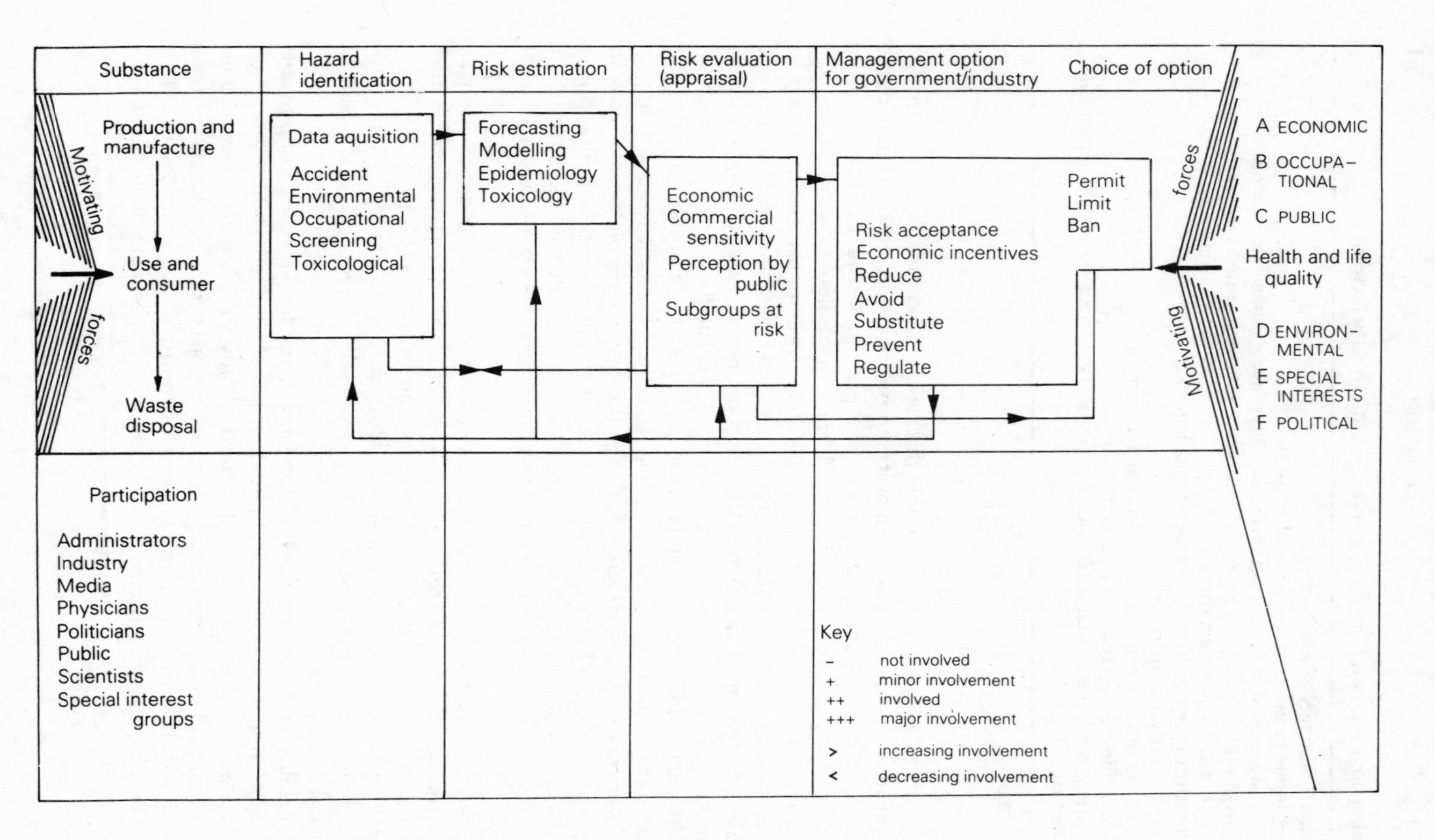

Figure 15.1 The risk assessment process.

Environmental health impact assessment

A schema for environmental health impact assessment is shown in Table 15.2. It is clear that some extension of established EIA procedures would be required to accommodate environmental health considerations if this proposal were adopted. The initial stages involving the identification of direct and indirect environmental change, steps 1 and 2, are already well established in EIA. Various methods exist to carry out these analyses, and it should be noted that these stages of the process are common to all types of impacts, not specific to the consideration of potential health impacts.

Table 15.2 Proposed environmental health impact assessment process.

Steps to be taken	*Tools to be used*
Step 1 Assessment of primary impacts on environmental parameters	Regular EIA process
Step 2 Assessment of secondary or tertiary impacts on environmental parameters resulting from the primary ones	Regular EIA process
Step 3 Screening of impacted environmental parameters of recognized health significance (EH factors)	Epidemiological knowledge
Step 4 Assessment of the magnitude of exposed population for each group of EH factors	Census, land-use planning
Step 5 Assessment of the magnitude of risk groups included in each group of exposed population	Census
Step 6 Computation of health impacts in terms of morbidity and mortality	Results from risk assessment studies
Step 7 Definition of acceptable risks (or of significant health impacts)	Assessment of trade-off between human and economic requirements
Step 8 Identification of efficient mitigation measures to reduce significant health impacts	Abatement of EH factors' magnitude, reduction of exposure, reduction of exposed populations, protection of risk groups
Step 9 Final decision Yes, if public health authorities are satisfied with proposed mitigation measures to control significant health impact No, if significant health impact was assessed and if doubt remains on the efficiency of proposed mitigation measures	

The health component of EIA, in effect, starts with the third step in the process, namely the screening of environmental parameters to identify those of health significance. Only environmental parameters which have a health significance, called, for convenience, environmental health parameters, are considered in the subsequent stages of the analysis. Epidemiological studies provide the most important means of assessing the health significance of an environmental parameter. Although epidemiology is a well-established science, the results of which should be used, research shows that epidemiological studies have not been a component of EIA in the past (CEMP/ISS 1986). None of the 13 EIAs reviewed in this report contained epidemiological data. A checklist of disease agents and environmental health factors which could be used in this stage of the assessment is included in Table 15.3

The fourth stage of the analysis involves an assessment of the extent to which the project will increase exposure of the population to environmental health parameters, that is, to increase contacts between human beings and infectious or offensive agents. In many respects, it is perhaps more important to assess impacts upon exposure rather than to identify the increase in the magnitude of an environmental health parameter *per se*. Such increases have an effect upon health, only if a sensitive population is exposed to them. Although most present EIAs include some measure of increased levels of pollutants, such as sulphur dioxide (SO_2) or nitrogen oxides (NO_x), and model their likely distribution,

Table 15.3 Checklists for use in environmental health impact assessment.

Disease agents:
- Biological:
  - Parasites, helminths, protozoa, bacteria, mycobacteria, *Rickettsia*, viruses
- Physical:
  - Noise, vibration, inert dust, ionizing radiation, non-ionizing radiation, excessive temperature or humidity
- Chemical:
  - Toxic chemicals, heavy metals, organics, inorganics, fermentable organics

Environmental health factors
- Primary:
  - Urban air pollution, indoor air pollution, improper solid waste disposal, improper liquid waste disposal, lack of proper drainage, toxic wastes, radioactive leakage, noise level, unstable structures (accidents)
- Secondary:
  - Increase of vector population resulting from increase in food supply, habitat, or reproduction sites

Exposure pathway of toxic chemicals:
- Food, drinking water, air breathed, skin contact

Exposed populations for hazardous industrial plants:
- Workers, their families, surrounding population, consumers of products

Risk groups
- Infants, children of pre-school age, pregnant women, elderly people, handicapped people, persons suffering from specific chronic diseases, persons with specific genetic deficiencies, workers in hazardous plants

these are rarely expressed in terms of health implications. The model projections need to be used in combination with a census and details of land-use distribution in order to assess the population exposed to change. A checklist of exposure pathways for toxic chemical hazards and exposed populations with respect to hazardous industrial plants is included in Table 15.3. Similarly, it will be necessary to determine the extent of the population particularly sensitive to the likely change, that is, the risk groups. Identification of risk groups is the fifth stage in the process. A checklist of potential risk groups is included in Table 15.3.

Impact on human health will be expressed in terms of changes in mortality (death rates) and morbidity (the incidence of particular diseases). In order to project the consequences of environmental health parameters in these terms, it is essential to have an adequate understanding of the dose–response relationship involved. At present, however, knowledge is far from complete and remedying this deficiency is an important area for research. It may be that monitoring the health effects of developments will provide important new data on such dose–response relationships. The importance of this factor can be seen from Figure 15.2. This figure shows two contrasted dose–response curves: a simple linear curve in which the response is directly proportional to the dose, for example, biological effects of ionizing radiation are recognized as being proportional to the cumulative dose (in rems) absorbed, and a non-linear curve. In the latter case, incremental changes have varying effects upon response, depending upon the level of the dose. At present, it is only possible to use the best available knowledge, which may be far from complete, to translate environmental health effects into likely mortality and morbidity impacts. Current understanding of long-term chronic exposures is particularly inadequate.

In any decision-making process, there are trade-offs between human health and economic considerations. It may be that low levels of risk can be accepted or that health risks can be accommodated through special monitoring and health care provisions for those in the risk groups. Defining the acceptability of the

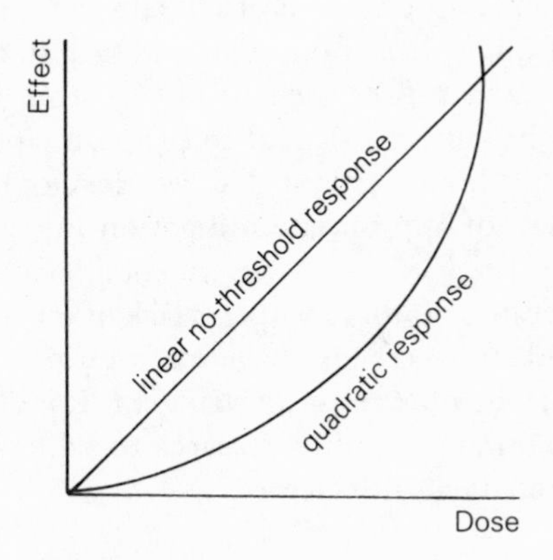

Figure 15.2 Dose–response curves.

risks involved is the seventh stage of the process. The availability of standards, particularly those developed by the WHO for health hazards, may help place decisions on trade-offs on a more rational basis.

When a significant health or safety hazard has been identified, it may be possible to suggest mitigation measures which would reduce its significance. This constitutes the penultimate stage. These reductions may be achieved in a number of ways. First, the impact may be reduced by changes which affect the magnitude of the environmental health factors. For example, proper collection, treatment and final disposal is a way to mitigate the health impact resulting from increased production of hazardous wastes. Secondly, exposure may be reduced through modifications which affect the major pathways. Thus, if there is significant evidence that a specific food is the major exposure pathway of a toxic chemical coming from an industrial source, the relevant crop could be forbidden in the polluted area. Crop selection is an alternative approach. Whereas, for example, potatoes which are eaten cooked may be irrigated with raw sewage, lettuce which is eaten raw could not. Finally, it may be possible to reduce the exposed population, particularly in relation to risk groups. Site selection is an important consideration in determining the size of the exposed population. Thus, hazardous industrial plants should be situated far from important concentrations of people such as high-density urban areas. Serious consideration should be given to alternative site selection, yet this is one aspect which has proved a weakness of many EIAs in the past. If no alternative is available, medical screenings should prevent sensitive persons from working in specific industries with high levels of exposure. Some examples of the interrelationships between environmental health factors, exposure, risk groups and mitigating measures are presented in Table 15.4.

The final stage of the process involves the decision on whether to proceed with the proposal. It is clear that if public health authorities are not satisfied with the mitigation measures proposed to abate a recognized health hazard, they have no choice other than to forbid the project. However, a minimal health hazard will always be accepted. For example, despite the major contribution of immunization techniques to the control of infectious diseases, there remains a very small health hazard in vaccination. It should be noted that the purpose of environmental health impact assessment is not to stop development, but to identify any potential health hazards and to promote appropriate mitigation measures to reduce them. In most if not all cases, development projects will be approved after inclusion of appropriate mitigation measures. Where there is doubt concerning the level of an environmental health parameter which constitutes a hazard, because of, for example, a lack of scientific evidence, public health authorities are likely to err on the side of caution and ask for a level of performance which may be unnecessarily stringent. It is clearly in the interests of all, therefore, to encourage scientific research to achieve more accurate risk assessment of environmental health factors.

Table 15.4 Examples of environmental health factors and related disease agents, exposure, risk groups and mitigation measures.

Air pollution
Factors and their effects:
inert dust (irritation of respiratory tracts)
pathogens on aerosols (respiratory diseases)
gaseous or suspended particulate toxic chemicals (carcinogenic effects)
oxygen deficit (asphyxia)
Exposure:
people breathe indoor and urban air (pollution of the higher atmosphere is not a health problem)
Risk groups:
people with chronic respiratory diseases
Mitigation:
abatement of emissions at source
dispersion of pollutants in the higher atmosphere
reduction in exposure of risk groups

Solid wastes improperly disposed
Factors and their effects:
inert materials such as stone, glass and metal (injury hazard)
toxic materials (human ingestion through water or food)
organic fermentation products (favours growth of pathogens)
food residues (increase population of disease animal vectors such as flies and rats)
Exposure:
contact with disease vectors
contact with toxic materials
consuming contaminated food or water
Risk groups:
children playing on discharge sites
garbage collection workers
consumers of water from aquifers contaminated by leachate
people within dispersal range of vectors
Mitigation measures:
proper selection of disposal sites
fencing of disposal sites
burying disposed waste under soil cover

Health considerations of an agricultural irrigation development project

Consideration of a hypothetical agricultural irrigation development project shows the way in which the approach to assessment, described in the previous section, can be applied. Throughout this example, no attempt has been made to identify all of the potential environmental health impacts. Rather, the ramifications of some of the more readily apparent environmental changes are followed sequentially through the assessment process.

PRIMARY IMPACTS ON ENVIRONMENTAL PARAMETERS

Of the wide range of potential effects of such developments, certain major changes appear to be more or less independent of site location. Invariably, the groundwater level will rise, which will lead to a consequent increase in the extent of wet areas, marshes and standing water. Thus, groundwater and surface water resources will become more readily accessible. In view of the intensification of agriculture associated with such schemes, groundwater is likely to become polluted by fertilizers and pesticides. Finally, irrigation schemes lead to increases in the salt content of the soil.

SECONDARY IMPACTS ON ENVIRONMENTAL PARAMETERS

A number of secondary changes are likely to occur. From considerations of environmental health, the most significant relate to increases in the extent of wet area. These are likely to lead to an extension of the breeding habitat of disease vectors. Water snails are likely to increase in numbers, and these areas will also serve as breeding sites for many insects, particularly biting insects such as mosquitoes.

SCREENING ENVIRONMENTAL PARAMETERS FOR POSSIBLE HEALTH EFFECTS

Certain primary effects give rise to impacts which adversely affect human health. Water quality is an important consideration in human health. Pollution by agro-chemicals is likely to be particularly significant. Epidemiological studies provide some guidance on possible significance. Nitrate in drinking water, for example, has been identified as a source in methaemolglobinaemia in babies ('blue baby' syndrome) as well as being more tentatively linked to the production of N.nitroso compounds which are known carcinogens. Many pesticides are highly toxic even at low concentrations.

Similarly some secondary changes have adverse health implications. Both rising water tables and increasing wet areas are likely to lead to a higher incidence of disease in an area if they result in an extension in the range or greater numbers of disease vectors. Thus, increased snail populations will be significant if these are *Bulinus* snails (intermediate host of the parasite which causes schistosomiasis). Changes in mosquito populations could have serious consequences on the incidence of malaria if *Anopheles* mosquitoes are present. *Anopheles* mosquitoes are intermediate hosts of the malaria parasite.

Not all of the effects are necessarily adverse. Increased accessibility of groundwater and surface water resources may increase drinking-water supply in rural areas where this might otherwise be a problem. Rising water tables may have either beneficial or adverse effects depending upon location. Thus, while it will facilitate the pumping of drinking water, it will decrease the efficiency of water filtration by the soil and subsurface strata. At a certain level, this will result in poor water being pumped from an unconfined aquifer. Other impacts, such as the increased salinity of the soil which affects agricultural production, may be important, but not directly significant for human health.

ASSESSMENT OF THE MAGNITUDE OF EXPOSED POPULATIONS

The populations exposed to the environmental health impacts associated with elevated groundwater levels, groundwater pollution, and increased accessibility of water resources are rural populations who take their drinking water from the unconfined aquifer inside the irrigated area. The magnitude of the exposed population depends upon the local settlement pattern and may be affected by the implementation of rural settlement policies.

People dwelling within the flying range of *Anopheles* mosquitoes constitute the exposed population. However, this represents only exposure to mosquito bites. Exposure to malaria only exists if the mosquitoes are infected with the parasite. This will almost certainly be the case if malaria is endemic in the area. An alternative source may be migrant workers, perhaps associated with the development in question, from malaria-infected areas.

ASSESSMENT OF THE MAGNITUDE OF RISK GROUPS

The sensitivity of a population to the two diseases which may increase as a result of the development proposals may show considerable regional variation, dependent upon the past history of the area. Thus, the sensitivity of populations to the parasitic diseases, malaria and schistosomiasis, will be greater if they are imported into an area previously free of disease (for example, by migrant workers) than if they occur in an area where part of the population has acquired a degree of immunity.

In the case of nitrate contamination of groundwater supplies, censuses can be taken to determine the extent of the risk group. There is a general consensus that infants and in particular bottle-fed babies are especially at risk when the concentration of nitrates in drinking water is in excess of 200 mg/l. Some consider that risks of methaemoglobinaemia exist even in the range 100 to 200 mg/l. The WHO recommended nitrate guideline value currently stands at 10 mg NO_3-Nl (WHO 1984), while the EEC drinking water directive has a guide value of 25 mg NO_3/l and an imperative value of 50 mg NO_3/l (Council of the European Communities 1980). The EC imperative value is equivalent to 11.3 mg NO_3-N/l.

COMPUTATION OF HEALTH IMPACTS IN TERMS OF MORBIDITY AND MORTALITY

Mathematical models have been developed to compute, for example, the extension of a malaria epidemic. These models could be used to assess the likely disease incidence amongst the affected population. Alternative models could be used to assess the consequences of development upon other diseases. The practical value of predicting the extension of malaria or other epidemics, however, must be called into question. If there is a direct risk of a parasitic disease epidemic which can be identified at the project assessment stage, there can be little justification in proceeding with the development as proposed. Either the project must be stopped or efficient mitigation measures must be sought and adopted before the project can be allowed to proceed.

DEFINITION OF ACCEPTABLE RISK

A project which will result in a significant increase in the *Anopheles* mosquito populations may be acceptable in a country free of malaria and where no labour force has moved in from infected countries. Of course, a project which would result in a significant increase in snail populations could not be accepted in an area where schistosomiasis is endemic.

IDENTIFICATION OF MITIGATING MEASURES

Proper engineering design of irrigation facilities and the selection of appropriate irrigation techniques will decrease the impact of the project upon mosquito breeding areas and snail populations. The adoption of appropriate agricultural practices will reduce the exposure of workers to the disease vectors. Similarly, suitably planned settlements as well as adequate clean water supplies and sanitation will decrease the exposure of the population to parasitic diseases. These measures will also have the advantage of, concurrently, increasing their health status in relation to enteric diseases. Other ways of controlling either the disease vector, for example through pesticide spraying, or the disease, through an immunization programme, might also be implemented. Derban (1984) describes the measures adopted retrospectively to mitigate the effects of water-related parasitic diseases for the Volta Dam project,

An adequate supply of clean water would overcome the problems caused by groundwater resources polluted by agro-chemicals. This may be piped from an unpolluted source. Alternatively, the contaminated supply may be blended with unpolluted water or treated to reduce the concentration of pollutants. High costs may make treatment impracticable. The supply of bottled water specifically for the risk group has been adopted as a means for overcoming the problem of nitrate contamination of drinking water.

Conclusions

WHO has recommended that studies of environmental health impacts should be a component of the assessment for all development projects. There is a need for further research and methodological development, however, before this can be achieved effectively. One significant advance would be the systematic analysis of proposals following the procedure outlined in this paper. A study of 13 development projects (CEMP/ISS 1986) has indicated that medical experts are rarely involved in EIA. This situation must be redressed if environmental health impact assessment is to become a reality. Clearly, public health authorities and expertise should be involved at a very early stage in the process.

The purpose of environmental health impact assessment is not to prevent economic developments; indeed, these are often a vital means for raising health standards. The process should aim at identifying whether there are significant health hazards resulting from a project. Whether a health hazard should be considered 'significant' is clearly a subjective notion which involves amongst other things the need to balance the protection of human life with the need for economic development. When a significant health hazard is identified, public health authorities will either request mitigation measures or oppose a particular project. A positive approach to environmental health impact assessment, therefore, should also include the identification of mitigation measures to reduce health hazards to an acceptable level.

One problem is likely to remain. Descriptions of the environmental health implications of development proposals in terms of prognoses for morbidity and mortality rates are likely to be particularly sensitive. It may prove impossible to include these kinds of data in an assessment intended for broad public distribution.

WHO sees environmental health impact assessment as a promising tool to achieve better environmental health planning and management. In a forthcoming document, WHO will be issuing guidelines stressing that health impacts should become a routine feature of project appraisal, that medical personnel should be involved from the outset to determine the scope of an EIA and that common health indicators such as local life expectancies should be more widely used in establishing baseline environmental conditions.

16 *Environmental impact assessment and bilateral development aid: an overview*

W. V. KENNEDY

Introduction

The Organization for Economic Co-operation and Development (OECD) has been giving attention to environmental impact assessment (EIA) since about 1974. That attention has been centred primarily in the Environment Committee where, as in other OECD committees, representatives of the 24 Member countries meet to discuss common problems, the international economic implications of those problems and solutions to them. The Environment Committee has passed two recommendations related to EIA which call upon member countries to establish procedures and methodologies for assessing the environmental impacts of significant public and private projects and to exchange information on matters which could help them better forecast the environmental effects of such projects (OECD 1974, 1979).

Since the passage of these recommendations most OECD Member countries have established some type of EIA system. No two countries, however, have adopted an identical approach with the result that the scope and requirements of national EIA processes vary greatly. None the less, OECD Member countries are gaining experience in assessing the environmental impacts of their domestic activities, in both the public and private spheres. The application of EIA to development aid activities, however, has been much less common. Thus, whereas large industrial and infrastructural projects within OECD countries are often routinely submitted to EIA, when included as part of a foreign aid programme in a developing country, they are not. There are probably several reasons for this, a main one being that aid policy and foreign environments have not been subject to the same internal political pressures as domestic environment issues. The situation, however, is changing and the need to inject an environmental component into aid programmes is being recognized increasingly.

Within the OECD this need was recognized in the creation of the *ad hoc* Group on Environmental Assessment and Development Assistance which had its first meeting in October 1983. This group represented a co-operative effort between the Environment Committee and the Development Assistance Committee working together for the first time. It comprised representatives from Member countries' environmental and developmental administrations.

Early on, the *ad hoc* group decided to use the term 'environmental assessment' rather than 'environmental impact assessment' or EIA. Although in some national contexts this term is used to describe a preliminary evaluation carried out to determine the need for a full EIA, no such distinction is intended here and the terms 'environmental assessment' and 'EIA' are used interchangeably throughout this paper.

At the first meeting, agreement of the *ad hoc* group was reached upon a programme of work centred around four objectives. First, it was necessary to identify the types of development aid most in need of EIA. Secondly, the constraints faced by developing countries in assessing the environmental impacts of aid projects and programmes had to be identified. Thirdly, the experience of aid agencies in actually carrying out assessments had to be examined. Finally, based on the first three objectives, recommendations regarding the kind of procedures, organization and resources needed to ensure that aid proposals are adequately assessed for their environmental impact have had to be made. The following discussion is based primarily on the final report of the *ad hoc* group which was submitted to, and approved by, both the Environment Committee and the Development Assistance Committee of the OECD in late 1985 and early 1986 respectively and subsequently published as an OECD environment monograph (OECD 1986a).

Types of projects most in need of EIA

There are a number of reasons for first identifying the types of aid projects and programmes most in need of EIA. For example, some activities because of their very nature and size pose greater environmental threats if implemented than others. In addition, the amount and kind of information needed to assess adequately potential impacts can vary greatly depending on project type, size and location. Lastly, the human and financial resources needed for carrying out EIAs are often limited and detailed assessment cannot be performed for all projects.

In order to come up with a list of the most important aid project types, the group looked first at how OECD countries had tackled this problem regarding their domestic activities. It was found that, where EIA systems exist, the types of projects covered are determined in one of three ways, namely inclusion in a list, through the use of screening criteria, or by a combination of the two.

Some OECD countries, such as the Netherlands, have established a positive list of specific project types which must always be submitted to environmental assessment. Lists can also be made up of project types which do not require assessment, for example, the 'categorical exclusions' specified by a number of US federal agencies under the National Environmental Policy Act.

Other OECD countries, for example, Canada and Australia, have established screening criteria or guidelines which are applied to projects on a case-by-case basis to determine which ones should undergo an assessment. The number and type of criteria vary from country to country, but they usually attempt to determine the significance of such issues as changes in the natural, physical, or social environment; pollution levels; cumulative effects; endangered species; sensitive ecosystems; and the level of public controversy.

In addition to the types of projects covered by the EIA systems of OECD member countries, the group also examined the steps taken by various developing countries to identify project and programme types in need of EIA. Lastly, it investigated the types of project which have been assessed by both bilateral and multilateral lending institutions.

The results of this research and the corresponding discussions within the group led to an agreement on seven types of aid projects and programmes most in need of EIA. These are listed in Table 16.1. This list of project types, together with issues which should be considered when carrying out EIAs make up the first recommendation of the group which was endorsed by the Environment Committee at ministerial level on 18 June 1985 and officially adopted by the OECD Council two days later.

Constraints in developing countries to carrying out EIA

In work on the second objective the group was able to draw upon the experience of many of its members concerning assessments in the Third World. In addition, it also undertook a general literature search and contacted directly the governments of a number of developing countries. The countries for which information was gathered were Brazil, Ecuador, Indonesia, Kenya, Malaysia, Mexico, Rwanda, Somalia, Sri Lanka, Sudan, Tanzania and Thailand.

The constraints identified were a general lack of political will or awareness of the need for environmental assessment; insufficient public participation; lacking or inadequate legislative frameworks; lack of an institutional base; insufficient skilled manpower; lack of scientific data and information; and insufficient financial resources. The extent to which any or all of these constraints are operative in the Third World varies from region to region and from country to country.

It was found, for example, that generally speaking, each of these constraints would apply to most African countries. In South-East Asia, on the other hand, insufficient data and skilled manpower appear to be, for the most part, bigger constraints than legislative or institutional frameworks which are, in many countries, already in place. On the whole, it was found that South-East Asia and the Pacific tended to be the most advanced region in the developing world regarding the establishment of and experience with environmental assessment. One study revealed, for example, that 66% of the countries in that region have passed legislation requiring EIA for certain types of projects, compared to 57% in Latin America and 41% of the countries in Africa and the Middle East (Sammy 1982).

Table 16.1 Development aid projects most in need of an environmental assessment.

Renewable resource use	
Substantial changes in land use; Introduction of forestry; Introduction of arable or bush fallow farming;	New lands development; Resettlement projects; Rural development projects.
Farming and fishing practices	
Large scale mechanization; Introduction of new crops; Conversion of land to agricultural production; Large scale chemical control of disease vectors and pests.	Fertilizer use; Pesticide use; Major change in farming and fishing practice.
Exploitation of hydraulic resources	
Substantial changes in water use; Water resources development; Dams and impoundments; River basin and water management;	Potable water supply; Irrigation projects; Hydro-electric power; Drainage projects.
Large infrastructure	
Major infrastructure; Road and penetration road building;	Ports; Road improvements.
Industrial activities	
Industrial plants; Chemical plants; Industrial processes generating toxic waste.	Power plants; Pulp mills.
Extractive industries	
Forest industries;	Mining.
Waste Management and disposal	
Waste management; Sewerage and sewage treatment plants.	Cattle dips.

Source: OECD (1986b)

OVERCOMING CONSTRAINTS

Developing countries, on their own and together with help from OECD Member countries, are beginning to take steps to overcome these constraints. A number of examples can be cited. Thus, the Canadian government has initiated an 'Environmental Manpower Development' project in Indonesia to help that country overcome the lack of trained individuals to undertake environmental assessment. Through a combination of training courses at Indonesian universities and environmental centres together with a programme to send Indonesians to Canadian universities, a large number of government officials, academics and local consultancy firms are acquiring the necessary skills for conducting EIAs.

In addition, the United States Agency for International Development

(USAID) together with UNESCO's Man and the Biosphere Programme has completed over 40 'environmental profiles' of developing countries. These reports are the first step in a process to develop better information for missions, host country officials and others detailing the environmental situation in specific countries as well as beginning to identify the most critical areas of concern.

Personnel in developing countries also have increasing opportunities to gain practical experience of EIA. Although not widespread, there is growing evidence that, where aid agencies carry out an EIA, they are involving host country officials and others in scoping sessions, in the collection of baseline data and in the actual preparation of assessments.

Aid agency experience with environmental assessment

Nine delegations from OECD countries – Australia, Canada, Denmark, France, the Netherlands, Norway, Japan, the United Kingdom and the United States – prepared a total of 16 case studies on environmental assessments which had been carried out by their aid agencies in Asian, African and Latin American countries. USAID is the only agency with a legal requirement for carrying out EIA. As a result, it has the greatest experience and submitted six case studies illustrating a wide range of project types as well as approaches to EIA.

As the other agencies have no legal requirement to conduct EIAs their cases represent, for the most part, *ad hoc* or special situations in which the assessment was carried out either because of obvious potential environmental effects or because the host country specifically requested it. In one case (a hydroelectric power project in Indonesia) the assessment was undertaken on the initiative of an engineering consultancy appointed to carry out an engineering feasibility study. The terms of reference issued by the Canadian aid agency made no reference to EIA, The consultant firm, however, which had conducted EIAs on similar types of project within Canada, suggested that an EIA be included and the aid agency agreed.

The case studies covered six of the seven types of projects and programmes identified above as being most in need of environmental assessment. The following summaries provide basic information on the case studies such as their size, location and major design elements on a country-by-country and project-by-project basis.

THE UNITED STATES

Sri Lanka's Mahaweli Development Program: this project focuses on the construction of four new dams and associated downstream works along the Mahaweli River to bring 117000ha of undeveloped land under irrigated cultivation. When completed, the project is expected to more than double the country's total electricity generating capacity; increase food production by 547000 tons annually; provide storage water to irrigate an additional 121000ha;

and to create significant employment opportunities through construction work, farming activities and off-farm employment.

Rural development in Peru's Upper Huallaga Valley: this development, begun in 1981, is designed to provide an array of services aimed at generating alternative sources of income for farmers presently engaged in illegal coca production. In its final form the project consisted of developing and applying agricultural production packages, such as road repairs, drilling of wells, crop improvements and strengthening the public sector agricultural support services in co-ordination with the Peruvian coca-eradication programme.

Rural electrification in Dominica: the installation of a 1500 kW diesel engine generator and the extension of electrical transmission lines from an existing hydroelectric facility on the west coast to the east coast of this Caribbean island are the major elements of this project.

Refugee settlement in Somalia: as an alternative to the establishment of refugee camps, this project aims at the development of a settlement programme, including agriculture and animal husbandry skills training and infrastructure development.

Meat processing factory in Thailand: the construction of a moderately sized plant north-west of Bangkok serving 2000 farmers is part of a programme for providing extension services to small and medium-sized livestock suppliers. The development is intended to stimulate economic growth. The facility will be constructed and operated by a Thai firm under licence to a US company. The plant will be built on an industrial estate of 2.59 km^2. An existing natural lagoon 100 m from the plant site intercepts drainage from the area before it reaches the Chao Phrava River.

The Central Selva Natural Resources Management Project: the original aim of this project involved only the construction of a north–south road in a remote jungle valley of eastern Peru. Through the environmental assessment process, however, the scope was enlarged to include a resource management project. The project area, the valley floor, for which the assessment was carried out covered 95 000 ha.

JAPAN

Coal-fired power stations and integrated steel mill in Singapore: this project concerns the planning and construction of two coal-fired power stations (one with three 250 MW units and one with two 350 MW units) and a steel mill with an anticipated annual production of 1 million tons. Both projects are expected to be completed by 1990.

Metropolitan Manila outer major roads project: this proposal involves the improvement of 18.2 km of two existing roads and the construction of 21 km of new road to the south-east of Manila. Their purpose is to strengthen connections between new industrial and housing areas which are being developed in the greater metropolitan area.

Bangkok solid waste management study: this study consisted of two components: Phase I, the formulation of a master plan, and Phase II, a feasibility study. In order to determine an optimum master plan, seven types of basic combinations of functional elements, such as solid waste collection, transport, intermediate treatment and final disposal, were formulated and later developed into 30 master plan alternatives. A further evaluation reduced the number to three alternatives which were subjected to environmental assessments as well as economic, financial and technological analyses.

FRANCE

Rice production in north Cameroun: the project aims at irrigating 5000 ha of an area near the Logone River in Cameroun, followed by 7000 ha in an area further north, for rice production.

DENMARK

A case study on cattle dips in Kenya: following the construction of a large number of cattle dips (tanks of acaricides in which cattle are immersed) to help eradicate tick-borne diseases during the 1970s, problems related to the correct concentration of acaricide and timing of the dips were encountered. This project aims at optimizing the use of dips and includes provisions for the training of advisory staff; the supply of acaricides to 3400 dips and the establishment of efficient control of the acaricide. This is supplemented by support for the maintenance and repair of dips together with the supply of water for them.

UNITED KINGDOM

The influence of EIA on the development of more target-specific tsetse fly control techniques: this project concerns three tsetse fly control campaigns in Zimbabwe, Botswana and Somalia carried out between 1975 and 1984. In contrast to earlier tsetse control techniques which involved extensive bush clearing, game destruction, or the application of mammalian-toxic insecticides such as dieldrin, these new programmes used newer insecticidal methods which were expected to be less environmentally harmful.

CANADA

The Sentani Hydroelectric Project in Indonesia: the purpose of this project, located in the province of Irian Jaya in the extreme north-east of the country, is to substitute water-generated electricity for the present, more expensive, diesel-generated supply for the provincial capital and surrounding region. The scheme will utilize the outflow from Lake Sentani and divert it through a power station. The power station is small and would have a generating capacity of 6 MW.

THE NETHERLANDS

Environmental assessment in Dutch bilateral assistance to rural development in Colombia: this project is divided into two subsets. The first comprises three integrated rural development projects, and the second consists of projects for the provision of long-term technical support. The three rural development projects were carried out in different ecological zones to test and demonstrate sustainable land-use systems appropriate for small landholders.

NORWAY

Road construction in Kenya: unlike the other case studies, this describes an environmental assessment carried out after the project had been constructed. An international trunk road connecting Kapenguria in Kenya with Juba in the Sudan was constructed in 1975. In 1982 the Norwegian Agency for International Development commissioned an impact study on the road in order to outline a generalized approach for road impact assessments; to test alternative methods for impact assessment; and to carry out a case-study description of the road impacts which occurred in the study area.

AUSTRALIA

The Northern Upland Development Project in Thailand: this project, an ongoing programme which began in 1967, aims at replacing a shifting form of agriculture with a stable, permanent rain-fed system. For this purpose, natural vegetation was cleared and the soil cultivated to final seed-bed condition. Support services including an agronomic and soils research station and training programme were established. By 1986, some 25000 ha will have been developed. No separate formal environmental assessment was conducted, but comparable aspects were built into the programme development and review.

Characteristics of the case studies

In analysing the case studies, the *ad hoc* group looked at five main issues. First, the form of the EIA was considered. The case studies showed that two general approaches had been taken. Either a special environmental impact statement or report was prepared or information on the environmental aspects of the project or programme was incorporated as part of the feasibility study and other project planning documents. No link was found between the form of the assessment and either its content or its effects on any project decisions. In other words, there was no apparent advantage of one form over another in terms of assessing environmental impacts or of aiding the decision.

Secondly, the way the EIA was prepared, in particular who had been responsible for its preparation and how it was carried out, were determined. In approximately half of the cases the documentation was prepared by outside consultants, either individuals or consultancy firms, and in the remainder by aid agency officials. Whether prepared 'in-house' or by others, an interdisciplinary team generally prepared a more comprehensive EIA than a single individual undertaking the task alone. Regardless of who actually prepared the assessment, all case studies showed evidence of co-ordination between those responsible for preparing the assessment and host country officials. Indeed, in some cases it was the host country government itself which requested the aid agency to undertake an assessment. This is a particularly interesting finding, as it contradicts a widely held attitude in development aid circles that the biggest constraint to carrying out EIA in developing countries is the attitude of the host countries. Although inter-agency co-operation with host country officials is a more or less standard feature of the case studies, the incorporation of public participation in project planning is much less common.

It appears from the case-study material that the main reason for failing to provide for public participation is that aid agencies in general view such matters as the responsibility of the host government. In the few cases where public involvement did take place it was carried out in a much more informal way than has been the case with environmental assessments carried out in OECD Member countries. The case studies which showed the most evidence of public participation were those for the Sentani Hydroelectric Project in Indonesia and the Central Selva Natural Resources Management Project in Peru.

The consultant team which prepared the assessment for the Sentani project interviewed a number of local residents to obtain information on their needs and their views on the proposed development. As a result of these efforts design changes were made to the project which eliminated the need for resettlement.

In the early stages of the Central Selva project a number of Peruvian and international environmental organizations were opposed to USAID'S approach. They viewed the project as a forced resettlement scheme being carried out without regard to the land tenure and rights of the local people. As a result of a meeting between representatives of the groups and USAID officials, USAID reviewed the preliminary environmental and social impact studies done by individual consultants and discussed plans for more detailed investigations. As a consequence, the organizations encouraged USAID to stay involved with the project, to carry out the planned examinations and to try to influence the decisions of the Peruvian government towards sustainable development. This was done and involved, amongst other things, a presentation to the Peruvian President outlining the findings and conclusions of the environmental and social studies.

Thirdly, in reviewing the content of the EIAs, the description of the present state of the environment; the number and types of alternatives considered; the identification and assessment of environmental impacts; and mitigating measures were considered by the group. The projects described in the case studies differed so widely that it was difficult to make any generalized statements or conclusions. For example, the extent to which the existing environment was described in the case studies varied greatly. For some projects extensive descriptions of the natural and man-made environmental setting were provided. For others, only those environmental factors which would be directly affected by the project, such as air and water quality, were described.

Regarding alternatives, most of the case studies concerned a single, site-specific development proposal without any indication that alternatives, either sites or different means of achieving development, were considered. In such cases, the EIA served as a mitigation plan for limiting negative impacts rather than as a decision-making tool for comparing alternative means of achieving development goals. In some instances, however, particularly those related to highway construction, alternatives were assessed and compared on the basis of their potential environmental impacts.

The actual number and type of environmental impacts which were assessed seemed to be dependent upon the characteristics of the project itself and the way in which the EIA preparer viewed the situation. In other words, there was very little evidence that EIA guidelines or checklists related to the project type were used. In two cases, a scoping process took place to identify the most significant impacts. In one of these cases, the proposal to construct electrical transmission lines in Dominica, the scoping meeting took place between USAID's regional environmental management specialist and cabinet-level officials, including the Prime Minister.

Almost all the case studies showed evidence of the adoption of measures to mitigate the negative effects associated with the proposal. In some instances, mitigation was achieved by design changes, such as the inclusion of waste-water treatment facilities in the steel production facility in Singapore. In other cases, it was in the form of a mitigation programme such as the creation of a wildlife reserve to compensate for the conversion of wildlife areas to agricultural production in Sri Lanka.

Fourthly, the results of the EIA, in terms of its effect on the outcome of the project or programme, were determined. In no case did the environmental assessment result in a decision to halt the project or programme. In several cases, however, the assessment brought about project or programme design changes for reducing negative environmental impacts. The hydroelectric project in Indonesia, for example, included design features to mitigate the impact on the natural environment and tribal lifestyles while maintaining adequate energy production.

Finally, the constraints encountered in undertaking the assessments were analysed. The most frequently mentioned hindrance was the lack of baseline data. This was overcome by sending assessment teams into the project areas to carry out the necessary field studies. Other constraints mentioned included the lack of trained host country counterparts, budgetary restrictions and inappropriate training. An example of the latter can be found in the meat processing project where EIA, coming late in the decision-making process, had little or no effect upon project design. Notable by its absence was the constraint of lack of interest in, or the support of, the assessment by host government officials. Indeed in several cases, the very initiation of the assessment itself was the result of a request from the host country.

As a general, concluding observation it may be said that, in comparison to assessments of similar projects appraised under domestic EIA requirements, such as in the USA and Canada, the content of the aid EIAs was not as comprehensive. The factors responsible for a successful assessment, however, appear to be more or less the same.

Key factors for a successful EIA

Based on the case studies, five 'key factors' for a successful EIA which seem to have relevance wherever EIA is applied were identified. These relate to timing, personnel, scoping, information and monitoring.

TIMING

All of the case studies pointed to the need to integrate environmental assessment at an early stage of project planning. Where it is seen as an extra or as an 'add-on' to projects which already have been determined on the basis of their engineering, technical and economic feasibility, it can perhaps suggest mitigation measures, but can have no real effect on the project design. When integrated early in project planning it can result in projects with built-in mitigation which is designed to minimize negative effects and maximize benefits.

PERSONNEL

The success of an environmental assessment is very much dependent on the individual, or team, responsible for preparing it. In view of the great diversity of project and programme types to which assessment has been and can be applied, it is difficult to determine an ideal profile for an 'EIA preparer' which would fit every situation. As the case studies indicate, some types of project can be assessed adequately by a single person with the right qualifications and experience working together with host government officials and local experts over a short time period. Other projects demand interdisciplinary teams of experts to carry out extensive field investigations and data gathering. In both cases, the need could conceivably be met from within aid agencies themselves. A more likely situation, however, is one in which the developer will have to approach private consultants or consultancy firms for help. In those situations, it is necessary that terms of reference be prepared in such a way as to ensure that the individual or group brings sound environmental knowledge and experience to the job.

SCOPING

A crucial task in carrying out environmental assessment is to identify, early in project planning, the most significant, serious, environmental impacts associated with a project and the reasonable alternatives available for constructing the project in an environmentally sound manner. Scoping is a procedure for accomplishing these tasks. An early meeting of the donor agency, host government officials, environmental experts and other interested parties to determine the scope of the project can result in quicker, less expensive and more efficient environmental assessments.

INFORMATION

The need for reliable data and information is a common theme in case studies. Where an adequate data base is missing it becomes particularly important to work closely with local universities, research institutes and the affected public to obtain an insight into existing environmental conditions. The time and expense involved in 'starting from scratch' makes it advisable to tie data gathering to the major environmental impacts identified during scoping.

MONITORING

An important lesson to be learned from experience with environmental assessment is the need for monitoring of environmental impacts. Although as yet it is not required by any aid agency, most are coming to see the need for auditing completed projects not only as a sound management measure, but also as a means of testing the accuracy of the environmental assessments. Knowledge of, for example, which impacts proved to be significant and which did not can result in the improved scoping of future projects.

Time and costs

It is difficult, if not impossible, to draw conclusions regarding the average time and cost involved in carrying out an environmental assessment for an aid project or programme. The main reason for this is that, to a large extent, the time and money needed for an assessment vary with the size, type and location of the project itself. The assessment of a multi-million dollar hydroelectric project, for example, is a much more extensive undertaking than one for the improvement of a 10-km length of existing roadway. Other factors play a role in determining the time and costs involved. First, the amount of information which is readily available versus that which must be obtained through field studies and other research is a major determinant. Secondly, whether the assessment is contracted out to an interdisciplinary team, is put together by a consultancy firm, or is done 'in-house' by a single individual affects the resources required. Finally, the point in time and the way in which the assessment is carried out is important. For example, whether an assessment is undertaken as part of engineering–economic feasibility studies early in project or programme planning or as an additional report initiated separately from and after all other planning studies have been completed affects costs and time schedules.

As the projects described in the case studies vary greatly in terms of their size, type and location and, as the assessments also differ greatly in terms of their scope and the way in which they were prepared, an averaging of time and cost would be of little value. The following statistics, however, throw some light on the order of magnitude involved.

The most expensive and time-consuming assessment was that for the Mahaweli Development Program which was conducted at a cost of \$775 000 and carried out over a 95 person-month period between August 1979 and October 1981. The quickest assessment was that for the rural electrification scheme in Dominica which was conducted in 10 days; as the assessment was carried out by USAID's environmental management specialist for the Caribbean, its cost was subsumed under the aid mission's personnel and administrative expenditures.

The figures for the other case studies which are available fall somewhat between these two extremes. The cost of the Central Selva Natural Resources Management Project EIA was $250 000 and took four months to complete. Assessment of the rice production project in Cameroun cost $80 000 and was completed by a six-person team in one month. The assessment for the Sentani Hydroelectric Project was $100 000 and involved five people working for several months.

Looking at the actual time and cost in isolation may not give a true picture of the situation. For example, expressing the environmental assessment as a percentage of total project costs gives a different perspective. The cost of the assessment for the Mahaweli project, for example, while three times greater than that for the Central Selva project, represented only 0.08% of the total project budget. The figure for the Central Selva project was 1.4%. As a general rule it can be said that the size of a project and the percentage of total project costs devoted to an environmental impact assessment are inversely related.

Similarly, the consideration of time periods in isolation can also be misleading. Eight months spent preparing an assessment, for example, need not lead to delays if integrated with other planning and feasibility studies. Half that time can mean delay if it comes after all other planning studies have already been completed. Although the case studies did not provide actual figures on time and cost saving obtained by carrying out an assessment, there were indications that, for some projects, this was indeed the situation.

EIA procedures

With completion of the work directed towards the first three objectives, the *ad hoc* group addressed the fourth issue, namely the kinds of processes, procedures, organization and resources needed for the environmental assessment of development projects and programmes. This activity culminated in the drafting of a second recommendation detailing the measures required to facilitate the environmental assessment of aid projects and programmes (OECD 1986b). This recommendation, which was adopted by the Council in 1986, outlines a suggested approach for an assessment process to be adopted by aid agencies as well as proposing measures for improving the capability of developing countries to carry out environmental assessments on their own.

Generally speaking, the results of the work indicate that EIA needs to be seen as a comprehensive process integrated at an early stage of project and programme planning. In addition, it should be co-ordinated with the host country government and the people likely to be affected. The results of an EIA should be reflected in the implementation of the activity and should be followed up by monitoring and a post-development audit.

Conclusion

Though some aid agencies are relatively enlightened in the way they view environmental assessment, at present, in none of them is this comprehensive approach in operation. It is, of course, not necessary that all aid agencies adopt identical processes and procedures for EIA. However, common goals and principles such as those outlined in the second recommendation should be agreed upon.

In this regard, the recommendation also urges aid agencies to adopt a specific, clearly stated policy towards EIA. Only through a specific policy can those responsible for implementation be given clear directions on carrying out environmental impact assessments. When followed by the commitment of skilled personnel and money one will hopefully begin to see more evidence of development which ensures economic growth while preventing environmental degradation and protecting the long-term productivity of the natural resources on which development depends.

17 *Fitting USAID to the environmental assessment provisions of NEPA*

J. A. J. HORBERRY

Introduction

The United States Agency for International Development (USAID) administers the US government's development assistance, economic security assistance and 'Food for peace' programmes. It is the most important bilateral aid agency and has considerable resources to meet its obligations under these programmes. Its total budget for the years 1977–82 is shown in Table 17.1

USAID is a federal agency and, as such, has a wide accountability for its activities. Thus, it is accountable via its administrator to the Executive. Similarly, Congress not only passes the legislation that spells out its mandate, but also monitors the agency's compliance with that mandate during the annual budget appropriations process. USAID is also accountable to the Judiciary in the same way as other federal agencies and, therefore, can be taken to court if it fails to uphold federal legislation applicable to its activities.

Like any bilateral development assistance organization, USAID is not only a creation, but also an agent of its government. Unlike most bilateral agencies, however, USAID serves a government that passed strong environmental protection legislation in the early 1970s.

Not only did the US government adopt a vigorous domestic environmental protection policy, but its agencies found themselves answerable to the courts and public interest groups for the implementation of that policy. After USAID was challenged in the courts for failing to comply with the provisions of the National Environmental Policy Act (NEPA), it revised its procedures, receiving forceful mandates from the President and Congress. By 1978, USAID was preparing the only enforceable and systematic environmental assessments of projects in the development assistance community.

Table 17.1 Total US agency for international development commitments (US$ millions).

Fiscal year:	1977	1978	1979	1980	1981	1982
Amount:	3181	4082	3848	4062	4209	4990

While any bilateral agency responds to the policies and laws of its government, USAID's capacity to review and mitigate the potential environmental damages of its projects stems from the focus and strictness of its accountability to the various parts of government and some if its constituents. Bilateral agencies do not respond equally to all government policies nor do all of their efforts to implement change in their funding programme have equal success. Minimizing environmental damage within an aid programme is not an easy task, given the tendencies of bureaucracies to resist change, to avoid delays in spending money and to maintain smooth relationships with recipient governments. The story of USAID's environmental assessment policy illustrates how a planning reform can overcome resistance and take hold in relation to the political and financial structure of the organization.

EIA is but one of USAID's policies for environmental and natural resource aspects of development assistance. Its policy has three goals. First, USAID aims to assist the less developed countries, not only in building the institutional and scientific capacity required for identifying, assessing and solving their critical environmental and natural resource problems, but also with establishing programmes to address the management of natural resources. Secondly, it attempts to ensure the environmental soundness and long-term sustainability of USAID assistance programmes and projects. Finally, it seeks to promote environmentally sound development projects funded by multilateral and bilateral development assistance organizations (USAID 1983a).

EIA, however, is the aspect of USAID's environmental policy that came first, is most formal, had the most effect on the funding programme of the agency and facilitated the introduction of other environmental activities. USAID's environmental staff tend to play down its significance, partly because they are aware that NEPA-style assessments may not be the most effective tools for environmental planning or the management of development projects, and partly because they feel that overall sensitivity to environmental issues within the agency has improved so that assessments are often redundant.

Similarly, the agency probably would not have committed as many resources to environmental activities, nor would recipients have co-operated in assessment, had not the initial reform been mandatory. While it may be true that some USAID missions now integrate environmental planning on a routine basis and, consequently, do not need to carry out assessments, it is unlikely that this would be the case if the threat of the assessment requirement was not there.

How aid works

There are two major components of USAID's budget, namely the development assistance account and the economic security assistance (formerly the security-supporting assistance) programme which grants financial aid to promote US security interests in selected nations. The main development assistance programme of USAID, about $1350 million in fiscal year 1984, comprises six functional areas: agriculture, rural development and nutrition; population planning; health, education and human resources; energy; private voluntary organizations and selected development activities; and science and technology (USAID 1983b). The relative allocation to different areas between 1980 and 1983 is shown in Table 17.2. The economic security assistance programme, while comprising mainly cash and commodity import support, includes some traditional capital projects, particularly in the Middle East. Between 1975 and 1979, 15 per cent of the economic security assistance programme was for aid (USAID 1980).

The broad policy objectives that have shaped the development assistance

Table 17.2 International Development Cooperation Agency for international development program trends: Fiscal years 1980–3 (US$ millions).

	FY 1980 actual	*FY 1981 actual*	*FY 1982 estimated*	*FY 1983 estimated*
Functional Development Assistance:				
Agriculture, rural development and nutrition	630.8	652.6	700.5	700.0
Population planning	184.9	189.9	211.0	201.0
Health	129.9	143.3	133.7	114.1
Education and human resources development	97.8	102.7	103.8	116.4
Energy, private voluntary, organizations and selected development activities	119.8	112.9	138.1	156.7
Science and technology	—	11.9	10.0	10.0
Subtotal, functional accounts	1163.4	1213.3	1297.1	1298.2
(Grants, included above)	(726.6)	(826.3)	(906.3)	(905.8)
(Loans, included above)	(436.7)	(387.0)	(390.9)	(392.5)
Sahel Development Program	76.5	95.6	96.2	93.8
American schools and hospitals abroad	25.0	20.0	20.0	7.5
International disaster assistance	55.9	51.5	73.2	25.0
Miscellaneous prior year accounts	—	2.1	1.8	—
Subtotal, functional and other	1320.8	1382.5	1488.3	1424.5
Operating expenses	273.0	302.8	333.0	377.0
Foreign service retirement fund	26.7	27.8	33.6	35.4
Total, aid development assistance	1620.5	1713.2	1854.9	1836.0
Economic support fund	2158.1	2199.3	2564.0	2886.0
Total, Agency for International Development	3778.7	3912.5	4418.9	4722.9

Source: United States Agency for International Development, *Congressional Presentation, Fiscal Year 1983, Main Volume,* 1982. These totals do not include contributions to multilateral organizations.

programme over the past decade are set out in the 1973 'new direction' and 1978 'basic human needs' amendments to the Foreign Assistance Act. These were designed to reorient assistance towards alleviating poverty and inequitable income distribution, as well as meeting the basic needs of the poor, with an emphasis on food production and rural development. Additional amendments call attention to environment and natural resources; appropriate technology; women in development; energy; and assisting the private sector (Hough 1982). Thus USAID's legislative mandate makes it clear that its task is to carry out specific development assistance programmes according to technical criteria, in addition to the political and economic interests underlying USAID's overall programme.

The development assistance account is appropriated for the functional sectors listed above. Regional bureaux receive allocations to be distributed among the missions according to their applications for specific proposals. Following the 'new direction' and 'basic human needs' amendments, most of the development assistance account goes to the poorest people in the poorest countries. It is made available at highly concessionary rates. For the fiscal year 1984, 42% of USAID's development assistance was allocated to countries with per capita incomes under $375, and 75.5% to those with under $795 (USAID 1983b). The grant element of all US official development assistance was 93.4% in 1981 (OECD 1982).

The organization is more decentralized than other development assistance agencies, with approximately 42% of its total staff based in overseas missions (Fig. 17.1). Missions enjoy considerable discretion over their activities. Since 1979, mission directors have had the authority to approve proposals having a total value of up to $5 million over the life of the project (Mickelwait 1979).

Formal procedures for development, review and authorization of projects which govern the responsibilities of missions and headquarters exist to ensure that a project is adequately prepared before approval. In brief, missions either receive requests for assistance from recipient governments or identify projects jointly with the government during, for instance, the course of preparing an annual report on the development strategy for a particular country. The mission then prepares a project identification document for each proposed development and submits them to both the regional bureau and the policy bureau in Washington for review. When project identification documents are approved, they are incorporated into the mission's contribution to USAID's annual budget proposal which is submitted to the Office of Management and Budget, and ultimately, to Congress for the approval of funds. Meanwhile, once the project identification document is approved, the mission and regional bureau can start to prepare a project paper presenting the full analysis and design of a project. Final approval of the project paper by Washington allows a project agreement to be drawn up between USAID and the recipient government. It usually takes at least two years from the project identification document stage to implementation and, sometimes, considerably longer.

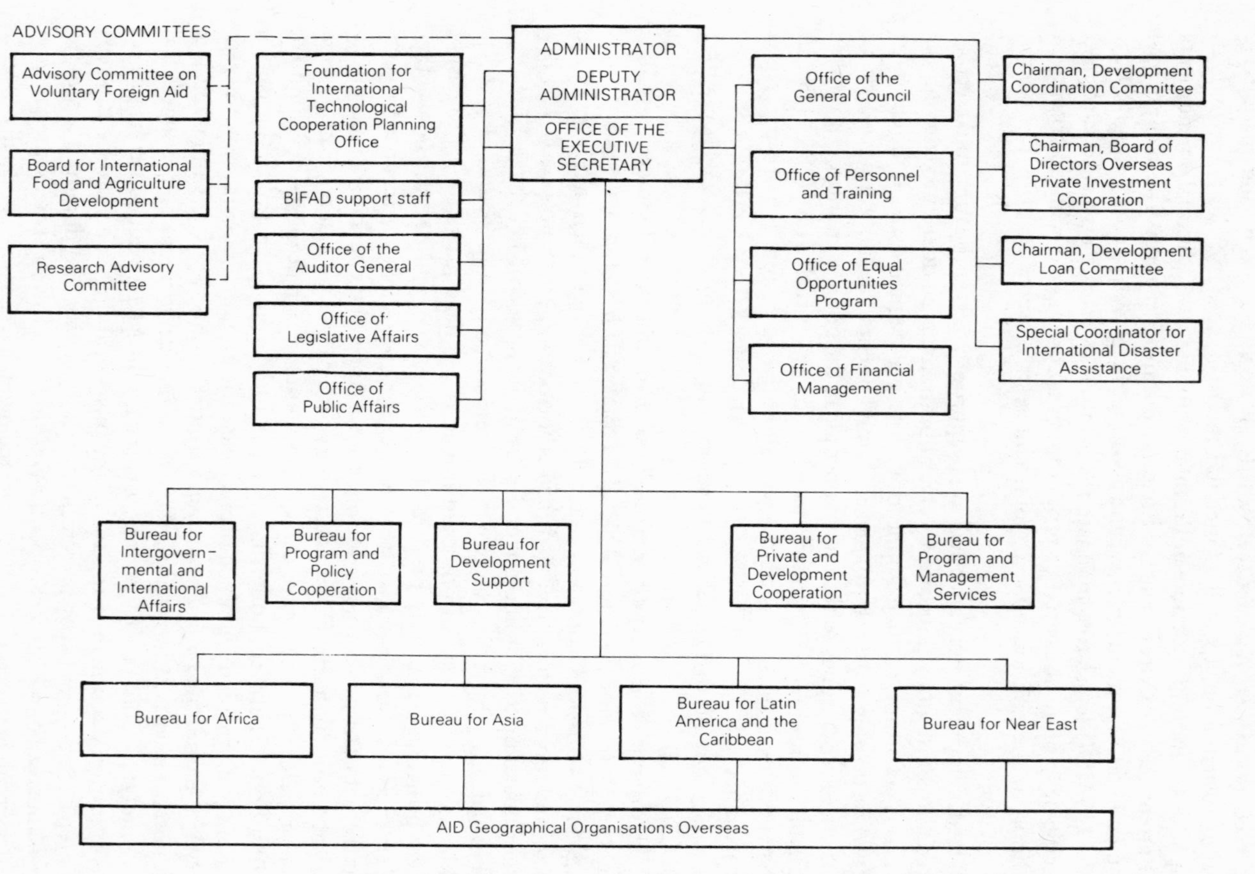

Figure 17.1 Administrative structure of USAID

NEPA and development assistance

Soon after NEPA was enacted, USAID, being a federal agency responsible for projects with possible environmental impacts, albeit overseas, adopted procedures for carrying out environmental assessments (USAID 1970, 1971). However, these procedures were very limited, only applying to traditional engineering and industrial projects. In view of USAID's changing mandate over the period, it is important to note that these provisions did not apply to a range of its main activities, for example, in financing commodities such as pesticides and in the provision of technical assistance. The main administrative change was the establishment of a special USAID committee, the Committee on Environment and Development to oversee the response to NEPA, in May 1971.

In fact, USAID responded slowly and reluctantly to NEPA, believing that the Act was domestic in intent and that the United States should not enforce such legislation beyond its own territory. However, in 1971, the Council on Environmental Quality (CEQ) proposed amendments to the existing regulations that would extend NEPA to development assistance to the fullest extent possible in an effort to persuade USAID to comply (Horberry 1984a).

At the same time, the Center for Law and Social Policy, a public interest lobbying organization involved in ensuring that federal agencies responded to NEPA, began to apply pressure on USAID to implement the Act. However, USAID appeared to be stalling and seemed unprepared to take any firm action, particularly in relation to one of the Center's main concerns, namely pesticide use.

In April 1973, the Center made a formal objection to USAID's response to NEPA on a number of grounds. First, the environmental procedures that USAID had adopted only applied to capital projects which no longer formed a major part of USAID's programme. Secondly, USAID had not applied any environmental procedures to the financing of commodities such as pesticides or to technical assistance. Thirdly, contrary to NEPA practice, there had been no public input into a review of pesticide use. Finally it was pointed out that in the two years that had elapsed since USAID had adopted its limited environmental procedures, it had not carried out a single EIS (Horberry 1984a).

In 1974, both the Center and CEQ intensified attention on the pesticide programme, but with little response from USAID. By 1975, the environmental organizations concerned with the implementation of NEPA came to the realization that USAID would not comply voluntarily and decided to file suit. In mid-1975, a consortium of environmental organizations brought a suit against USAID based on the grounds discussed above.

Once the suit was filed, USAID settled quickly. The agency was quite shocked to discover that litigation had, in fact, been initiated and proved willing to negotiate with the environmental lobbies and CEQ. In December 1975, the parties concluded a settlement. USAID agreed to prepare new regulations for implementing NEPA, to undertake an environmental impact statement (EIS) on its pesticide programme, and to produce interim and, eventually, final pesticide regulations. These new environmental procedures, issued in June 1976, also required an initial environmental examination (IEE) for all projects to determine the need for a formal environmental assessment or, in the case of a capital project, an EIS (USAID 1978).

The settlement also prompted the agency to issue an environmental policy determination that undertook to 'assist in developing the indigenous capabilities of developing countries, and to assess and mitigate the effects of proposed USAID projects in conjunction with the host government' (USAID 1978). In addition, amendments to the Foreign Assistance Act contained a new mandate, Section 118, calling upon USAID to 'furnish assistance . . . for developing and strengthening the capacity of less developed countries to protect and manage their environment and natural resources'.

Soon after 1978, the Congress amended Section 102 of the Act, directing USAID to include environmental and natural resources in the list of 'critical problems' to be addressed, and Section 118, obliging the agency to consider the environmental impacts of all its development assistance activities. Finally, President Carter reaffirmed the extra-territoriality of NEPA, already fairly clear to the agency, in Executive Order 12114 (President's Office 1979). Although the original suit had cited the failure to adopt adequate environmental impact assessment procedures, the net effect of the 1975 and 1978 policy determinations and the congressional mandates of 1977 and 1978 went beyond this to embrace a much wider environmental and natural resource management policy.

Congress also mandated USAID to address the problems of deforestation and soil erosion within the part of its budget allocated to agriculture, nutrition and rural development. The 1978 policy statement lists new categories of assistance which USAID expected to provide, namely reforestation, watershed protection, wildlife protection, improvement in the physical environment, environmental education and institution strengthening. It also committed the agency to training its own personnel, drawing upon the expertise of other federal agencies, and to co-operating with non-governmental organizations and other international donors (USAID 1978).

Thus, USAID mandates and policy statements now combine its responsibilities under NEPA with a more development orientated mandate, in line with the 'new directions' and 'basic human needs' provisions discussed earlier. It is likely that policy makers within USAID found a mandate to include consideration of the environment and natural resources more palatable than a precise commitment to carry out assessments. Many staff members, especially in the field, probably felt that NEPA-style EISs were not suited to the agency's goals and operating style, and would be hard to implement in developing countries. Attention to severe environmental degradation and natural resource management problems in developing countries, however, was included in USAID's main policy statement of that period (USAID 1978).

At first, the bulk of the agency's effort was devoted to fulfilling the assessment requirements. USAID Environmental Coordinator estimated that during the period 1976–9, 90 per cent of environmental staff effort was taken up with assessments, dropping to 60 per cent during the period 1979–81 and to 20 per cent since 1981 (Horberry 1984a).

Implementing the environmental assessment regulations required considerable reform within USAID. The agency had to recruit staff members with appropriate experience and to identify eligible consulting firms and contractors qualified to prepare assessments. At this time, however, there was a shortage of US consultancy firms with both environmental assessment and overseas development experience (Horberry & Johnson 1981). Initially, three contractors were appointed to provide environmental services. Not all of their efforts, however, were successful. In 1979, a further six were appointed with better prospects of success. Finally, USAID had to educate its regional bureaux and missions about the regulations and had to prepare procedural and technical guidelines.

The quality of the first generation of environmental assessments was not high and, certainly, not of much value in improving the preparation of the projects concerned.

> At first the assessments took on some of the characteristics of the early NEPA impact statements – separate documents prepared by a visiting team, performed often after the project planning was well under way, and highly duplicative of material presented elsewhere in the project documentation.
> (McPherson 1982)

From a study of the environmental policies, procedures and performance of USAID, Blake *et al.* (1980) reported four main conclusions about the practical implications on these events. First, the EIS on pest management resulted in significant changes in USAID's operations. Secondly, several other assessments had positive effects on the design of projects. Thirdly, the new procedures had increased the sensitivity of USAID staff members to environmental problems, brought about environmental training and demonstrated the needs for technical guidance.

Finally, a number of difficulties associated with the initial period of implementation were identified. Within the agency, there was poor knowledge about the procedural and technical requirements. Inexperienced contractors were often used to prepare assessments which, frequently, were reviewed inadequately. IEEs were more procedural than substantive and provided an inadequate basis for judging the need for further assessment. Field staff had shown resistance to the new procedures, while Washington had provided poor guidance to the field staff. Excessive emphasis was placed on the procedural requirements which led to unnecessary length and poor focus on potential problems. There was inadequate integration with project design and insufficient attention to mitigation measures. Finally, assessments were often produced too late to contribute much to the formulation of the proposal.

Horberry (1984a) reviewed some twenty early environmental assessments to see whether specific environmental problems had been identified and whether changes in project design had been recommended. It was clear, that most assessments of that period were isolated from the project design process, presented no analysis of alternative design features or implementation measures and contained much irrelevant, descriptive material. What was not clear was what potential problem triggered the decision that assessment was needed. Subsequent interviews with the regional environmental officers confirmed that few of the early assessments influenced project design. Many assessments had simply concluded that no problems were likely and others were merely produced to satisfy the procedural requirements (Horberry 1984b).

The recipient governments were indifferent to the environmental procedures so long as the studies did not consume their funds. Any resistance on their part, however, could be overcome by arguing that US law required the assessments before funds could be released. Mission staff, whether they approved of assessments or not, had no choice but, at least, to comply with the procedures. Similarly, Washington staff may not all have welcomed the new policy, but had mandatory regulations to back up the necessary reforms. Once staff members were in place and had gained some experience, however, it was clearly in their interest to improve the implementation of the regulations.

The first steps towards improving the assessment procedures were better technical guidance, programmatic environmental assessments for individual classes of projects, and the preparation of design criteria for specific types of project. Some efforts were made to increase the participation of officials in recipient countries as opportunities presented themselves. For instance, environmental consultants were placed in USAID missions in Indonesia and the Philippines and local organizations were encouraged to co-operate in carrying out assessments. In India, the mission identified and evaluated a number of consulting organizations for possible collaboration (Blake *et al*. 1980).

Subsequently, like many other federal agencies, USAID revised its formal procedures for preparing assessments in line with the revisions to CEQ guidelines, aimed at introducing flexibility and reducing delays (USAID 1981). Some categories of project were eliminated, and a scoping mechanism and design criteria for projects likely to have significant effects were introduced.

Involvement with two types of project has enabled USAID to improve acceptance of environmental assessments within the agency while, at the same time, gaining valuable practical experience. The first includes the capital projects in the Middle East under the economic security assistance programme, such as the Maquarin Dam and Jordan Valley Irrigation System in Jordan, and the Alexandria and Greater Cairo Wastewater System in Egypt. The scale and complexity of these assessments has necessitated hiring large, experienced consultancy firms. As a consequence, agency staff have become better at ensuring that recipients participate in scoping the assessment, that consultants focus on alternative design options and that the assessment results in an agreement between USAID and the recipient to avoid or mitigate potential problems.

The other source of experience involved environmental assessments for large multi-donor projects in which USAID had only a minor financial stake. These have met with mixed fortunes. Thus, in the case of the vast Senegal River Basin Development neither USAID nor the consultancy firm had the necessary experience to resolve the environmental problems (Gannett, Fleming, Corrdry & Carpenter Inc. 1981). Assessment of the Mahaweli development scheme in Sri Lanka, however, was much more successful, at least, judging by the environmental mitigation plan which was adopted for the scheme.

USAID agency staff confirm the benefits of the new regulations in the design and management of projects and in avoiding serious mistakes since 1980. They also report that, in recent years, complying with the regulations has become a positive contribution to project planning. Missions now understand the assessment requirements and have better guidelines to assist them in carrying out the various stages of the assessment process. There is now greater emphasis on effects that cause significant harm to the environment. Similarly, host country officials are increasingly involved in scoping the assessment. In addition, there is a greater inclination on the part of the agency staff to anticipate recurrent problems and address them at an early stage to avoid the need for more detailed investigation later (Horberry 1984a).

Looking at the whole picture, USAID's current environmental activities bear the hallmark, not of an unwanted domestic legal requirement, but rather an integrated process that addresses the resources and management strategies on which so many of the rural poor in developing countries depend (USAID 1983b). The 1981 amendments to the Foreign Assessment Act and USAID's 1983 policy statement reflect this more balanced approach. The result is a relatively well integrated assessment procedure and a more positive emphasis on directing some of the available assistance resources to environmental and natural resource programmes of relevance to the recipient government (USAID 1983a). The current situation appears to owe more to a development mandate than to NEPA. Without NEPA, its supporters and their commitment to enforcing its provisions, however, it is unlikely that the necessary reforms would have come about. These pressures ensured: that the Congress would develop the agency's mandate; that the necessary expertise would be put in place; that the missions would appoint environmental officers, however inexperienced and uninterested; and that many recipient governments would co-operate in preparing assessments.

Accountability, organizational response and environmental planning

The agency's environmental assessment procedures stemmed from domestic legislation to which it was held accountable by environmental groups and the Judiciary. The congressional mandate came later and was supported by the same environmental groups who saw it as a means of strengthening their cause. The importance of these two factors, operating together, can be gauged by comparing environmental assessment with other initiatives, namely the 'women in development' policy and procedures designed to incorporate social considerations into assessment through the use of social soundness analysis.

The pervasive objective of developing its programme and ensuring the necessary disbursement of funds, in order to sustain or expand the agency, is as true of USAID as of any other government organization. What is distinctive about USAID is its degree of decentralization which is designed to ensure that there are sufficient staff in the field to manage the programme. There is a two-part relationship in the administrative system which not only holds the agency accountable for its activities but also offers professional incentives to staff. The first concerns the Washington staff and the need to satisfy Congress and other government agencies. The other concerns the mission staff responsible for developing proposals for the programme and achieving the desired disbursement.

Both Tendler (1975) and Mickelwait (1979) found that there was conflict between being accountable to Congress for the selection and packaging of proposals and the task of managing and implementing effective projects in the field. Washington staff devote their energies to satisfying Congress, thereby withholding from the missions the authority for making decisions about designing, reviewing and, most importantly, modifying projects. Mission staff face keen incentives to generate more projects and achieve greater disbursement, but experience delays and obstructions in the project review process. They attribute these delays to congressional requirements, initiated by special interests, which provide few benefits for the projects themselves. Consequently, their energies are diverted away from project implementation to packaging proposals.

The implications of these observations for individual policy objectives are somewhat disheartening. Policies are translated into requirements for project selection and packaging by the Washington staff. These requirements are imposed on field staff who realize that it is not the effect on project design that is important, but rather the need to assure Congress that policy objectives have been taken into account before funds are committed. Thus, analysis of project impacts or certain design features plays an important advocacy role. Only rarely do policy objectives become positive influences on the design and implementation of projects and even more rarely do they provide opportunities to develop the funding programme. The acid test of the assessment procedures, in the eyes of agency staff, is the extent to which assessments facilitate spending money.

To put it another way, USAID's need to fend off criticism and to satisfy numerous special interests in order to be seen to comply with the mandate from Congress conflicts with the kind of organizational functions needed for its primary task. Many of the policies that the agency is required to promote stem from domestic concerns and, while they may be quite desirable in themselves, are enforced in ways counterproductive to the business of generating and implementing aid projects. Mickelwait (1979), for instance, concludes that the 'new direction' goals cannot be achieved without greater decentralization of both the review and approval of projects. Furthermore, it is considered that a shift of accountability from project proposals to the effects of projects after they have been implemented is required.

Of course, different policy objectives fare differently in the face of various demands of Congress, the Administration, the public and factions within the agency, as can be seen from a comparison of EIA with the 'women in development' policy and the social soundness requirements. These policy objectives are both similar to and different from environmental assessment in important ways. Although the 'women in development' policy came about as a result of a strong outside lobby, there has never been a requirement for analysis of all proposals with respect to its provisions. There are requirements for social soundness analysis, a policy initiative that had its origins within USAID, but it lacks a strong external lobby or active constituency.

The 'women in development' programme started as a relatively symbolic response to a strong outside lobby and has reached a stage where it is helping to facilitate expenditure. In 1974, the agency set up an office specifically to deal with this issue and a commitment has been made to allocate a proportion of the budget to further these ends.

However, it has never become a routine aspect of project preparation and analysis. An active lobby has ensured that USAID's Washington staff respond to the issue and exploit its potential for enhancing the agency's activities. The groups that support the policy, however, have not been able to influence the day-to-day concerns of the field staff.

Social soundness analysis was one of the procedures developed for the 'new directions' programme. Agency staff perceived the need for analysing the socioeconomic organization, culture and attitudes of project beneficiaries and enlisted anthropologists to prepare the procedures and methodology. However, the natural constituency of such a procedure, the anthropological community, has always been ambivalent about alliances with government agencies and was still wary after some unfortunate experiences with USAID during the Vietnam War. Thus, no external lobby materialized to force compliance with these provisions at the crucial stage when staff were being asked to incorporate another facet into assessments. No effective constituency has since emerged.

The comparison of these three policies illustrates that two factors were important for the success of environmental assessment. First, EIA procedures had external legitimacy and, once the court case was settled, there could be no real argument by mission staff against extending the project assessment to include environmental issues. The other is that there was a well-organized, experienced and successful lobbying community to make sure that USAID did not let the matter rest with inadequate regulations and perfunctory assessments. The environmental lobbies had both legitimacy and clout behind their efforts to hold USAID accountable for environmental assessment.

In the absence of strong domestic environmental policy goals, however, the pressure for environmental planning to be extended to foreign assistance would have barely existed. Accountability only penetrates as far as the allocation of funds and the submission of project proposals. There is very little opportunity to probe the effects of projects after implementation. In addition, without the need for economic appraisal, there is no incentive to pay attention to even the direct economic costs of environmental damage.

Conclusions

This analysis shows that environmental considerations would, certainly, have played a role in project review in the absence of a rigorous domestic environmental policy. However, the accountability for performing such a review would have been shallow and would not have brought about genuine reform in how missions identify and prepare project proposals. In addition, there would have been only marginal effects on how projects are assessed. USAID's Washington staff members would have enforced the mandate if they expected Congress to apply environmental criteria to the budgetary process.

In contrast to the situation with capital projects, assessment of proposals in the environmental sector, technical assistance and institution building, probably, would have fared better. Environmental and natural resource management projects fit the orientation of the development assistance programme related to rural development, health and the provision of basic human needs. The missions, in their efforts to match the characteristics of their proposals with the criteria of the development assistance programme, might have found such projects a desirable means of extending their 'portfolio' of activities. The major constraints facing these activities were the low level of demand for such projects from recipient governments and agency staff's general lack of experience and expertise related to this sector. As a result of USAID's compliance with NEPA, however, mission staff are now much more knowledgeable about environmental aspects and have stimulated greater demand for assistance in this area from recipient governments.

Indeed, there has been some convergence between the orientation of assessments towards environmental mitigation, and natural resource projects. A recent case study demonstrates how environmental policy mechanisms are supposed to benefit project design and to promote sustainable natural resource utilization.

The Peruvian government proposed an extensive highway and land colonization development, the Pichas Palcazu Project, in the forested foothills of the Andes. The project enjoyed the personal favour of the Peruvian President and a group of donor institutions were eager to participate. USAID considered providing $22 million for one component of the project in the Palcazu Valley. The government's main emphasis was on building the highway, but USAID, partly at the suggestion of environmental organizations, became concerned about the possible environmental effects of the colonization and cultivation plans.

The environmental assessment, carried out after the project identification document had been produced, became a central input during detailed formulation of the proposals. Using environmental and socioeconomic analyses of the proposed project and comparing other colonization experience in Peru, the consultants concluded that the project would not be sustainable and made a set of proposals indicating how the project could be redesigned. However, the Peruvian government would not accept serious modifications to its highway proposals. Consequently, USAID tried to adapt the colonization plans to ensure that the project would be ecologically sound, using the land suitability analysis carried out as part of the environmental assessment. Accordingly, the development has been renamed the Central Selva Natural Resources Management Project.

There has since been a long delay in executing the project, which has undermined the support of the Peruvian government. In addition, highway construction has also fallen behind schedule, which might make it necessary to modify the colonization plans once more. However, according to USAID's regional environmental specialist at that time, this is how environmental assessment should work (Horberry 1984a). The recipient government proposes a project and the subsequent assessment comes early enough to contribute to the detailed formulation of the proposals. The assessment is organized around testing the feasibility of the original proposal and suggesting alternatives that would avoid the environmental difficulties that have become evident. Thus, the final shape of the project is negotiated with the host government so that its interests and the environmental management priorities can be reconciled.

References

Chapter 1

Ahmad, Y. J. & G. K. Sammy, 1985. *Guidelines to environmental impact assessment in developing countries*. London: Hodder & Stoughton.

Alexander, C. & M. L. Manheim, 1962. *The use of diagrams in highway route location: An experiment*. Department of Civil Engineering Publication No. 161, Cambridge, Mass: Massachusetts Institute of Technology.

Anderson, F. R. 1973. *NEPA in the courts: a legal analysis of the National Environmental Policy Act*. Baltimore: Johns Hopkins University Press.

Bingham, G. 1986. *Resolving environmental disputes: a decade of experience*. Washington DC: The Conservation Foundation.

Bisset, R. 1978. Quantification, decision making and environmental impact assessment in the United Kingdom. *Journal of Environmental Management*, **7**, 43–58.

Caldwell, L. K., R. V. Bartlett, D. E. Parker, & D. L. Keys, 1982. *A study of ways to improve the scientific content and methodology of environmental impact analysis*. Advanced Studies in Science, Technology and Public Affairs. School of Public and Environmental Affairs, Indiana University, Bloomington.

Canter, L. W. 1977. *Environmental impact assessment*. New York: McGraw-Hill.

Canter, L. W. 1984. Environmental impact studies in the United States. In *Perspectives in environmental impact assessment*, B. D. Clark, A. Gilad, R. Bisset & P. Tomlinson (eds), 15–24. Dordrecht: Reidel.

Carpenter, R. A. (ed.) 1983. *Natural systems for development: what planners need to know*. London: Macmillan.

Catlow, J. & C. G. Thirlwall, 1976. *Environmental impact analysis*. DoE Research Report no. 11, Department of the Environment, London.

Cheremisinoff, P. N. & A. C. Morresi, 1977. *Environmental assessment and impact statement handbook*. Ann Arbor, Mich: Ann Arbor Science.

Clark, B. D., R. Bisset & P. Wathern, 1980. *Environmental impact assessment*. London: Mansell.

Clark, B. D., R. Bisset, and P. Wathern, 1981a. The British experience. In *Project appraisal and policy review*, T. O'Riordan & W. D. Sewell (eds), 125–53. Chichester: Wiley.

Clark, B. D., K. Chapman, R. Bisset & P. Wathern, 1976. *Assessment of major industrial applications: a manual*. DoE Research Report no. 13, Department of the Environment, London.

Clark, B. D., K. Chapman, R. Bisset & P. Wathern, 1978. *US environmental impact assessment: a critical review*. DoE Research Report no. 26, Department of the Environment, London.

Clark, B. D., K. Chapman, R. Bisset, P. Wathern & M. Barret, 1981b. *A manual for the assessment of major developments*. London: HMSO.

Collins, J. 1986. *Integrating environmental impact assessment into planning processes in Cheshire*. Paper to CEMP Conference on the EEC Environmental Assessment Directive, 30–31 January 1986, London.

Commission of the European Communities 1980. Draft directive concerning the assessment of the environmental effect of certain public and private projects 7973/80:/ Com (80313 Final). *Official Journal, c169, 9.7.80*, 14–17.

Council on Environmental Quality 1976. *Environmental impact statements: an analysis of six years' experience by seventy federal agencies*. Washington DC: Council on Environmental Quality.

Council on Environmental Quality 1978. National Environmental Policy Act – Regulations. *Federal Register*, **43**, 55978–56007.

Council of the European Communities 1985. On the assessment of the effects of certain public and private projects on the environment. *Official Journal, L175, 28.5.85*, 40–8.

Culhane, P. J. & T. V. Armentano, 1982. State-of-the-art scientific understanding of acid deposition in environmental assessments of fossil fuel powerplants. In *A study of ways to improve the scientific content and methodology of environmental impact analysis*, L. K. Caldwell, R. V. Bartlett & D. L. Keys (eds), 455–75. Advanced Studies in Science, Technology and Public Affairs, School of Public and Environmental Affairs, Indiana University, Bloomington.

Davies, G. S. & F. G. Muller, 1983. *A handbook on environmental impact assessment for use in developing countries*. Report submitted to United Nations Environment Programme, Nairobi.

Dee, N., J. K. Baker, N. L. Drobny, K. M. Duke, I. Whitman & D. C. Fahringer, 1973. An environmental evaluation system for water resource planning. *Water Resources Research*, **9**, 523–35.

Department of Housing and Urban Development 1981. *Areawide environmental assessment: a guidebook*. Washington DC: Department of Housing and Urban Development.

Dooley, J. E. & R. W. Newkirk, 1976. *Corridor selection methods to minimize the impact of an electricity transmission line*. Toronto: MacLaren.

Environmental Protection Agency 1975. *Manual: review of federal actions impacting the environment*. Washington DC: Environmental Protection Agency.

Environmental Protection Agency 1980. *Evaluation of EPA's EIA program for wastewater treatment facilities* Washington DC: Environmental Protection Agency.

Environment Canada 1974. *An environmental assessment of Nanaimo port alternatives*. Ottawa: Environment Canada.

Everitt, R. R. 1983. Adaptive environmental assessment and management: some current applications. In *Environmental impact assessment*, PADC Environmental Impact Assessment and Planning Unit (eds), 293–306. The Hague: Nijhoff.

Fairfax, S. K. 1978. A disaster in the environment movement. *Science*, **199**, 743–8.

Fairfax, S. K. & H. M. Ingram, 1981. The United States experience. In *Project appraisal and policy review*, T. O'Riordan & W. D. Sewell (eds), 29–45. Chichester: Wiley.

Federal Environmental Assessment Review Office 1978. *Guide for environmental screening*. Ottawa: Federal Environmental Assessment Review Office.

Foster, B. J. 1984. Environmental impact assessment in the UK. *Zeitschrift für Umweltpolitik*, **7**, 389–404.

Heer, J. E. & D. J. Hagerty, 1977. *Environmental assessments and statements*. New York: Van Nostrand Reinhold.

Holland, M. C. 1985. Judicial review of compliance with the National Environmental Policy Act, an opportunity for the rule of reason. *Boston College Environmental Law Review*, **12**, 743–90.

Holling, C. S. 1978. *Adaptive environmental assessment and management*. Chichester: Wiley.

Horberry, J. A. J. 1984. *Development assistance and the environment: a question of accountability*. Ph.D. Thesis, Massachusetts Institute of Technology, Cambridge.

in't Anker, M. C. & M. Burggraaff, 1979. Environmental aspects in physical planning. *Planning and Development in the Netherlands*, **11**, 128–45.

Jain, R. K., L. V. Urban and G. S. Stacey, 1977. *Environmental impact assessment*. New York: Van Nostrand Reinhold.

Kennedy, W. V. 1984. US and Canadian experience with environmental impact assessment: relevance for the European Communities? *Zeitschrift für Umweltpolitik*, **7**, 339–66.

Kennedy, W. V. 1986. *Environmental impact assessment and highway planning: a comparative case study analysis of the United States and the Federal Republic of Germany*. Berlin: Edition Sigma.

Klennert, K. (ed) 1984. *Environmental impact assessment (EIA) for development*. Feldafing: Deutsche Stifting für Internationale Entwicklung.

Krauskopf, J. M. & D. C. Bunde, 1972. Evaluation of environmental impact through a computer modelling process. In *Environmental impact analysis: philosophy and methods*, R. B. Ditton & T. I. Goodale (eds), 107–26. University of Wisconsin Sea Grant Program, Madison.

Lee, N. & C. Wood, 1978. Environmental impact assessment of projects in EEC countries. *Journal of Environmental Management*, **6**, 57–71.

Legore, S. 1984. Experience with environmental impact assessment procedures in the USA. In *Planning and ecology*, R. D. Roberts & T. M. Roberts (eds), 103–28. London: Chapman & Hall.

Leistritz, F. L. & B. L. Ekstrom, 1986. *Social impact assessment – an annotated bibliography*. New York: Garland.

Leopold, L. B., F. E. Clark, B. B. Hanshaw & J. R. Balsley, 1971. *A procedure for evaluating environmental impact*. US Geological Survey Circular 645, Department of Interior, Washington DC.

Liroff, R. A. 1985. NEPA litigation in the 1970s: a deluge or a dribble. In *Enclosing the environment*, C. Kury (ed.), 162–77. University of New Mexico School of Law, Albuquerque.

MacDougall, E. B. 1975. The accuracy of overlay maps, *Landscape Planning*, **2**, 23–30.

McHarg, I. 1968. *A comprehensive route selection method*. Highway Research Record 246, Highway Research Board, Washington DC.

Munn, R. E. 1979. *Environmental impact analysis. Principles and procedures*, 2nd edn, SCOPE Report no. 5. Chichester: Wiley.

Nijkamp, P. 1980. *Environmental policy analysis: operational methods and models*. Chichester: Wiley.

Nuclear Regulatory Commission 1976. *Generic environmental statement on the use of plutonium in mixed oxide fuel in light water reactors*. Washington DC: Nuclear Regulatory Commission.

Schaenam, P. S. 1976. *Using an impact measurement system to evaluate land development*. Washington DC: The Urban Institute.

Solomon, R. C., B. K. Colbert, W. J. Hanson, S. E. Richardson, L. W. Canter, & E. C. Vlachos, 1977. *Water resources assessment methodology (WRAM): impact assessment and alternative evaluation*. Technical Report no. Y-77-1, US Army Corps of Engineers, Vicksburg, Miss.

Sorensen, J. C. 1971. *A framework for identifying and control of resource degradation and conflict in the multiple use of the coastal zone*. Masters Thesis, Department of Landscape Architecture, University of California.

Steinitz, C., P. Parker, & L. Jordan, 1976. Hand drawn overlays: their history and prospective uses, *Landscape Architecture*, **66**, 444–55.

United Nations Environment Programme 1980. *Guidelines for assessing industrial environmental impact and environmental criteria for the siting of industry*. Industry and Environment Office, United Nations Environment Programme, Paris.

Walker, B. H., G. A. Norton, G. R. Conway, H. N. Comins, & M. Birley, 1978. A procedure for multidisciplinary ecosystem research: with reference to the South African Savanna Ecosystem Project. *Journal of Applied Ecology*, **15**, 481–502.

Wathern, P. 1984. Methods for assessing indirect impacts. In *Perspectives on environmental impact assessment*, B. D. Clark, A. Gilad, R. Bisset & P. Tomlinson (eds), 213–31. Dordecht: Reidel.

Wathern, P., S. N. Young, I. W. Brown, & D. A. Roberts, 1986. Ecological evaluation techniques. *Landscape Planning*, **12**, 403–20.

Wathern, P., S. N. Young, I. W. Brown, and D. A. Roberts, 1987. Assessing the environmental impacts of policy: a generalised framework for appraisal. *Landscape and Urban Planning*, **14**, 321–30.

Chapter 2

Anon. 1975. *Georges Bank Conference: marine environmental assessment needs on the Georges Bank related to petroleum exploration and development. Proceedings of a conference and workshop*. New England Natural Resources Center, Bentley College, Waltham, Mass., May 1975.

Beanlands, G. E. & P. N. Duinker, 1983. *An ecological framework for environmental assessment in Canada*. Institute for Resource and Environmental Studies, Dalhousie University, Halifax, NS and Federal Environmental Assessment Review Office, Hull, Que.

Beaufort Sea Environmental Assessment Panel 1982. *Beaufort Sea hydrocarbon production proposal: interim report of the environmental assessment panel*. Hull, Que.: Federal Environmental Assessment Review Office.

Cairns, J. 1975. Critical species, including man, within the biosphere. *Naturwissenschaften*, **62**, 193–9.

Council on Environmental Quality 1980. *Environmental quality – the Eleventh Annual Report of the Council on Environmental Quality*. Washington DC: Council on Environmental Quality.

Eberhardt, L. L. 1976. Quantitative ecology and impact assessment. *Journal of Environmental Management*, **4**, 27–70.

Fritz, E. S., P. J. Rago, & I. D. Murarka, 1980. *Strategy for assessing impacts of power plants on fish and shellfish populations FWS/OBS–80/34*. Fish and Wildlife Service, Department of the Interior, Ann Arbor, Mich.

Hirsch, A. 1980. The baseline study as a tool in environmental impact assessment. In *Proceedings of the Symposium on Biological Evaluation of Environmental Impacts FWS/OBS-80/26*, 84–93. Council on Environmental Quality and Fish and Wildlife Service, Department of the Interior, Washington DC.

Imperial Oil Ltd., Aquitaine Co. of Canada and Canadian-Cities Service Ltd. 1978. *Summary: environmental impact statement for exploratory drilling in Davis Strait region*.

McMichael, D. F. 1975. Tiering the EIS to the decision-making process. In *The EIS Technique*, 50–2. The Australian Conservation Foundation, Melbourne.

States, J. B., P. T. Haug, T. G. Shoemaker, L. W. Reed, and E. B. Reed, 1978. *A systems approach to ecological baseline studies FWS/OBS-78/21*. Fish and Wildlife Service. Department of the Interior, Fort Collins, Colo.

Truett, J. C. 1978. Ecosystem process analysis – a new approach to impact assessment. In *Proceedings of the Symposium Energy/Environment '78*, 70–5. Society of Petroleum Industry Biologists, Los Angeles, Calif.

United States Atomic Energy Commission 1973. *Preparation of environmental reports for nuclear power plants*. U.S. Atomic Energy Commission Directorate of Regulatory Standards 1973. Regulatory Guide 4.2, USAEC Regulatory Guide Series, Washington DC.

Walsh, D. A. 1983. *Baseline studies for environmental impact assessment: a Labrador case study*. Masters Dissertation, Institute for Resource and Environmental Studies, Dalhousie University, Halifax, NS.

Chapter 3

Andrews, R. N. L. 1973. *Approaches to impact assessment: comparison and critique*. Paper presented to short course on Impact Assessment in Water Resource Planning, Ann Arbor, Mich., 9 June.

Beanlands, G. E. & P. N. Duinker, 1982. *An ecological framework for environmental impact assessment in Canada*. Institute for Resource and Environmental Studies, Dalhousie University, Halifax, NS.

Bisset, R. 1978. Quantification, decision-making and environmental impact assessment in the United Kingdom. *Journal of Environmental Management*, **7**, 43–58.

Bisset, R. 1984a. Methods for assessing direct impacts. In *Perspectives on environmental impact assessment*, B. D. Clark, A. Gilad, R. Bisset & P. Tomlinson (eds), 195–212. Dordecht: Reidel.

Bisset, R. 1984b. Post-development audits to investigate the accuracy of environmental impact predictions. *Zeitschrift für Umweltpolitik*, **7**, 463–84.

Caldwell, L. K., R. V. Bartlett, D. E. Parker, & D. L. Keys, 1982. *A study of ways to improve the scientific content and methodology of environmental impact analysis*. Advanced Studies in Science, Technology and Public Affairs, School of Public and Environmental Affairs, Indiana University, Bloomington.

Canter, L. W. 1983. Methods for environmental impact assessment: theory and application (emphasis on weighting-scaling checklists and networks). In *Environmental impact assessment*, PADC Environmental Impact Assessment and Planning Unit (eds), 165–233. The Hague: Nijhoff.

Clark, B. D., R. Bisset, and P. Wathern, 1980. *Environmental impact assessment*. London: Mansell.

Collins, J. P. & E. A. Glysson, 1980. Multivariate utility theory and environmental decisions. *Journal of the Environmental Engineering Division, Proceedings of the American Society of Civil Engineers, 106, no. EE4*, 815–30.

Dee, N., J. K. Baker, N. L. Drobny, K. M. Duke, I. Whitman, & D. C. Fahringer, 1973. Environmental evaluation system for water resource planning. *Water Resources Research*, **9**, 523–35.

Doremus, C., D. C. McNaught, P. Cross, T. Fuist, E. Stanley, & B. Youngberg, 1978. An ecological approach to environmental impact assessment. *Environmental Management*, **2**, 245–8.

Environment Canada 1982. *Review and evaluation of adaptive environmental assessment and management*. Vancouver: Environment Canada.

Everitt, R. R. 1983. Adaptive environmental assessment and management: some current applications. In *Environmental impact assessment*, PADC Environmental Impact Assessment and Planning Unit (eds), 293–306. The Hague: Nijhoff.

Fritz, E. S., P. J. Rago, & I. D. Murarka, 1980. *Strategy for assessing impacts of power plants on fish and shellfish populations. FWS/OBS–80/34.* Fish and Wildlife Service, Department of the Interior, Washington DC.

Gallopin, G., T. R. Lee, & M. Nelson, 1980. The environmental dimension in water management: the case of the dam at Salto Grande. *Water Supply and Management*, **4**, 221–41.

Gilliland, M. W. & P. G. Risser, 1977. The use of systems diagrams for environmental impact assessment: procedures and an application. *Ecological Modelling*, **3**, 188–209.

Hollick, M. 1981. The role of quantitative decision-making methods in environmental impact assessment. *Journal of Environmental Management*, **12**, 65–78.

Holling, C. S. (ed). 1978. *Adaptive environmental assessment and management*. Chichester: Wiley.

Keeney, R. L. & H. Raiffa, 1976. *Decisions with multiple objectives: preferences and value tradeoffs*. New York: Wiley.

Keeney, R. L. & G. A. Robilliard, 1977. Assessing and evaluating environmental impacts at proposed nuclear power plant sites. *Journal of Environmental Economics and Management*, **4**, 153–66.

Kirkwood, C. W. 1982. A case history of nuclear power plant site selection. *Journal of the Operational Research Society*, **33**, 353–63.

Lavine, M. J., T. Butler, & A. H. Meyburg, 1978. Bridging the gap between economic and environmental concerns in EIA. *EIA Review*, **2**, 28–32.

Leopold, L. B., F. E. Clark, B. B. Hanshaw, & J. R. Balsley, 1971. *A procedure for evaluating environmental impact*. US Geological Survey Circular 645, Department of the Interior, Washington DC.

Longley, W. L. 1979. An environmental impact assessment procedure emphasizing changes in the organization and function of ecological systems. In *Proceedings of the Ecological Damage Conference*, 355–77. Society of Petroleum Industry Biologists, Los Angeles.

Mekong Secretariat 1982. *Nam Pong Environmental Management Research Project, Phase III*. Mekong Secretariat, United Nations Economic and Social Commission for Asia and the Pacific, Bangkok.

Mongkol, P. 1982. A conceptual development of quantitative environmental impact assessment methodology for decision-makers. *Journal of Environmental Management*, **14**, 301–7.

Odum, H. T. 1971. *Environment, power and society*. New York: Wiley Interscience.

Odum, H. T. 1972. Use of energy diagrams for environmental impact statements. In *Proceedings of the Conference Tools of Coastal Management*, 197–231. Marine Technology Society, Washington DC.

Sanders, F. S. *et al*. 1980. *Strategies for ecological effects assessment at DOE energy activity sites. ORNL/TM-6783*. Oak Ridge National Laboratory, Oak Ridge, Tenn.

Solomon, R. C., B. K. Colbert, W. J. Hanson, S.E. Richardson, L. W. Canter & E. C. Vlachos, 1977. *Water resources assessment methodology (WRAM) – impact assessment and alternative evaluation. Technical Report no. Y-77-1*. Army Corps of Engineers, Vicksburg, Miss.

Sondheim, M. W. 1978. A comprehensive methodology for assessing environmental impact. *Journal of Environmental Management*, **6**, 27–42.

Sonntag, N., R. R. Everitt, & M. J. Staley, 1980. Integration: a role for adaptive environmental assessment and management. In *Proceedings of the Symposium on Effects of Air Pollutants on Mediterranean and Temperate Forest Ecosystems, Report PWS–43*, P. Miller (ed.), Pacific Southwest Forest and Range Experiment Station, Forest Service, Department of Agriculture, Berkeley, Calif.

Srivardhana, R. 1983. *The Nam Pong case study: some lessons to be learned.* Working paper of the East West Environment Policy Institute, East West Center, Hawaii.

Staley, M. J. 1978. *Report of a simulation modelling workshop on the environmental effects of Albert oil sands development.* Vancouver: Environmental and Social Systems Analysts Ltd. (ESSA).

State of Texas General Land Office 1978. *Activity Assessment Routine:* Ecological Systems Component (User's Manual). State of Texas General Land Office, Austin, Tex.

Truett, J. C. 1978. Ecosystem process analysis – a new approach to impact assessment. In *Proceedings of the Symposium Energy/Environment '78*, 70–5. Society of Petroleum Industry Biologists, Los Angeles.

Uys, P. 1981. *Modelling the environmental impact of energy supply in South Africa – a preliminary report.* Research Report 25, Department of Applied Mathematics, University of Pretoria.

von Neumann, J. & O. Morgenstern, 1953. *Theory of games and economic behaviour*, 3rd edn. Princeton, N.J: Princeton University Press.

Ward, D. V. 1978. *Biological environmental impact studies: theory and methods.* New York: Academic Press.

Wathern, P. 1984. Methods for assessing indirect impacts. In *Perspectives on environmental impact assessment*, B. D. Clark, A. Gilad, R. Bisset & P. Tomlinson (eds), 213–31. Dordrecht: Reidel.

Yapijakis, C. 1983. *A comprehensive methodology for project appraisal and environmental protection in multinational water resources development.* Paper presented to the Symposium on Environmental Impact Assessment: Current Status and Future Prospects, Crete, April 10–17.

Chapter 4

Beanlands, G. E. & P. N. Duinker, 1982. *An ecological framework for environmental impact assessment in Canada.* Institute for Resource and Environmental Studies, Dalhousie University, Halifax, NS.

Bisset, R. 1980. Methods for environmental impact analysis: recent trends and future prospects. *Journal of Environmental Management*, **11**, 27–43.

Caldwell, L. K. 1982. *Science and the National Environmental Policy Act.* Alabama: University of Alabama Press.

Canter, L. W. 1977. *Environmental impact assessment.* New York: McGraw-Hill.

Canter, L. W. 1983. *Environmental impact assessment: current status and future directions.* Paper to the symposium on Environmental Impact Assessment, Crete, April 1983.

Clark, B. D., R. Bisset, & P. Wathern, 1980. *Environmental impact assessment.* London: Mansell.

Council on Environmental Quality 1978. National Environmental Policy Act, implementation of procedural provisions, final regulations. *Federal Register,* **43**, 55978–56007.

Council on Environmental Quality 1982. *Summary of the Inquiry on the Implementation of the CEQ-regulations of 1978.* Washington DC: Council on Environmental Quality.

Curtis, F. A. 1983. Integrating environmental mediation. *EIA Impact Assessment Bulletin*, **2**, 17–25.

Elliot, M. L. 1981. Pulling the pieces together: amalgamation in environmental impact assessment. *Environmental Impact Assessment Review*, **2**, 11–38.

Environmental Resources Ltd. 1981. *Scoping guidelines and methodologies*. EIA publication series nos 2, 3 and 4, The Hague: Ministries of Health and Environmental Protection and of Culture, Social Welfare and Recreation.

Environmental Resources Ltd. 1984. *Prediction in environmental impact assessment*. E.I.A. publication series no. 17, The Hague: Ministries of Housing, Physical Planning and Environment and of Agriculture and Fisheries.

Environmental Resources Ltd. 1985. *Handling uncertainty in prediction*. E.I.A. publication series no. 18, The Hague: Ministries of Housing, Physical Planning and Environment and of Agriculture and Fisheries.

Environmental Resources Ltd. 1986. *Management of risks and uncertainties in environmental policy*. The Hague: Ministry of Housing, Physical Planning and Environment.

Fischhoff, B. & P. J. M. Stallen, 1985. *Alternative strategies for risk assessment*. Internal report, Studiecentrum voor Technologie en Belied, TNO, Apeldoorn, The Netherlands.

Friesema, H. P. 1982. The scientific content of environmental impact statements: workshop conclusions. In *A study of ways to improve the scientific content and methodology of environmental impact analysis*, L. K. Caldwell, R. V. Bartlett, D. E. Parker & D. L. Keys (eds), 479–96. Bloomington Advanced Studies in Science, Technology and Public Affairs, School of Public and Environmental Affairs, Indiana University.

Harter, P. J. 1982. Negotiating regulations: a cure for the malaise? *Environmental Impact Assessment Review*, **3**, 75–91.

Hickling, A. 1974. *Managing Decisions*. Rugby, War.: Mantac.

Hickling, A. 1975. *Aids to strategic choice*. Vancouver Centre for Continuing Education, University of British Columbia.

Hickling, A., R. Hartman, & J. G. Meester, 1976. *Werken met strategische keuze, een toepassing in de Ruimtelijke Planning*. Alphen aan de Rijn, The Netherlands: Samson uitgeverij.

Hobbs, B. F. 1985. Choosing how to choose: comparing amalgamation methods for environmental impact assessment. *Environmental Impact Assessment Review*, **5**, 301–19.

Holling, C. S. (ed.) 1978. *Adaptive environmental assessment and management*. Chichester: Wiley.

De Jongh. P. E. 1983. *Results and conclusions of recent Dutch studies on EIA methodologies and predictive methods*. Paper to the symposium on Environmental Impact Assessment, Crete, April 1983.

De Jongh, P. E. 1984. *From science to decision-making, the Dutch study on predictive methods*. Paper to the Dutch–US Seminar on Environmental Management, Washington DC, April 1984.

De Jongh, P. E. 1985a. *Environmental impact assessment: methodologies, prediction and uncertainty*. Paper to the Congress of the International Association of Impact Assessment, Utrecht, June 1985.

De Jongh, P. E. 1985b. *Some remarks on technical aspects of training in EIA with emphasis on ecological impacts. Report on the Seminar on Training for Environmental Assessment*. European Institute for Public Administration, Maastricht, September 1985.

Magness, T. H. 1984. *NEPA, the regulations of the US Council on Environmental Quality and the heart of EIA*. Paper to the Dutch–US Seminar on Environmental Management, Washington DC, April 1984.

Ministry of Housing, Physical Planning and Environment 1984. *Brochure on environmental impact assessment in the Netherlands*. The Hague: Ministry of Housing, Physical Planning and Environment.

Ministry of Housing, Physical Planning and Environment 1985. *Indicative environmental programme for 1986 to 1990*. The Hague: Ministry of Housing, Physical Planning and Environment.

Otway, H. J. & M. Peltu, 1985. *Regulating industrial risks, science, hazards and public protection*. Sevenoaks: Butterworth.

Rau, J. G. & D. C. Wooten, 1980. *Environmental impact analysis handbook*. New York: McGraw-Hill.

Ruckelshaus, W. D. 1983. Science, risk and public policy. *Science*, **221**, 1026–8.

Susskind, L. & C. Ozawa, 1983. Mediated negotiation in the public sector. *American Behavioural Scientist*, **27**, 255–79.

Susskind, L., L. Bacow & M. Wheeler, 1983. *Resolving environmental regulatory disputes*. Cambridge, Mass.: Schenkman.

Susskind, L., J. R. Richardson, & K. J. Hildebrand, 1978. *Resolving environmental disputes approaches to intervention, negotiation and conflict resolution*. EIA Project Paper, Laboratory of Architecture and Planning, Massachusetts Institute of Technology.

Voogd, J. H. 1982. *Multicriteria evaluation for urban and regional planning*. Dissertation, Eindhoven Technical University, Eindhoven, The Netherlands.

Ward, D. V. 1978. *Biological environmental impact studies, theory and methods*. New York: Academic Press.

Wondolleck, J. 1985. The importance of process in resolving environmental disputes. *Environmental Impact Assessment Review*, **5**, 341–56.

Chapter 5

Andrews, R. N. L. 1976. *Environmental policy and administrative change*. Lexington, Mass.: Lexington Books.

Atkisson, A. A., M. E. Kraft, & L. L. Philipson, 1985. *Risk analysis methods and their employment in governmental risk management*. Technical Report no. PRA 85–1398–1 to the National Science Foundation, J. H. Wiggins, Redondo Beach, Calif.

Beanlands, G. E. 1984a. *Environmental health impact assessment*. Paper to CEMP conference on Environmental Health Impact Assessment, Adana, Turkey, 16 November–5 December 1984.

Beanlands, G. E. 1984b. *Selected EIA procedures: Canada*. Paper to CEMP conference on Environmental Health Impact Assessment, Adana, Turkey. 16 November–5 December 1984.

Beanlands, G. E. & P. N. Duinker, 1983. *An ecological framework for environmental impact assessment in Canada*. Institute for Resource and Environmental Studies, Dalhousie University, Halifax, NS.

Caldwell, L. K., R. V. Bartlett, D. E. Parker and D. L. Keys, 1982. *A study of ways to improve the scientific content and methodology of environmental impact analysis*. Advanced Studies in Science, Technology and Public Affairs, School of Public and Environmental Affairs, Indiana University, Bloomington.

Canter, L. W. 1983. *Risk assessment*. Paper to the American Society for Public Administration Region VII Meeting, Oklahoma City, 5–7 October 1983.

Clark, B. D. 1984a. *Basic concepts of environmental impact assessment and environmental health impact assessment*. Paper to CEMP conference on Environmental Health Impact Assessment, Adana, Turkey, 26 November–5 December 1984.

Clark, B. D. 1984b. *Selected EIA procedures: the European Economic Community directive on environmental assessment*. Paper to CEMP conference on Environmental Health Impact Assessment, Adana, Turkey, 26 November–5 December 1984.

Cohen, A. V. & B.G. Davies, 1981. The wider implications of the Canvey Island study. In *Technological risk assessment*, R. F. Ricci (ed.), 101–10. Boston: Nijhoff.

Conservation Foundation 1984. *State of the environment: an assessment at mid-decade*. Washington DC: The Conservation Foundation.

Covello, V. T. & J. P. Fiksel, 1985. *The suitability and applicability of risk assessment methods for environmental applications of biotechnology*. Report no NSF/PRA 8502286, National Science Foundation, Washington DC.

Covello, V. T. & J. Menkes, in press. *Risk assessment and risk management methods: the state of the art*. Washington DC: National Science Foundation.

Covello, V. T. & J. Mumpower, 1985. Risk analysis and risk management: an historical perspective. *Risk Analysis*, **5**, 103–20.

Council on Environmental Quality 1985. *Draft Amendment to 40 C.F.R. 1502.22 'Worst case analysis'*. Council of Environmental Quality memo, Washington DC.

Dooley, J. E., in press. Risk theory and the environmental assessment process. In *Environmental impact assessment, technology assessment, and risk analysis: contributions from the psychological and decision sciences*, V. T. Covello (ed.). Heidelberg: Springer.

Environmental Protection Agency 1984. *Risk assessment and risk management*. Washington DC: Environmental Protection Agency.

Fischhoff, B., S. Lichtenstein, P. Slovic, S. L. Derby & R. L. Keeney, 1981. *Acceptable risk*. Cambridge: Cambridge University Press.

Giroult, E. 1984. *The health component of environmental impact assessment*. Paper to the CEMP conference on Environmental Health Impact Assessment, Adana, Turkey, 26 November–5 December 1984.

Haemisegger, E. R., A. D. Jones, & F. L., Reinhardt, 1985. EPA's experience with assessment of site-specific environmental problems: a review of IEMD's geographic study of Philadelphia. *Journal of Air Pollution Control Association*, **35**, 809–15.

Hattis, D. & J. A. Smith, 1985. *What's wrong with quantitative risk analysis?* Paper to conference on Moral Issues and Public Policy Issues in the Use of the Method of Quantitative Risk Assessment, Georgia State University, Atlanta, 26–7 September 1985.

Holling, C. S. 1978. *Adaptive environmental assessment and management*. Chichester: Wiley.

Kleindorfer, P. & T. H. Yoon, 1984. *Toward a theory of strategic problem formulation*. Paper to the Fourth Annual Strategic Management Society, Philadelphia, Pa.

Martin, J. E. 1984. *Methods for environmental health impact assessment*. Paper to the CEMP conference on Environmental Health Impact Assessment, Adana, Turkey, 26 November–5 December 1984.

Mason, R. & I. Mitroff, 1981. *Challenging strategic planning assumptions*. Chichester: Wiley.

Munn, R. E. (ed.) 1979. *Environmental impact assessment: principles and procedures*, 2nd edn, SCOPE report 5. Chichester: Wiley.

National Research Council 1982. *Risk assessment in the federal government: managing the process*. Washington DC: National Research Council.

Northwest Coalition for Alternatives to Pesticides 1985. *Comments on the Council on Environmental Quality Draft Amendment to 40 C.F.R. 1502.22 'Worst case analysis'. Statement*. Eugene, Oreg.

O'Riordan, T. 1979. EIA and RA in a management perspective. In *Energy risk management*, G. Goodman & W. D. Rowe (eds) London: Academic Press.

O'Riordan, T. 1982. Risk-perception studies and policy priorities. *Risk Analysis*, **2**, 95–100.

O'Riordan, T. in press. The impact of EIA on decision making. In *Environmental impact assessment, technology assessment, and risk analysis: contributions from the psychological and decision sciences*. V. T. Covello (ed.). Heidelberg: Springer.

Otway, H. & K. Thomas, 1982. Reflections on risk perception and policy. *Risk Analysis*, **2**, 69–82.

Popper, F. J. 1983. LP/HC and LULUs: the political uses of risk analysis in land use planning. *Risk Analysis*, **3**, 255–63.

Reeve, M. 1984. Scientific uncertainty and the National Environmental Policy Act – the Council on Environmental Quality's regulation 40 C.F.R. section 1502.22. *Washington Law Review*, **60**, 101–16.

Spangler, M. 1982. The role of interdisciplinary analysis in bridging the gap between the technical and human sides of risk assessment. *Risk Analysis*, **2**, 101–14.

Starr, C. 1985. Risk management, assessment, and acceptability. *Risk Analysis*, **5**, 97–102.

Susskind, L. E. 1985. The siting puzzle; balancing economic and environmental gains and losses. *Environmental Impact Assessment Review*, **5**, 157–63.

Vari, A., J. Vecsenyi, & Z. Paprika, 1985. *Supporting problem structuring in high level decisions: the case of the siting of a hazardous waste incinerator*. Paper to the tenth SPUDM conference, Helsinki, Finland, August 1985.

Vlachos, E., in press. Assessing long-range cumulative impacts. In *Environmental impact assessment, technology assessment, and risk analysis: contributions from the psychological and decision sciences*. T. Covello (ed.). Heidelberg: Springer.

Wilson, R. & E. Crouch, 1982. *Risk–benefit analysis*. Cambridge, Mass.: Ballinger.

Chapter 6

Cheshire County Council 1986. *Mersey Marshes local plan: written statement and proposals map*. Chester: Cheshire County Council.

Collins, J. 1986. *Integrating environmental impact assessment into the planning process in Cheshire*. Paper to the CEMP conference on the EEC Environmental Assessment Directive, London, January 1986.

Commission of the European Communities 1980. Draft directive on the assessment of the environmental effects of certain public and private projects. *Official Journal*, **c169**, 14–17.

Council of the European Communities 1985. On the assessment of the effects of certain public and private projects on the environment. *Official Journal*, **L175**, 40–8.

Foster, B. J. 1983. *Land-use and plan-making: the role of EIA in forward planning*. Paper to the symposium on Environmental Impact Assessment, Crete, April 1983.

Hall, R. 1977. MEIRS – a method for evaluating the environmental impacts of general plans. *Water, Air and Soil Pollution*, **7**, 251–60.

Healy, P. 1986. *Local plans in British land use planning*. Oxford: Pergamon.

Holling, C. S. (ed.) 1978. *Adaptive environmental assessment and management*. Chichester: Wiley.

Housing and Urban Development (U.S. Department of) 1978. *New Castle County I–95/Route 40 growth corridor: final area-wide environmental impact statement*. Washington DC: Department of Housing and Urban Development.

Housing and Urban Development (U.S. Department of) 1981. *Areawide environmental impact assessment: a guidebook*. Washington DC: Department of Housing and Urban Development.

in't Anker, M. C. & M. Burggraaff 1979. Environmental aspects in physical planning. *Planning and Development in the Netherlands*, **11**, 128–45.

Jones, M. G. 1981. Environmental impact assessment at the planning level: two systems. In *Environmental impact assessment*, Project Appraisal for Development Control (eds), 63–94. The Hague: Nijhoff.

Lee, N. & C. Wood, 1978. EIA – a European perspective. *Built Environment*, **4**, 101–10.

Lee, N. & C. Wood, 1980. *Methods of environmental impact assessment for use in project appraisal and physical planning*. Occasional paper 7, Department of Town and Country Planning, University of Manchester.

Lee, N. & C. Wood, 1983. The effects of population and industrial change on environmental quality in major conurbations. *Journal of Environmental Management*, **16**, 91–107.

Lee, N., C. Wood & V. Gazidellis, 1985. *Arrangements for environmental impact assessment and their training implications in the European Communities and North America: country studies*. Occasional paper 13, Department of Town and Country Planning, University of Manchester.

Lichfield, N., P. Kettle & M. Whitbread, 1975. *Evaluation in the planning process*. Oxford: Pergamon.

Local Government Operational Research Unit 1976. *Development plan evaluation and robustness*. Research Report 5, Department of the Environment, London.

Lyddon, W. D. C. 1983. Land use and environmental planning for the development of North Sea gas and oil: the Scottish experience. *Environmental Impact Assessment Review*, **4**, 473–92.

McAllister, D. M. 1980. *Evaluation in environmental planning, assessing environmental, social, economic and political trade-offs*. Cambridge, Mass.: MIT Press.

McElligott, C. R. 1978. The HUD areawide environmental impact statement: an innovation. *Environmental Impact Assessment Report (1978)*, 18–22.

McHarg, I. 1969. *Design with nature*. New York: Natural History Press.

Merrill, F. 1981. Areawide environmental impact assessment guidebook. *Environmental Impact Assessment Review*, **2**, 204–8.

O'Riordan, T. & W. R. D. Sewell (eds) 1982. *Project assessment and policy review*. Chichester: Wiley.

Roberts, M. 1974. *Introduction to town planning techniques*. London: Hutchinson.

Rodgers, J. L. 1976. *Environmental impact assessment, growth management and the comprehensive plan*. Cambridge, Mass.: Ballinger.

Williams, R. H. (ed.) 1984. *Planning in Europe*. London: Allen & Unwin.

Wood, C. & N. Lee, 1978. *Physical planning in the member states of the European Community*. Occasional paper 2, Department of Town and Country Planning, University of Manchester.

Wood, C. 1976. *Town planning and pollution control*. Manchester: Manchester University Press.

Chapter 7

Andrews, W. H., G. E. Madsen & G. J. Lagaz, 1974. *Social impacts of water resource development and their implications for urban and rural development: a post-audit analysis of the Weber Basin Project in Utah*. Research monograph 4, Institute for Social Science Research on Natural Resources, University of Utah, Logan.

Bisset, R. 1981. Problems and issues in the implementation of EIA audits. *Environmental Impact Assessment Review*, **1**, 379–96.

Bisset, R. 1984. Post development audits to investigate the accuracy of environmental impact predictions. *Zeitschrift für Umweltpolitik*, **7**, 463–84.

Bureau of Land Management 1975. *1974 Barstow–Las Vegas Motorcycle Race: Evaluation Report*. Department of Interior, Sacramento, Calif.

Carley, M. 1982. Lessons for impact monitoring. *IAIA Bulletin*, **1**, 76–9.

Environmental Protection Agency 1980. *Wisconsin Power Plant Impact Study*. Duluth, Minn.: Environmental Protection Agency.

Gilmore, J. S., D. M. Hammond, J. M. Uhlmann, K. D. Moore & D. C. Coddington, 1980. The impacts of power plant construction: a retrospective analysis. *Environmental Impact Assessment Review*, **1**, 417–20.

Gore, K. L., J. M. Thomas & D. G. Watson, 1979. Quantitative evaluation of environmental impact assessment based on aquatic monitoring programs at three nuclear power plants. *Journal of Environmental Management*, **8**, 1–7.

Green, R. H. 1979. *Sampling design and statistical methods for environmental biologists*. Chichester: Wiley.

Harvey, T. 1981. Environmental interventions: the monitoring paradigm. I. The monitoring concept and practices of descriptive monitoring. *The Environmentalist*, **1**, 283–91.

Institute for Environmental Studies 1977. *Documentation of environmental change related to the Columbia Electric Generating Station*. IES Report 82, Institute of Environmental Studies, University of Wisconsin-Madison, Madison, Wis.

Johnson, W. C. & S. P. Bratton, 1978. Biological monitoring in UNESCO biosphere reserves with special reference to the Great Smoky Mountains National Park. *Biological Conservation*, **13**, 105–15.

Leistritz, F. L. & K. C. Maki, 1981. *Socio-economic effects of large-scale resource development projects in rural areas: the case of McLean County, North Dakota*. Department of Agricultural Economics, North Dakota State University, Fargo, N. Dak.

Marmer, G. J. & A. J. Policastro undated. *Evaluation of utility monitoring and pre-operational hydrothermal modelling at three nuclear power plant sites*. Argonne National Laboratory, Argonne, Ill.

Murarka, I. P. 1976. *An evaluation of environmental data relating to selected nuclear power plant sites: Prairie Island generating site*. ANL/E156 (Prairie Island). Argonne National Laboratory, Argonne, Ill.

Murdock, S. H., F. L. Leistritz, R. R. Hamm & S.-S. Hwang, 1982. An assessment of socio-economic assessments: utility, accuracy, and policy considerations. *Environmental Impact Assessment Review*, **3**, 333–50.

Power Station Impacts Research Team 1979. *The socio-economic effects of power stations on their localities*. Department of Town and Country Planning, Oxford Polytechnic, Oxford.

Skalski, J. R. & D. H. McKenzie, 1982. A design for aquatic monitoring programs. *Journal of Environmental Management*, **14**, 237–51.

Tomlinson, P. (ed.) 1987. Environmental audits special edition. *Environmental Monitoring and Assessment*, **8** (3), 183–261.

Chapter 8

Beanlands, G. E. *et al.* (eds) 1986. *Cumulative environmental effects: a bi-national perspective.* Canadian Environmental Assessment Research Council, Hull, Que.

Beanlands, G. E. & P. Duinker, 1983. *An ecological framework for environmental impact assessment.* Institute for Resource and Environmental Studies, Dalhousie University, Halifax, NS.

Beaufort Sea Environmental Assessment Panel 1984. *Beaufort Sea hydrocarbon production and transportation. Final Report of the Environmental Assessment Panel.* Federal Environmental Assessment Review Office, Hull, Que.

Bisset, R. 1980. Problems and issues in the implementation of EIA audits. *Environmental Impact Assessment Review*, **1**, 379–95.

Boothroyd, P. & W. E. Rees, 1984. *Impact assessment from pseudo-science to planning process.* Discussion paper no. 3, School of Community and Regional Planning, University of British Columbia, Vancouver.

Caldwell, L. K., R. V. Bartlett, D. E. Parker & D. L. Keys, 1982. *A study of ways to improve the scientific content and methodology of environmental impact analysis.* Advanced Studies in Science, Technology and Public Affairs, School of Public and Environmental Affairs, Indiana University, Bloomington.

Case, E. S., P. Z. R. Finkle & A. R. Lucas (eds) 1983. *Fairness in environmental and social impact assessment processes.* The Canadian Institute of Resources Law, Calgary, Alta.

Clark, S. D. (ed.) 1981. *Environmental assessment in Australia and Canada.* Westwater Research Centre, University of British Columbia, Vancouver.

Clark, W. C. & R. E. Munn (eds) 1985. *Sustainable development of the biosphere.* Cambridge: Cambridge University Press.

Cornford, A. B., J. O'Riordan & B. Sadler, 1985. Planning assessment and implementation: a strategy for integration. In *Environmental protection and resource development: convergence for today.* B. Sadler (ed.), 47–76. Calgary, Alta: University of Calgary Press.

Dorcey, A. H. J. & B. R. Martin, 1985. *Impact assessment, monitoring and management: a case study of the Utah and Amax mines.* Report to Environment Canada, Hull, Que.

Erickson, D. 1985. *A comparative analysis of two social impact assessment studies.* Background paper no. 2, Canadian Environmental Assessment Research Council, Hull, Que.

Everitt, R. R. & N. C. Sonntag, 1985. *Follow-up study of environmental assessments for selected frontier oil and gas projects.* Report to Environment Canada, Hull, Que.

Fee Yee Consulting Ltd. 1985. *A downstream perspective: Dene concerns with the environmental assessment, monitoring and surveillance of the Norman Wells project.* Report to the Dene Nation, Yellowknife, N.W.T.

Fenge, T., I. Fox, B. Sadler & S. Washington, 1985. A proposed port on the North Slope of Yukon: the anatomy of conflict. In *Environmental protection and resource development: convergence for today.* B. Sadler (ed.). Calgary, Alta: University of Calgary Press.

Hecky, R. E., R. W. Newbury, R. A. Bodacy, K. Patalas & D. M. Rosenberg, 1984. Environmental impact prediction and assessment: the South Lake experience. *Canadian Journal of Fisheries and Aquatic Science*, **41**, 579–90.

Holling, C. S. (ed.) 1978. *Adaptive environmental assessment and management.* Chichester: Wiley.

Jakimchuk, R. D., C. D. Schick, L. G. Sopuck & Y. K. Olyn, 1985. *Follow-up to environmental assessments for selected projects in Canada.* Report to Federal Environmental Assessment Review Office, Hull, Que.

Janes, S. H. & W. A. Ross, 1985. *Follow-up study to the Banff Highway Twinning Project, Alberta*. Report to Environment Canada, Hull, Que.

Kiell, D. J., J. L. Barnes & E. L. Hill, 1985. *Follow-up to environmental impact assessment: Hinds Lake, Upper Salmon and Cat Arm hydro-electric developments in Newfoundland*. Report to Environment Canada, Hull, Que.

Larminie, T. G. 1984. Control, consultation or incentives for environmental management. *Industry and Environment, 1984 (July/August/September)*, 12–14.

McCallum, D. 1985. *Environmental follow-up to federal projects*. Draft report to Environment Canada, Hull, Que.

Moncrieff, I., L. Torrens & E. Mackintosh, 1985. *An environmental performance audit for selected pipeline projects in southern Ontario*. Report to Environment Canada, Hull, Que.

Munn, R. E. (ed.) 1979. *Environmental impact assessment: principles and procedures*. Chichester: Wiley.

Munro, D. A., T. J. Bryant & A. Matte-Baker, 1986. *Learning from experience: a state of the art review and evaluation of environmental impact assessment audits*. Canadian Environmental Assessment Research Council, Hull, Que.

O'Riordan, T. 1976. Policy making and environmental management: some thoughts on processes and research ideas. *Natural Resources Journal*, **16**, 55–72.

O'Riordan, T. & W. R. D. Sewell (eds) 1981. *Project appraisal and policy review*. Chichester: Wiley.

Rosenberg, D. M. *et al*. 1981. Recent trends in environmental assessment. *Canadian Journal of Fisheries and Aquatic Science*, **38**, 591–624.

Rowsell, J. A. & P. Seidl, 1985. *The effectiveness of the environmental assessment and review process as applied to the Sarnia Montreal pipeline*. Report to Environment Canada, Hull, Que.

Ruggles, P. 1985. *Follow-up ecological studies at the Wreck Cove hydroelectric development, Nova Scotia*. Report to Environment Canada, Hull, Que.

Sadler, B. (ed.) 1980. *Public participation in environmental decision making: strategies for change*. Environmental Council for Alberta, Edmonton, Alta.

Sadler, B. 1981. The regulation of the Red Deer River: conflict and choice. In *Water problems and policies*, W. D. R. Sewell & M. L. Barker (eds) Department of Geography, University of Victoria, Victoria, BC.

Sadler, B. 1983. Fairness and existing processes: a policy review. In *Fairness in environmental and social impact assessment processes*, E. S. Case, P. Z. R. Finkle & A. R. Lucas (eds), 101–7. The Canadian Institute of Resource Law, Calgary, Alta.

Sadler, B. 1984. *Energy development on the Arctic frontier of Canada: an analysis of project decision making*. Paper to the international seminar on Environmentally Sound Development in the Mining and Energy Industries, Chania, Greece, October 1984.

Sadler, B. 1985. *Process evaluation in EIA: a review of Canadian experience*. Paper to the sixth international Seminar on Environmental Impact Assessment, Aberdeen, Scotland.

Sadler, B. 1986a. Impact assessment in transition: a framework for redeployment. In *Integrated approaches to resource planning*, R. Lang (ed.) 99–129. University of Calgary, Calgary, Alta.

Sadler, B. 1986b. Environmental conflict resolution in Canada. *Resolve*, **18**, 1–4.

Sadler. B. (ed.) 1987. *Audit and Evaluation in Environmental Assessment and Management: Canadian and International Experience* 2 vols. Hull, Que.: Environment Canada.

Spencer, R. B. 1985. *Shakwak follow up investigation*. Report to Environment Canada, Hull, Que.

Wiebe, J. D., E. H. Hustan & S. Hum, 1984. *Environmental planning for large scale development projects*. Vancouver: Environment Canada.

Zallen, M., J. McDonald & P. Richwa, 1985. *Follow up review of projected and residual impacts within the Coquihalla Valley, British Columbia*. Report to Environment Canada, Hull, Que.

Chapter 9

Canter, L. W. 1983. *A review of recent research on the utility of environmental impact assessment*. Paper to symposium on Environmental Impact Assessment, April 1983, Chania, Crete.

Cook, P. L. 1979. *Costs of environmental impact statements and the benefits they yield to improvements to projects and opportunities for public involvement*. Paper to the Economic Commission for Europe Seminar on Environmental Impact Assessment, Villach, Austria.

Council on Environmental Quality 1976. *Environmental impact statements: an analysis of six years' experience by seventy federal agencies*. Washington DC: Council on Environmental Quality.

Dean, F. E. 1979. *The use of environmental impact analysis by the British gas industry*. Paper to the symposium on Practices in Environmental Impact Assessment, European Commission, Brussels.

Lee, N. 1984. *Training for environmental impact assessment, main report*. Report to the Commission of the European Communities, Brussels.

Lee, N. & C. M. Wood, 1979. *Methods of environmental impact assessment for use in project appraisal and physical planning*. Occasional paper 7, Department of Town and Country Planning, University of Manchester.

Lee, N. & C. M. Wood, 1985. Training for environmental impact assessment within the European Economic Community. *Journal of Environmental Management*, **21**, 271–86.

Lee, N., C. M. Wood & V. Gazidellis, 1985. *Arrangements for environmental impact assessment and their training implications in the European Communities and North America: country studies*. Occasional paper 13, Department of Town and Country Planning, University of Manchester.

Lee, N. 1987. *Environmental impact assessment: a training guide*, Occasional paper 18, Department of Town and Country Planning, University of Manchester.

Wood, C. M. 1985. The adequacy of training for EIA in the USA. *Environmental Impact Assessment Review*, **5**, 321–37.

Wood, C. M. & V. Gazidellis, 1985. *A guide to training materials for environmental impact assessment*, Occasional paper 14, Department of Town and Country Planning, University of Manchester.

Wood, C. M. & N. Lee (eds) 1987. *Environmental impact assessment: five training case studies*, Occasional paper 19, Department of Town and Country Planning, University of Manchester.

Chapter 10

Ackerman, B. & R. Stewart, 1985. Reforming environmental law. *Stanford Law Review*, **37**, 1333–65.

Albany Law Review 1982. Symposium on the New York State Environmental Quality Review Act, *Albany Law Review* **46**(4), 1097–306.

Anderson, F. 1973. *NEPA in the courts*. Baltimore: Johns Hopkins University Press.

Andrews, R. 1973. After mammoth: *Friends of Mammoth* and the amended California Environmental Policy Act. *Ecology Law Quarterly*, **3**, 349–89.

Andrews, R. N. L. 1976. *Environmental policy and administrative change*. Lexington, Mass.: Lexington Books.

Axelrod, R. 1981. The emergence of cooperation among egoists. *American Political Science Review*, **75**, 306–18.

Bass, R. 1983. CEQA 1983: alive! but how well? *California Regulatory Law Reporter*, **3(4)**, 3–6.

Belsky, M. 1984. Environmental policy law in the 1980s: shifting back the burden of proof. *Ecology Law Quarterly*, **12**, 1–88.

Berzok, L. 1986. The role of impact assessment in environmental decision making in New England: a ten year retrospective. *Environmental Impact Assessment Review*, **6**, 103–33.

Caldwell, L. K. 1982. *Science and the National Environmental Policy Act*. Alabama: University of Alabama Press.

Caldwell, L. K., R. V. Bartlett & D. L. Keys, 1982. *A study of ways to improve the scientific content and methodology of environmental impact analysis*. Advanced Studies in Science, Technology and Public Affairs, School of Environmental Affairs, Indiana University, Bloomington.

CEQA/Housing Task Force 1984. *Recommendations to improve implementation of the California Environmental Quality Act. A report to the Governor*. Business and Transportation Agency, Sacramento, Calif.

Diver, C. 1982. Engineers and entrepreneurs: the dilemma of public management. *Journal of Policy Analysis and Management*, **1**, 402–6.

Eastman, J. 1981. Probable future projects: their role in environmental assessments under the California Environmental Quality Act. *Santa Clara Law Review*, **21**, 727–49.

Fairfax, S. 1978. A disaster in the environmental movement. *Science*, **199**, 743–8.

Fulton, W. 1985. The environmental push and pull. *California Lawyer*, **5(3)**, 51–4.

Futrell, W. 1985. NEPA, environment's 'quiet teenager', an error to assume all is well. *Environmental Forum*, **39** (*Jan. 1985*), 43–4.

George, A. 1972. The case for multiple advocacy. *American Political Science Review*, **66**, 751–85.

Hawkins, K. 1984. *Environment and enforcement*. Oxford: Clarendon Press.

Heclo, H. 1974. *Social policy in Britain and Sweden*. New Haven, Conn.: Yale University Press.

Heclo, H. 1978. Issue networks and the executive establishment. In *The New American Political System*, A. King (ed), 87–124. Washington DC: American Enterprise Institute.

Herson, A. 1986. Project mitigation revisited: most courts approve findings of no significant impact justified by mitigation. *Ecology Law Quarterly*, **13**, 51–72.

Hill, N. 1975. What is going on in Sacramento? *California EIR Monitor*, **2(7)**, 1–9.

Hill, N. 1983. Amendments to guidelines adopted. *California EIR Monitor*, **10(1)**, 1–10.

Kaufman, F.-X., G. Majone & V. Ostrom (eds) 1986. *Guidance, control and evaluation in the public sector*. Berlin: Walter de Gruyter.

Landau, M. 1969. Redundancy, rationality, and the problem of duplication and overlap. *Public Administration Review*, **29**, 346–58.

Landau, M. 1973. On the concept of a self-correcting organization. *Public Administration Review*, **33**, 533–42.

Latin, H. 1985. Ideal versus real regulatory efficiency: implementation of uniform standards and 'fine-tuning' regulatory reforms. *Stanford Law Review*, **37**, 1267–1332.

Lee, N. 1982. The future development of environmental impact assessment. *Journal of Environmental Management*, **14**, 71–90.

Lewis, E. 1980. *Public entrepreneurship*. Bloomington: Indiana University Press.

Lindblom, C. 1965. *The intelligence of democracy*. New York: The Free Press.

Liroff, R. 1976. *A national policy for the environment*. Bloomington: Indiana University Press.

Macaulay, S. 1963. Non-contractual relations in business: a preliminary study. *American Sociological Review*, **28**, 55–67.

Mazmanian, D. & J. Nienaber, 1979. *Can organizations change?* Washington DC: Brookings Institution.

Mnookin, R. & W. Kornhauser, 1979. Bargaining in the shadow of the law: the case of divorce. *Yale Law Journal*, **88**, 950–97.

Murchison, K. 1984. Does NEPA matter? An analysis of the historical development and contemporary significance of the National Environmental Policy Act. *University of Richmond Law Review*, **18**, 557–614.

Nevins, H. 1984. The application of emergency exemptions under CEQA: loopholes in need of amendment? *Pacific Law Journal*, **15**, 1089–1126.

Pearlman, K. 1977. State environmental policy acts: local decision making and land use planning. *AIP Journal*, **43**, 42–53.

Perlstein, J. 1981. Substantive enforcement of the California Environmental Quality Act. *California Law Review*, **69**, 112–88.

Planning and Conservation League Foundation 1985. *Citizen's guide to the California Environmental Quality Act*. Sacramento, Calif.: Planning and Conservation League Foundation.

Rabin, R. 1985. Some thoughts on the dynamics of continuing relations in the administrative process. *Wisconsin Law Review*, *1985*, 741–9.

Renz, J. 1984. The coming of age of state environmental policy acts. *Public Land Law Review*, **5**, 31–54.

Roberts, J. 1985. *CEQA update. An unofficial report from Sacramento*. Paper to the 1985 annual conference of the Association of Environmental Professionals, Sacramento, April 1985.

Rose, R. 1986. *Steering the ship of state: one tiller but two pairs of hands*. Studies in Public Policy 154, Centre for the Study of Public Policy, University of Strathclyde, Glasgow.

Rosenberg, D. M. *et al*. 1981. Recent trends in environmental impact assessment. *Journal of Fisheries and Aquatic Sciences*, **38**, 591–624.

Sax, J. 1973. The (unhappy) truth about NEPA. *Oklahoma Law Review*, **26**, 239–48.

Selmi, D. 1984. The judicial development of the California Environmental Quality Act. *UCD Law Review*, **18**, 197–286.

Seyman, J. 1986. *Cases and materials on CEQA*. Unpublished materials prepared for the Environmental Law Class, School of Environmental Law, University of California, Davis.

Sproul, C. 1986. Public participation in the Point Conception LNG controversy: energy wasted or energy well spent? *Ecology Law Quarterly*, **13**, 73–153.

State Bar of California 1983. *The California Environmental Quality Act: recommendations for legislative and administrative change*. Report of the Committee on the Environment to the

Assembly Committee on Natural Resources, December 1983, State Bar of California, San Francisco.

Stewart, R. 1975. The reformation of American administrative law. *Harvard Law Review*, **88**, 1667–1813.

Stewart, R. 1985. The discontents of legalism: interest group relations in administrative regulation. *Wisconsin Law Review, 1985*, 655–86.

Taylor, S. 1984. *Making bureaucracies think*. Stanford, Calif.: Stanford, University Press.

Vandervelden, M. 1984. Is the state environmental act an endangered species? *California Lawyer*, **4(3)**, 45–6 and 52.

Vettel, S. 1985. San Francisco's downtown plan: environmental and urban design values in central business district regulation. *Ecology Law Quarterly*, **12**, 511–66.

Vig, N. & M. Kraft (eds) 1984. *Environmental policy in the 1980s*. Washington DC: CQ Press.

Wandesforde-Smith, G. 1977. Projects, policies and environmental impact assessment: a look inside California's black box. *Environmental Policy and Law*, **3**, 167–75.

Wandesforde-Smith, G. 1981. The evolution of environmental impact assessment in California. In *Project appraisal and policy review*, T. O'Riordan and W. Sewell (eds) 47–76. Chichester: Wiley.

Wandesforde-Smith, G. 1986. Learning from experience, planning for the future: the parable (and paradox) of environmentalists as pin-striped pantheists. *Ecology Law Quarterly*, **13**, 715–58.

Wandesforde-Smith, G. & I. Moreira, 1985. Subnational government and EIA in the developing world: bureaucratic strategy and political change in Rio de Janeiro, Brazil. *Environmental Impact Assessment Review*, **5**, 223–38.

Wengert, N. 1955. *Natural resources and the political struggle*. New York: Random House.

Yngvesson, B. 1985. Re-examining continuing relations and the law. *Wisconsin Law Review, 1985*, 623–46.

Yost, N. 1974. NEPA's progeny: state environmental policy acts. *Environmental Law Reporter*, **3**, 50090–8.

Yost, N. 1985. NEPA – the law that works. *Environmental Forum*, **38** *(Jan. 1985)*, 41–2.

Chapter 11

Acre, R. M. 1986. *Los Estudios de Impacto Ambiental en la CEE y los proyectos de infraestructura viaria*. Paper presented to the Coloquio Nacional sobre Carretera y Medio Ambiente, Madrid, 24–26 February 1986.

Anon. 1978. Interview avec M. L. Dhoore. *Environmental Policy and Law*, **4**, 42–5.

Anon. 1980. Redrafting the EEC Commissions's draft directive on EIA. *ENDS*, **44**, 17–18.

Anon. 1985a. Bad environmental science – or impoverished scientists? *ENDS*, **126**, 2.

Anon. 1985b. Environmental assessment directive adopted at last. *ENDS*, **125**, 9–11.

Anon. 1986. Environment Minister encourages public to discuss draft of new environmental law. *International Environmental Reporter*, **9**, 124.

Brouwer, H. C. G. M. 1986. *Experience in implementing environmental impact assessment in the Netherlands*. Paper to the CEMP Conference on the EEC Environmental Assessment Directive, 30–1 January 1986, London.

Bunge, T. 1984. Zur Umweltvertraglichkeitsprufung in der Bundesrepublik Deutschland – eine Zwischenbilanz. *Zeitschrift für Umweltpolitik*, **4**, 405–24.

Catlow, J. & C. G. Thirlwall, 1976. *Environmental impact analysis.* DoE Research Report no. 11, Department of the Environment, London.

Clark, B. D., R. Bisset & P. Wathern, 1981a. The British experience. In *Project appraisal and policy review*, T. O'Riordan & W. D. R. Sewell (eds), 125–53. Chichester: Wiley.

Clark, B. D., K. Chapman, R. Bisset & P. Wathern, 1976. *The assessment of major industrial applications – a manual.* DoE Research Report no. 13, Department of the Environment, London.

Clark, B. D., K. Chapman, R. Bisset, P. Wathern & M. Barrett, 1981b. *A manual for the assessment of major developments.* London: HMSO.

Commission of the European Communities 1980. Proposal for a Council directive concerning the assessment of environmental effects of certain public and private projects. *Official Journal*, **c169**, 9.7.80., 14–17.

Council of the European Communities 1977. Second programme of action on the environment. *Official Journal*, **c134**, 13.6.77.

Council of the European Communities 1985. On the assessment of the effects of certain public and private projects on the environment. *Official Journal*, **L175**, 28.5.85., 40–8.

Dalas, W. G. 1984. Experiences of environmental impact assessment procedures in Ireland. In *Planning and ecology*, R. D. Roberts & T. M. Roberts (eds), 389–95. London: Chapman & Hall.

Department of the Environment 1986. *Implementation of the Directive on Environmental Assessment.* Consultation Paper, Department of the Environment, London.

Foster, B. J. 1984. Environmental impact assessment in the UK. *Zeitschrift für Umweltpolitik*, **4**, 389–404.

Fuentes, F. 1985. *Marco Legal y Administrativo de las Evaluaciones de Impacto Ambiental.* Madrid: Ministerio de Obras Publicas y Urbanismo.

Fuller, G. I. 1986. *The environmental assessment directive. Proposals for implementation.* Paper to the CEMP Conference on the Environmental Assessment Directive, 30–1 January 1986, London.

Haigh, N. 1983. The EEC directive on environmental assessment of development projects. *Journal of Planning and Environmental Law*, Sept. 1983, 585–95.

Haigh, N. 1984. *EEC environmental policy and Britain.* London: Environmental Data Services.

House of Lords 1981. *Environmental assessment of projects.* 11th Report of the Select Committee on the European Communities. London: HMSO.

Jones, M. G. 1980. Developing an EIA process for the Netherlands. *Environmental Impact Assessment Review*, **1**, 167–80.

Jones, M. G. 1983. Environmental impact assessment at the planning level: two systems. In *Environmental impact assessment*, PADC (eds), 63–94. The Hague: Nijhoff.

Kennedy, W. V. 1981. The German experience. In *Project Appraisal and Policy Review*, T. O'Riordan and W. R. D. Sewell (eds), 155–85. Chichester: Wiley.

Lee, N. & C. Wood, 1978. Environmental impact assessment of projects in EEC countries. *Journal of Environmental Management*, **6**, 57–71.

Lee, N. & C. Wood, 1985. Training for environmental impact assessment within the European Economic Community. *Journal of Environmental Management*, **21**, 271–86.

Macrory, R. & M. Lafontaine, 1982. *Public inquiry and enquête publique.* London: Environmental Data Services.

Milne, R. 1986. Strange bedfellows from Brussels. *New Scientist*, **55**.

Monbailliu, X. 1981. Role of environmental impact assessment in plans and policies. In *Environmental impact assessment*, Project Appraisal for Development Control (eds), 95–101. The Hague: Nijhoff.

Prieur, M. 1984. Les études d'impact en droit français. *Zeitschrift für Umweltpolitik*, **4**, 367–88.

Waldegrave, W. 1986. *The EEC Environmental Assessment Directive*. Keynote speech to the CEMP Conference on the Environmental Assessment Directive, 30–1 January 1986, London.

Wathern, P., I. W. Brown, D. A. Roberts & S. N. Young, 1983. *Some effects of EEC policy*. Report to the Directorate General for the Environment, Consumer Protection and Nuclear Safety, Commission of the European Communities, Brussels.

Wathern, P., S. N. Young, I. W. Brown & D. A. Roberts, 1986. The EEC less favoured areas directive: implementation and impact on upland land use in the UK. *Land Use Policy*, **3**, 205–12.

Young, S. N. & P. Wathern, 1984. The EEC shellfish directive in Wales. *Water Science and Technology*, **17**, 1199–1209.

Chapter 12

Bochniarz, Z. & A. Kassenberg, 1985. *Environmental protection by integrated planning*. In proceedings of the conference on Economic and Social Problems of Environmental Protection, Warsaw.

Enyedi, G. & V. Zentai, 1985. *Environmental policy in Hungary*. Report for the ECOI Project, Vienna Centre, Vienna.

Galperyn, Z. 1985. *Procedure for EIA review*. In proceedings of the course on Environmental Impact Assessment, Warsaw (in Polish).

Institute for Environmental Development 1983. *Environmental impact assessment system*. Final Report of the Project POL/RCE-003, Institute for Environmental Development, Katowice (in Polish).

Janikowski, R. & A. Starzewska, 1986. EIA project in Poland. *Environmental Impact Assessment Worldletter*, 1986 (May/June), 1–4.

Kotyczka, K. 1984. *Environmental policy of the GDR*. National report to the European Environmental Project, Vienna Centre, Vienna.

Kozlowski, S. 1985. *Problem of impact of large industrial developments upon the natural environment*. In proceedings of the course on Environmental Impact Assessment, Warsaw (in Polish).

Lisitsin, E. 1985. *Fundamentals of the environmental policies of European socialist countries: a public-policy approach*. Report to the ECOI Project, Vienna Centre, Vienna.

Madar, Z. 1985. *Main principles of the care of environment Czechoslovak Socialist Republic*. Report to the ECOI Project, Vienna Centre, Vienna.

Tomaszek, S. 1985. *EIA and land use planning in Poland*. In proceedings of the course on Environmental Impact Assessment, Warsaw (in Polish).

Chapter 13

ASEAN – UNEP – CDG 1985. *Report of Workshop on the Evaluation of Environmental Impact Assessment Applications in ASEAN Countries*. Bandung, Indonesia.

ESCAP 1985. *Environmental impact assessment – guidelines for planners and decision makers.* UN/ESCAP, Bangkok, Thailand.

Fowler, R. J. 1982. *Environmental impact assessment, planning and pollution measures in Australia.* Canberra: Australia Government Publishing Service.

Lohani, B. N. & N. Halim, 1982. Environmental impact identification and prediction. In *Environmental management for developing countries, Volume 3. Environmental assessment and management,* G. Tharun & R. Bidwell (eds), 25–55. Asian Institute of Technology, Bangkok, Thailand.

Nay Htun 1984. Development of UNEP guidelines for assessing industrial environmental impacts and environmental criteria for siting of industry. In *Perspectives on environmental impact assessment,* B. D. Clark, A. Gilad, R. Bisset & P. Tomlinson (eds), 253–63. Dordrecht: Reidel.

Snidvongs, K. 1985. Problems and prospects of environmental impact assessment in Thailand. In *Environmental impact assessment (EIA) for development. Proceedings of a joint DSE/UNEP international seminar,* K. Klennert (ed.), 152–70. Feldafing, Federal Republic of Germany: Deutsche Stiftung für Internationale Entwicklung.

United Nations Environment Programme 1982. *The use of environmental impact assessment for development project planning in ASEAN countries.* UNEP Regional Office for Asia and the Pacific, Bangkok, Thailand.

Chapter 14

Aquino, R. S. L. *et al.* 1984. *História das sociedades americanas.* Rio de Janeiro: Livraria Eu & Voce Editora.

Arocha, J. L. M. 1985. *Situación y experiencias en estudios de impacto ambiental en Venezuela.* Paper presented at Reunión de Expertos ECO-OPS/ORPALC-PNUMA, Caracas.

Azuela, A. 1982. *Sintesis de la legislación ambiental en Panamá, Republica Dominicana y Nicaragua.* Opiniones, Fascículos sobre Medio Ambiente no. 8. Madrid: CIFCA.

Balderiote, M. 1985. *Documento sobre situación y experiencias de EIA en Argentina.* Paper presented at Reunión de Expertos ECO-OPS/ORPALC-PNUMA, Buenos Aires.

Ballesteros, R. B. 1981. *La legislación ambiental en América Latina: visión comparativa.* Mexico: Universidad Autónoma Metropolitana.

Ballesteros, R. B. 1982. *El derecho ambiental en America Latina.* Opiniones, Fascículos sobre Medio Ambiente no. 1. Madrid: CIFCA.

Delgado, C. M. 1983. *La evalución de impacto ambiental en Venezuela.* Paper for Curso Avaliação de Impacto Ambiental. FEEMA/PNUMA, Rio de Janeiro.

Ferrari, R. P. 1983. *Informe.* Paper for Curso Avaliação de Impacto Ambiental. FEEMA/PNUMA, Rio de Janeiro.

Furtado, C. 1970. *Formação econômica da América Latina,* 2nd edn, Rio de Janeiro: Lia Editor.

Giglo, N. 1982. *Medio ambiente y planificación: las estrategias políticas a corto y mediano plazo.* Opiniones, Fascículos de Medio Ambiente no. 2. Madrid: CIFCA.

La Garza, M. F. 1984. *The environmental procedure in México.* Mexico: Secretaria de Desarollo Urbano y Ecología.

Lámbarri, M. A. 1985. *Situación y experiencias de la evaluación de impacto ambiental en Mexico.* Paper presented at Reunión de Expertos ECO-OPS/ORPALC-PNUMA, Mexico.

Olano, J. E. M. 1985. *Situación y experiencias de evaluación de impacto ambiental en el Peru*. Paper presented at Reunión de Expertos ECO-OPS/ORPALC-PNUMA. Lima.

ONERN 1978. *Oficina Nacional de Evalución de Recursos Naturales*. Lima: ONERN.

Pérez, A. D. 1981. *Algumas experiências sobre evalución de impactos ambientales y ubicación de industrias en Colombia*. Bogotá: INDERENA.

Pérez, A. D. 1985. *Situación y experiencias de evaluación de impacto ambiental en Colombia*. Paper presented at Reunión de Expertos ECO-OPS/ORPALC-PNUMA, Bogotá.

UNEP 1984. *Legislación ambiental en America Latina y Caribe*. Serie Documentos 2. Centro de Documentación e Información de la ORPALC, México.

Wandesforde-Smith, G., R. Carpenter & J. A. J. Horberry, 1985. EIA in developing countries: an introduction. *Environmental Impact Assessment Review*, **5**, 201–6.

Wandesforde-Smith, G. & I. V. D. Moreira, 1985. Sub-national government and EIA in the developing world: bureaucratic strategy and political change in Rio de Janeiro. *Environmental Impact Assessment Review*, **5**, 223–38.

Chapter 15

Centre for Environmental Management and Planning/Istitutio Superiore di Sanita 1986. *The health and safety component of environmental impact assessment – a case study analysis of environmental assessments of chemical industry projects*. Report to Regional Office for Europe, World Health Organization, Copenhagen.

Cohen, H. 1983. *Orientation of the future extension and application of knowledge in the field of environmental health WHO/EURO ICP/PPE/009(2)/8*. Paper to the WHO/EURO Working Group on Health and the Environment, 12–16 December, Vienna.

Council of the European Communities 1980. Directive on the quality of water intended for human consumption (80/778/EEC). *Official Journal*, **L229**, 30.8.80, 11–28.

D'Appolonia SA 1982. *The petro-chemical industry: a survey covering environmental impact, pollution control, health and safety*. Report to the Regional Office for Europe, World Health Organization, Copenhagen.

Derban, L. K. A. 1984. Health impacts of the Volta Dam, Ghana. In *Perspectives on environmental impact assessment*, B. Clark, A. Gilad, R. Bisset & P. Tomlinson (eds), 105–20. Dordrecht: Reidel.

Donaldson, R. J. 1984. Medical effects of atmospheric pollution and noise. In *Perspectives on environmental impact assessment*, B. Clark, A. Gilad, R. Bisset & P. Tomlinson (eds), 105–19. Dordrecht: Reidel.

Environmental Resources Ltd. 1983. *Environmental health impact assessment for irrigated development projects – guidelines and recommendations, final report*. Report to Regional Office for Europe, World Health Organization, Copenhagen.

Environmental Resources Ltd. 1985. *Environmental health impact assessment of urban development projects – guidelines and recommendations*. Report to Regional Office for Europe, World Health Organization, Copenhagen.

Gilad, A. 1984. The health component of the environmental impact assessment process. In *Perspectives on environmental impact assessment*, B. Clark, A. Gilad, R. Bisset & P. Tomlinson (eds), 93–104. Dordrecht: Reidel.

Parke, D. 1983. *Limits and constraints of scientifically-based environmental health risk assessment WHO/EURO ICP/PPE/009(2)/6*. Paper to WHO/EURO Working Group on Health and the Environment, 12–16 December, Vienna.

Robinson, J. D., M. D. Higgins & P. K. Bolyard, 1983. Assessing environmental impacts on health – a role for the behavioural science. *Environmental Impact Assessment Review*, **4**, 41–54.

World Health Organization 1982. *Rapid assessment of sources of air, water and land pollution.* Offset Publication no. 62, World Health Organization, Geneva.

World Health Organization 1983a. *Seventh programme of work covering the period 1984–89.* Health for All Series no. 8, World Health Organization, Geneva.

World Health Organization 1983b. *Selected techniques for environmental management, Training Manual EFP/83.50.* World Health Organization, Geneva.

World Health Organization 1983c. *Risk assessment and its use in the decision-making process for chemical control – report of a WHO/EURO consultation, 9–11 November, Ulm.* World Health Organization, Copenhagen.

World Health Organization 1983d. *Environmental pollution control in relation to development.* Technical Report Series no. 718, World Health Organization, Geneva.

World Health Organization 1984. *Guidelines for drinking water quality, Volume 1.* World Health Organization, Geneva.

Chapter 16

Organisation for Economic Co-operation and Development 1974. *Analysis of the environmental consequences of significant public and private projects. C(74)216.* Paris: Organisation for Economic Co-operation and Development.

Organisation for Economic Co-operation and Development 1979. *The assessment of projects with significant impact on the environment. C(79)116.* Paris: Organisation for Economic Co-operation and Development.

Organisation for Economic Co-operation and Development 1986a. *Environmental assessment and development assistance environment monograph No. 4.* Paris: Organisation for Economic Co-operation and Development.

Organisation for Economic Co-operation and Development 1986b. *Recommendation of the Council on Measures Required to Facilitate the Environmental Assessment of Development Assistance Projects and Programmes. C(86)26.* Paris: Organisation for Economic Co-operation and Development.

Sammy, G. K. 1982. *Environmental assessment in developing countries.* Ph.D. Thesis, University of Oklahoma, Norman.

Chapter 17

Blake, R. O., B. Lausche, S. J. Scherr, T. B. Stoel & A. T. Thomas, 1980. *Aiding the environment: a study of the environmental policies, procedures and performance of the United States Agency for International Development.* Washington DC: Natural Resources Defense Council.

Gannett, Fleming, Corrdry & Carpenter, Inc. 1981. *Assessment of environmental effects of proposed developments in the Senegal River Basin.* Harrisburg, Pa.: Gannett, Fleming, Corrdry & Carpenter.

Horberry, J. A. J. 1984a. *Development assistance and the environment.* Ph.D. thesis, Massachusetts Institute of Technology, Cambridge, Mass.

Horberry, J. A. J. 1984b. Status and application of environmental impact assessment for development. In *Environmental impact assessment (EIA) for development*, K. Klennert, (ed.), 269–377. Feldafing, Federal Republic of Germany: Deutsche Stiftung für Internationale Entwicklung.

Horberry, J. A. J. & B. Johnson, 1981. *The environmental performance of consulting firms in development aid*. London: International Institute for Environment and Development.

Hough, R. 1982. *Economic assistance and security: rethinking US policy*. Washington DC: National Defense University Press.

McPherson, P. 1982. Prepared statement of the Administrator, Agency for International Development. In *Review of the Global Environment 10 years after Stockholm. Hearings. US Congress House Subcommittee on Human Rights and International Organization*, p. 34. 97th Congress, 2nd Session. Government Printing Office, Washington DC.

Mickelwait, D. 1979. *New directions in development: a study of U.S. AID*. Boulder, Colo.: Westview.

Organization for Economic Co-operation and Development 1982. *Development co-operation: 1982 review*. Paris: Organization for Economic Co-operation and Development.

President's Office 1979. Environmental effects abroad of major federal actions. Executive Order 12114. *Federal Register*, **44**, 1957.

Tendler, J. 1975. *Inside foreign aid*. Baltimore: Johns Hopkins University Press.

United States Agency for International Development 1970. *Consideration of environmental aspects of US-assisted capital projects. Manual Circular 1221.2, 18 May 1970*. Washington DC: United States Agency for International Development.

United States Agency for International Development 1971. Environmental procedures. *Federal Register*, **36**, 22686.

United States Agency for International Development 1975. *Environmental policy determination PD-63*. Washington DC: United States Agency for International Development.

United States Agency for International Development 1978. Environmental procedures. *Federal Register*, **22(216)**.

United States Agency for International Development 1981. *Memorandum for the executive staff, mission directors and environmental officers, 22 January 1981*. Washington DC: United States Agency for International Development.

United States Agency for International Development 1983a. *Environmental and natural resource aspects of development assistance: policy determination PD-6*. Washington DC: United States Agency for International Development.

United States Agency for International Development 1983b. *Congressional presentation, fiscal year 1984*. Washington DC: United States Agency for International Development.

United States Congressional Budget Office 1980. *Assisting the developing countries: foreign aid and trade policies of the United States*. Washington DC: Government Printing Office.

Index